Lecture Notes on Data Engineering and Communications Technologies 186

Series Editor

The aim of the book series is to present cutting edge engineering approaches to data technologies and communications. It will publish latest advances on the engineering task of building and deploying distributed, scalable and reliable data infrastructures and communication systems.

The series will have a prominent applied focus on data technologies and communications with aim to promote the bridging from fundamental research on data science and networking to data engineering and communications that lead to industry products, business knowledge and standardisation.

Indexed by SCOPUS, INSPEC, EI Compendex.

All books published in the series are submitted for consideration in Web of Science.

Leonard Barolli
Editor

Advances on Broad-Band and Wireless Computing, Communication and Applications

Proceedings of the 18th International Conference on Broad-Band and Wireless Computing, Communication and Applications (BWCCA-2023)

Springer

Editor
Leonard Barolli
Department of Information and Communication Engineering
Fukuoka Institute of Technology
Fukuoka, Japan

ISSN 2367-4512 ISSN 2367-4520 (electronic)
Lecture Notes on Data Engineering and Communications Technologies
ISBN 978-3-031-46783-7 ISBN 978-3-031-46784-4 (eBook)
https://doi.org/10.1007/978-3-031-46784-4

This Springer imprint is published by the registered company Springer Nature Switzerland AG
The registered company address is: Gewerbestrasse 11, 6330 Cham, Switzerland

Paper in this product is recyclable.

Welcome Message of BWCCA-2023 International Conference Organizers

Welcome to the 18th International Conference on Broad-Band and Wireless Computing, Communication and Applications (BWCCA-2023), which will be held in conjunction with 3PGCIC-2023 International Conference from November 8 to November 10, 2023.

This International Conference is a forum for sharing ideas and research work in the emerging areas of broadband and wireless computing. Information networks of today are going through a rapid evolution. Different kinds of networks with different characteristics are emerging, and they are integrating into heterogeneous networks. For these reasons, there are many interconnection problems which may occur at different levels of the hardware and software design of communicating entities and communication networks. These kinds of networks need to manage an increasing usage demand, provide support for a significant number of services, guarantee their QoS, and optimize the network resources.

The success of all-IP networking and wireless technology has changed the ways of living the people around the world. The progress of electronic integration and wireless communications is going to pave the way to offer people the access to the wireless networks on the fly, based on which all electronic devices will be able to exchange the information with each other in ubiquitous way whenever necessary.

The aim of this conference is to present the innovative research and technologies as well as developments related to broadband networking, and mobile and wireless communications.

The organization of an International Conference requires the support and help of many people. A lot of people have helped and worked hard to produce a successful BWCCA-2023 technical program and conference proceedings. First, we would like to thank all authors for submitting their papers, Program Committee Members and reviewers who carried out the most difficult work by carefully evaluating the submitted papers.

We thank Web Administrators Co-Chairs and Finance Chair for their excellent work. We would like to express our gratitude to Prof. Makoto Takizawa, Hosei University, Japan, as Honorary Chair of BWCCA-2023 for his support and help. We give special thanks to the Keynote Speakers of BWCCA-2023 and the local arrangement team.

We are thankful to Daegu Convention & Visitors Bureau (DCVB) and Korea Tourism Organization (KTO) for providing us very good facilities to hold the conference and for the financial support.

We hope you will enjoy the conference proceedings.

BWCCA-2023 Organizing Committee

Honorary Chair

Makoto Takizawa	Hosei University, Japan

General Co-chairs

Hyunhee Park	Myongji University, South Korea
Tomoya Enokido	Rissho University, Japan
Hsing-Chung Chen	Asia University, Taiwan

Program Committee Co-chairs

Seunghyun Park	Hansung University, South Korea
Lidia Ogiela	AGH University of Science and Technology, Poland
Tetsuya Shigeyasu	Prefectural University of Hiroshima, Japan

International Advisory Committee

Fang-Yie Leu	Tunghai University, Taiwan
David Taniar	Monash University, Australia
Kangbin Yim	Soonchunhyang University, Korea

Publicity Co-chairs

Jinkyu Kang	Myongji University, South Korea
Keita Matsuo	Fukuoka Institute of Technology, Japan
Evjola Spaho	Polytechnic University of Tirana, Albania
Naohiro Hayashibara	Kyoto Sangyo University, Japan

Finance Chair

Makoto Ikeda	Fukuoka Institute of Technology, Japan

Web Administrator Chairs

Phudit Ampririt	Fukuoka Institute of Technology, Japan
Ermioni Qafzezi	Fukuoka Institute of Technology, Japan

Local Organizing Co-chairs

Seongah Jung	Kyungpook National University, South Korea
Jaehee Jung	Myongji University, South Korea

Steering Committee Chair

Leonard Barolli	Fukuoka Institute of Technology, Japan

Track Areas

1. Next Generation Wireless Networks

Chairs

Bhed Bista	Iwate Prefectural University, Japan
Szu-Yin Lin	National Ilan University, Taiwan
Sriram Chellappan	University of South Florida, USA

PC Members

Jiahong Wang	Iwate Prefectural University, Japan
Shigetomo Kimura	University of Tsukuba, Japan
Chotipat Pornavalai	King Mongkut's Institute of Technology Ladkrabang, Thailand
Danda B. Rawat	Howard University, USA
Gongjun Yan	University of Southern Indiana, USA
Vamsi Paruchuri	University of Central Arkansas, USA

Arjan Durresi	IUPUI, USA
Shih-Yi James Chien	National Sun Yat-sen University, Taiwan
Pei-Ju Lee	National Chung Cheng University, Taiwan
Chih-Hao Lin	Chung Yuan Christian University, Taiwan
Hao-Hsiang Ku	National Taiwan Ocean University, Taiwan
Jung-Bin Li	Fu Jen Catholic University, Taiwan
Thoshitha Gamage	Southern Illinois University, USA
Mukundan Sridharan	Samraksh Company, USA
Brijesh Chejerla	Florida Blue, USA
Srinivas Chakravarthi Thandu	Amazon, USA

2. Cloud and Service Computing

Chairs

Hwamin Lee	Soonchunhyang University, South Korea
Ramesh C. Hansdah	Indian Institute of Science, Bangalore, India
Baojiang Cui	Beijing University of Posts and Telecommunications, China

PC Members

Gang Wang	Nankai University, China
Jianxin Wang	Beijing Forestry University, China
Jie Cheng	Shandong University, China
Shaoyin Cheng	University of Science and Technology of China, China
Yan Zhang	Hubei University, China
Willy Susilo	University of Wollongong, Australia
Kamil Kluczniak	Wroclaw University of Technology, Poland
Francesco Palmieri	University of Salerno, Italy
Jian Shen	Nanjing University of Information Science and Technology, China
Jin Li	Guangzhou University, China
Fangguo Zhang	Sun Yat-sen University, China
Xinyi Huang	Fujian Normal University, China
Shengli Liu	Shanghai Jiao Tong University, China
Zhenjie Huang	Zhangzhou City University, China
Joseph K. Liu	Institute for Infocomm Research, Australia
Yong Yu	University of Wollongong, China
Ding Wang	Peking University, China

Tao Jiang	Xidian University, China
Jianfeng Wang	Xidian University, China
S. D. Madhu Kumar	NIT Calicut, India
Ashutosh Bhatia	BITS Pilani, Pilani Campus, India
Amulya Rathna Swain	KIIT, Bhubaneshwar, India
Yogesh Simmhan	IISc Bangalore, India
Soumya K. Ghosh	Indian Institute of Technology, India

3. Multimedia and Web Applications

Chairs

Yoshihiro Okada	Kyushu University, Japan
Chuan-Yu Chang	National Yunlin University of Science and Technology, Taiwan
Salem Alkhalaf	Qassim University, Saudi Arabia

PC Members

Kaoru Sugita	Fukuoka Institute of Technology, Japan
Tomoyuki Ishida	Fukuoka Institute of Technology, Japan
Makoto Nakashima	Oita University, Japan
Nobukazu Iguchi	Kinki University, Japan
Kenzi Watanabe	Hiroshima University, Japan
Nobuo Funabiki	Okayama University, Japan
Shinji Sugawara	Chiba Institute of Technology, Japan
Li-Wei Kang	National Yunlin University of Science and Technology, Taiwan
Chia-Hung Yeh	National Taiwan Normal University, Taiwan
Jun-Wei Hsieh	National Taiwan Ocean University, Taiwan
Wu-Chih Hu	National Penghu University of Science and Technology, Taiwan
Chien-Cheng Lee	Yuan Ze University, Taiwan
Muhammad Hussain	King Saud University, Saudi Arabia
Umair Azfar Khan	Habib University, Pakistan
Shigeru Takano	Kyushu University, Japan
Kosuke Kaneko	Kyushu University, Japan
Akira Haga	Kyushu University, Japan
Wei Shi	Kyushu University, Japan

4. Security and Privacy

Chairs

Tianhan Gao	Northeastern University, China
Masakatsu Nishigaki	Shizuoka University, Japan
Mohamed Abdur Rahman	Prince Mugrin University, KSA

PC Members

Nan Guo	Northeastern University, China
Zhenhua Tan	Northeastern University, China
Jian Xu	Northeastern University, China
Hiroaki Kikuchi	Meiji University, Japan
Takamichi Saito	Meiji University, Japan
Rashid Tahir	University of Prince Mugrin Madinah, Saudi Arabia
Syed Sadiq	University of Prince Mugrin Madinah, Saudi Arabia
Md. Mamunur Rashid (Mamun)	King's Business School, UK
Akhlaq Ahmad	Umm Al-Qura University Makkah, Saudi Arabia
Shyhtsun Felix Wu	University of California, Davis, USA
Zhen-Yu Wu	Penghu University of Science and Technology, Taiwan
Tsung-Chih Hsiao	Southeast University, China
Kuo-Kun Tseng	Harbin Institute of Technology, China
Akira Otsuka	Institute of Information Security, Japan
Naonobu Okazaki	University of Miyazaki, Japan
Masaki Shimaoka	Secom Co., Ltd., Japan
Davinder Kaur	IUPUI, USA

5. Network Protocols and Performance Analysis

Chairs

Tetsuya Shigeyasu	Prefectural University of Hiroshima, Japan
Ching-Feng Liang	Industrial Technology Research Institute, Taiwan
Vamsi Paruchuri	University of Central Arkansas, USA

PC Members

Xiaoyi Wang	Nokia Solutions and Networks, USA
Yu Sun	University of Central Arkansas, USA
Qiang Duan	Pennsylvania State University, USA
Han-Chieh Wei	Dallas Baptist University, USA
Masaaki Yamanaka	Japan Coast Guard Academy, Japan
Misako Urakami	Tokuyama College of Technology, Japan
Tomoya Kawakami	Nara Institute of Science and Technology, Japan
Masaaki Noro	Fujitsu Corp., Japan
Nobuyoshi Sato	Iwate Prefectural University, Japan
Phone Lin	National Taiwan University, Taiwan
Ray-Guang Cheng	National Taiwan University of Science and Technology, Taiwan
Shun-Ren Yang	National Tsing Hua University, Taiwan
Whai-En Chen	National Ilan University, Taiwan

6. Intelligent and Cognitive Computing

Chairs

Lidia Ogiela	AGH University of Science and Technology, Poland
Takahiro Uchiya	Nagoya Institute of Technology, Japan
Hai Dong	RMIT University, Australia

PC Members

Atsuko Mutoh	Nagoya Institute of Technology, Japan
Shinsuke Kajioka	Nagoya Institute of Technology, Japan
Ryota Nishimura	Tokushima University, Japan
Shohei Kato	Nagoya Institute of Technology, Japan
Francesco Pascale	University of Salerno, Italy
Jan Platoš	VŠB Technical University of Ostrava, Czech Republic
Pavel Krömer	VŠB Technical University of Ostrava, Czech Republic
Urszula Ogiela	AGH University of Science and Technology, Poland
Jana Nowaková	VŠB Technical University of Ostrava, Czech Republic

Chang Choi	Gachon University, South Korea
Hoon Ko	Chosun University, South Korea
Hae-Duck Joshua Jeong	Korean Bible University, South Korea
Pengcheng Zhang	Hohai University, China
Sajib Mistry	Curtin University, Australia
Tooba Aamir	RMIT University, Australia
Wei Du	Wuhan University of Technology, China
Wei Zhang	Macquarie University, Australia
Shang-Pin Ma	National Taiwan Ocean University, Taiwan

7. Distributed and Parallel Computing

Chairs

Naohiro Hayashibara	Kyoto Sangyo University, Japan
Omar Khadeer Hussain	University of New South Wales, Australia

PC Members

Sazia Parvin	Melbourne Polytechnic, Australia
Naeem Janjua	Edith Cowan University, Australia
Alireza Faed	Ryerson University, Canada
Adil Hammadi	Curtin University, Australia
Lucian Prodan	Polytechnic University Timisoara, Romania
Kanwalinderjit Kaur Gagneja	Florida Polytechnic University
Rohaya Latip	Universiti Putra Malaysia, Malaysia
Tomoya Enokido	Rissho University, Japan
Leonard Barolli	Fukuoka Institute of Technology, Japan
Minoru Uehara	Toyo University, Japan

8. IoT and Smart Environment

Chairs

Nadeem Javaid	COMSATS University Islamabad, Pakistan
Chun-Wei Tsai	National Chung Hsing University, Taiwan

PC Members

Zahoor Ali Khan	Higher Colleges of Technology, UAE
Umar Qasim	University of Engineering and Technology, Pakistan
Farookh Hussain	University of Technology Sydney, Australia
Elis Kulla	Fukuoka Institute of Technology, Japan
Keita Matsuo	Fukuoka Institute of Technology, Japan
Hsin-Hung Cho	National Ilan University, Taiwan
Fan-Hsun Tseng	National Taiwan Normal University, Taiwan
Hsin-Te Wu	National Penghu University of Science and Technology, Taiwan

9. Database, Data Mining and Big Data

Chairs

Antonio Esposito	University of Campania "Luigi Vanvitelli," Italy
Yao-Chung Fan	National Chung Hsing University, Taiwan

PC Members

Mehran Samavati	University of Sydney, Australia
Farshid Hajati	Griffith University, Australia
Jinnie Hee Yoon	Sejong University, South Korea
Elena Sitnikova	UNSW, Australia
Chen-Yi Lin	National Taichung University of Science and Technology, Taiwan
Lun-Chi Chen	National Center for High-Performance Computing (NCHC), Taiwan
Huan Chen	National Chung Hsing University, Taiwan
Luca Tasquier	University of Campania "Luigi Vanvitelli," Italy
Stefania Nacchia	University of Campania "Luigi Vanvitelli," Italy
Salvatore Augusto Maisto	University of Campania "Luigi Vanvitelli," Italy
Salvatore D'Angelo	University of Campania "Luigi Vanvitelli," Italy

10. Ubiquitous and Pervasive Computing

Chairs

Isaac Woungang	Toronto Metropolitan University, Canada
Asm Kayes	La Trobe University, Australia
Chyi-Ren Dow	Feng Chia University, Taiwan

PC Members

Evjola Spaho	Polytechnic University of Tirana, Albania
Makoto Ikeda	Fukuoka Institute of Technology, Japan
Elis Kulla	Fukuoka Institute of Technology, Japan
Admir Barolli	Aleksander Moisiu University of Durres, Albania
Donald Elmazi	Metropolitan University of Tirana, Albania
Alan Colman	Swinburne University of Technology, Australia
Iqbal H. Sarker	Swinburne University of Technology, Australia
Eric Pardede	La Trobe University, Australia
Syed Mahbub	La Trobe University, Australia
Patrick Hung	Ontario Tech University, Canada
Pei-Chun Lin	Feng Chia University, Taiwan
Zhang Kejun	Zhejiang University, China
Duc-Binh Nguyen	Thai Nguyen University of Information and Communications Technology (ICTU), Vietnam
Wei Lu	Keene State College, USA
Luca Caviglione	CNIT, Italy
Hamed Aly	Acadia University, Canada
Danda B. Rawat	Howard University, USA
Marcelo Luis Brocardo	University of Victoria, Canada
Glaucio Carvalho	Ryerson University, Canada

BWCCA-2023 Reviewers

Barolli Admir
Barolli Leonard
Bista Bhed
Chellappan Sriram
Chen Hsing-Chung
Cui Baojiang
Durresi Arjan
Enokido Tomoya
Fujisaki Kiyotaka
Fun Li Kin
Funabiki Nobuo
Gotoh Yusuke
Javaid Nadeem
Ikeda Makoto
Ishida Tomoyuki
Kamada Masaru

Kanzaki Akimitsu
Kayem Anne
Kohana Masaki
Kikuchi Hiroaki
Leu Fang-Yie
Maeda Hiroshi
Matsuo Keita
Nishigaki Masakatsu
Ogiela Lidia
Ogiela Marek
Okada Yoshihiro
Paruchuri Vamsi Krishna
Rahayu Wenny
Sakamoto Shinji
Shibata Yoshitaka
Shigeyasu Tetsuya
Saito Takamichi
Spaho Eviola
Taniar David
Uehara Minoru
Venticinque Salvatore
Wang Xu An
Woungang Isaac
Xhafa Fatos
Yi Liu
Yim Kangbin

BWCCA-2023 Keynote Talks

Secure Connected and Autonomous Vehicles: Research Trends and Challenges

Seunghyun Park

Hansung University, Seoul, Korea

Abstract. Automotives are evolving into large-scale smart devices connected to the Internet, and the self-driving functionalities in these vehicles are dramatically changing the mobility of users. However, new security threats not considered in existing vehicles are emerging behind this convenience. In particular, the diverse connectivity of automobiles, including the Internet, provides critical attack surfaces to which hackers can gain malicious access. Moreover, autonomous vehicles that take over control from humans require higher safety and security criteria because even minor security flaws or malfunctions can cause accidents with significant human casualties. In this talk, we will introduce the convenience of the recently evolving connected and autonomous vehicles (CAVs) and investigate the risks of cyberattacks accompanying such advantages. In addition, we analyze essential cybersecurity research trends to mitigate these cyber-risks and discuss the development direction of cybersecurity perspectives for CAVs.

Toward Human-Friendly Information System Design

Kaori Yoshida

Kyushu Institute of Technology, Kitakyushu, Japan

Abstract. As network technology advances, it is imperative to consider the end users' experiences when designing and implementing network infrastructure. The focus on human-friendly information system design is gaining popularity, a trend we should take seriously. With the increasing complexity of network infrastructure, it is easy to overlook the human characteristics of the design process. Designing systems that are easy to use and understand benefits end users. This talk provides an overview of information system design and its examples.

Contents

Resiliency of the Area-Segmentation in Vehicle Routing for Collecting the Disaster Information

Sanjukta Khwairakpam(✉), Masahiro Shibata, and Masato Tsuru

Computer Science and Systems Engineering, Kyushu Institute of Technology, Fukuoka, Japan
sanjukta.khwairakpam429@mail.kyutech.jp,
{shibata,tsuru}@csn.kyutech.ac.jp

Abstract. The emergency disaster response headquarters (HQ) is responsible for incrementally collecting the disaster information from the whole area of a region. Our previous work studied a routing problem for patrolling vehicles to monitor such information and bring it to HQ multiple times on the way. The optimal routes for vehicles can be found by a systematic search to minimize the average delay time for the information collection. Further, we have also studied an area-segmentation approach in which each vehicle collects the information only in one of sub-areas. However, in disaster situations, some parts of the vehicles' routes are likely to be damaged, incurring a considerable delay of information collection. In this paper, we investigate the resiliency of area-segmentations with different topologies, damaged locations, and delays incurred by damaged links. Our simulation results show that the area-segmentation approach is more robust than the whole-area approach in some cases by mitigating the risk of a large delay in information collection.

1 Introduction

The disaster management team takes adequate measures to protect human lives during or in the aftermath of the disaster. They urgently set up an on-site emergency disaster response headquarters (HQ) for disaster response and management only when a large-scale disaster happens. In the initial response and partial recovery for early and appropriate actions, the HQ should collect the information promptly and incrementally from the overall region that may be damaged. Patrolling vehicles equipped with mics, cameras, and other sensors often collect disaster information and travel all streets in an area affected by a disaster. During or in the aftermath of a disaster, since the high-speed telecommunication network infrastructures may be unavailable or degraded, each patrolling vehicle should move not to monitor the damaged information along streets but to bring it to HQ by itself.

Supported by NICT, Japan (No.22007) and JSPS KAKENHI Grant Number 21K17706.

L. Barolli (Ed.): BWCCA 2023, LNDECT 186, pp. 1–14, 2024.
https://doi.org/10.1007/978-3-031-46784-4_1

Therefore, we considered a routing problem for patrolling vehicles that start from HQ, cruise through streets, and end at HQ to monitor the information along the streets and bring it back to HQ [1], which is a kind of Arc Routing Problem (ARP) in Vehicle Routing Problem (VRP) [2,3]. We aim to minimize the average delay time of collecting the information along all streets in a region incrementally in which vehicles can return to HQ multiple times on the way to drop partially monitored information. To find good routes for this purpose, a key idea of the Chinese Postman Problem (CPP) [4] is borrowed. Our previous work studied a systematic search to find the optimal routes for vehicles to minimize the average delay time for the information collection [5,6]. An area-segmentation approach was also studied in which each vehicle collects the information only in one of sub-areas to reduce the complexity of finding the optimal routes.

However, in disaster situations, there is a high risk that some parts of the vehicular routes are damaged, which incurs a considerable delay of information collection. In this paper, therefore, we investigate the resiliency of information collection by area-segmentations in terms of mitigating the risk of a large delay of information collection. The investigation is performed based on various case-studies with different topologies, different damaged locations, and different delays incurred by damaged links. We also discuss a better segmentation design for resiliency to existence of damaged links.

There is extensive literature on using VRP and ARP for disaster response and recovery. For instance, in [7], VRP aids in a humanitarian relief effort by rescuing disaster victims and providing medical assistance. In this scenario, medical teams swiftly transported injured individuals to hospitals or evacuation centers for urgent care [8]. Using stochastic dynamic programming to select the shortest path is crucial in mitigating the risk of fatalities by ensuring short distances between casualties and the hospital [9].

Our setting is clearly different from them. The conventional VRP and ARP mainly consider delivering/collecting the necessary goods or conveying injured persons among two or more locations, and aim at minimizing the time and the cost taken to finish the task or at maximizing the effort of the task in a certain time duration. On the other hand, our target is to minimize the average delay time of collecting the information from the all streets incrementally.

The rest of this paper is organized as follows. The model is described in Sect. 2, the whole area approach and area-segmentation approach are proposed in Sect. 3. Through the experimental results are shown in Sect. 4 and finally, a conclusion is presented in Sect. 5.

2 Model

A street map in a region is considered as an undirected connected graph. The term "link" is used instead of "edge" and "node" instead of "vertex" on the graph. Each node is connected to neighboring nodes by links; the degree of a node is the number of those links. The emergency disaster response headquarters (HQ) is considered as a node in the graph representing a street map. We consider patrolling vehicles that start from HQ at the same time and bring monitored information when returning to HQ at possibly different times. A vehicle may return to HQ multiple times on the way to drop the monitored information to HQ. All links should be covered (i.e., monitored) by vehicles in total; A vehicle travels from HQ and finally returns to HQ along a "route" (or a travel route) that consists of a sequence of links. The term "length" of a path, a circuit, or a route on the graph is the number of links in the sequence of links of that. Note that a route can include the same link more than once and such a link is counted multiple times in "length". Let k be the degree of HQ. The vehicles need to return to HQ at least $\lceil k/2 \rceil$ times in total to cover those k links on their routes. Let R be the total number of returns of the vehicles to HQ.

The two assumptions of our model to make simple are as follows.

1. The unit time is taken for a vehicle to pass through one link in one direction between two nodes. Hence, the time taken for a vehicle's travel is the total number of links passed by that vehicle.
2. The unit amount of information is collected from one link. Since the information associated with a link is assumed to be unchanged, if a link is passed twice by the same or different vehicles, the same (duplicated) information is monitored.

We have the following two criteria focus on finding a good set of travel routes for vehicles:

1. Last Information Delay-time (LID): the time when the final vehicle brought the entire information of all links to HQ. To make "LID" small, the travel route from HQ to HQ should be as short as possible.
2. Information Delay-time Product (IDP): defined as

$$\int_0^{LID} (1 - u(t))\, dt$$

where $u(t)$ is the ratio of the amount of the information brought to HQ until time t to the amount of the entire information; t is the time taken from the start. Based on our assumption of unit information, $u(t)$ is the ratio of the number of different links passed by the vehicles returning to HQ before time t to the total number of links of the entire network. The IDP represents the average delay time for information collection. The small IDP benefits early decisions and responses by HQ. The shortest route is not always an excellent choice to make "IDP" small. Instead, a small IDP is realized by an appropriate number of multiple returns of the vehicles to HQ with balanced-length intervals [1].

3 The Whole Area Approach and Area-Segmentation Approach

1. Set the number R ($\geq \lceil k/2 \rceil$) of returns to HQ by letting k be the degree of HQ.

2. If $2R > k$, add virtual (duplicate) links to HQ. Then, randomly add $(2R - k)$ virtual links to HQ, and each link connecting to one of its k neighbors nodes makes a new extended graph.
3. Make an Euler graph from an extended graph. There will always be an even number of odd-degree nodes, say $2m$. Choose m pairs from a group of $2m$ nodes with odd degrees. For each pair of odd-degree nodes, add virtual (duplicate) links along a path connecting the pair nodes. Perform the pairing and path selection randomly as long as minimizing the total length of newly added virtual links and any connecting path that does not pass HQ, the value R is unchanged. The entire graph becomes an Euler graph by adding the virtual links, i.e., there are no odd-degree nodes.
4. We randomly discover an Eulerian circuit on the converted Euler graph. The Eulerian circuit comprises R sub-circuits, each passing through HQ only once.
5. Get each vehicle's travel route(s) to assign each sub-circuit. For example, for two vehicles, the R sub-circuits are separated into two groups: G_A and G_B. Vehicle A travels a route connecting all subcircuits in G_A in any order while vehicle B does the same in G_B. This process randomly groups sub-circuits in G_A and G_B while ensuring that the total lengths are balanced, resulting in slightly different lengths between the two routes. Obtained travel route(s) of vehicle(s) P depend on grouping and concatenation of sub-circuits.
6. Calculate LID and IDP of each P in Step 5, and choose the optimal P for the Eulerian circuit. Note that the IDP of P is for the original graph, not the converted Euler graph. We select the P with the smallest IDP and a small LID value as P^*, which is the best for an Eulerian circuit obtained in Step 4.
7. Search for candidates with good P^* across various options. Repeat Steps 1 to 6 while altering the selections for the number R, the converted Euler graph, and the Eulerian circuit. Ultimately, we will identify suitable candidates for ideal travel routes for small IDPs and LIDs.

The above procedure can find good travel routes that cover all links in a given (whole) area called "the whole-area approach". However, if the whole area is large, it may be a challenging and much effort because of the complexity of design (route optimization), management, and operation in practice. Thus, the area-segmentation approach is considered where each vehicle collects information from only one sub-area. This consists of the following steps; Assuming there are two vehicles and the HQ is not at the outer boundary of the whole area.

1. Divide the entire (whole) area into two sub-areas with approximately the same number of links to ensure the two vehicles' routes are of similar length. First, select an internal segmentation path (without loop) that connects a node on the external boundary, HQ, and another different node on the exterior boundary of the entire graph, which shares two subgraphs, say X and Y. Then, construct a segmented subgraph pair from (X, Y) to avoid duplicated links on the internal segmentation path in X and Y.
2. Use the whole-area approach to search for optimal single-vehicle routes for each sub-area. Note that each route is only suitable for its specific IDP and LID.

3. Choose optimal pairs of sub-area routes from the top m routes acquired in Step 2 by computing and comparing IDP and LID on the whole area for each pair across all possible combinations of two sub-areas, resulting in $m \times m$ total combinations.

4 Experimental Results

4.1 Grid Streets

As shown in Fig. 1(a), we consider a grid-like network with 85 links and 50 nodes as the street map without any damaged streets in a region in normal condition. HQ, denoted by a red dot with a degree of 4, is centered at a grid network. Let's imagine two vehicles starting and ending at HQ, with $R = 4$. In other words, the total return times of two vehicles to HQ four is four. We segment the whole grid area into two segments; each vehicle travels each segment. In our simulation, therefore, R = 2 is used in each segments; and R = 4 is used for the whole area. We compared the good routes for IDP and LID using a whole-area approach to investigate how area segmentation affects information collection performance. We examine two segmentation approaches by taking two examples: Pattern 1 and 2, illustrated in Fig. 1(b) and (c). We carried out 100,000 iterations for the entire (whole) area and each sub-area using the procedure outlined in Sect. 3, implemented in Python.

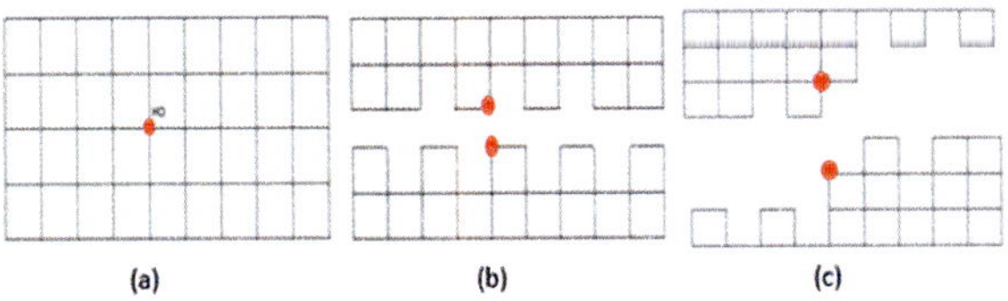

Fig. 1. (a) The whole area, (b) the segmented areas of Pattern 1 and (c) 2.

The optimal routes and five good cases of $u(t)$ are shown on the left and right sides of Fig. 2. The blue on the right side of Fig. 2 represents Case 1, which has the least IDP (= 36.59) and LID (=52). Our focus is on Case 1, where two vehicles left HQ simultaneously for the first round trip and returned at HQ at time 22, providing partial information of 0.26 (the information ratio is 22/85 = 0.26) and at time 26 with 0.29 fraction of information (the information ratio is (47−22)/85 = 0.29); in the second-round trip, both the vehicles return at the same time of 52 drop off all the remaining information to HQ. The first and second round trips of vehicles A and B are indicated by blue, yellow, green, and red, respectively, as shown in Fig. 2 (on the left).

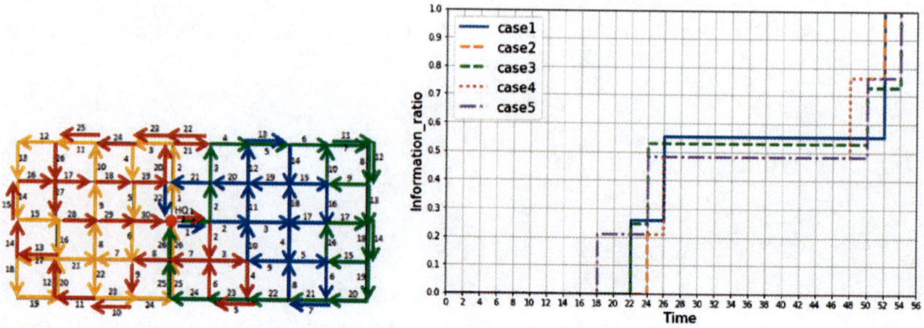

Fig. 2. One of the best routes (left) and the good five cases of $u(t)$ (right) in the whole area approach ($R = 4$, two vehicles).

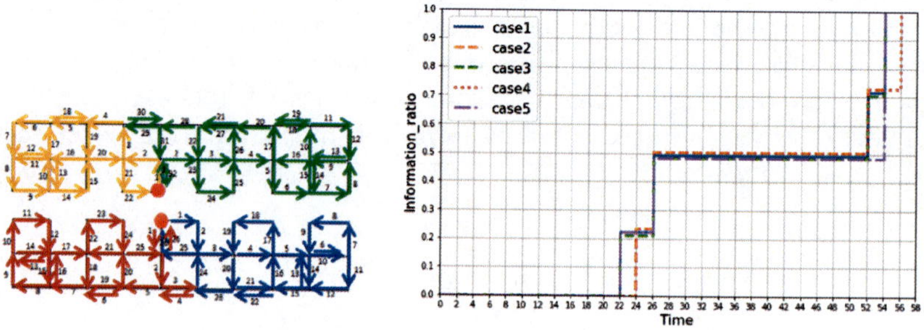

Fig. 3. One of the best routes (left) and the good five cases of $u(t)$ (right) in the segmentation approach Pattern 1 ($R = 4(2 + 2)$, two vehicles).

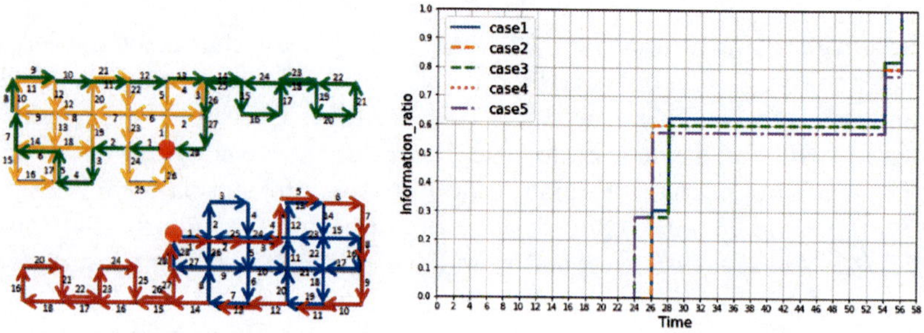

Fig. 4. One of the best routes (left) and the good five cases of $u(t)$ (right) in the segmentation approach Pattern 2 ($R = 4(2 + 2)$, two vehicles).

Pattern 1 segmented approach: Fig. 3 displays optimal routes (left) and five good cases of $u(t)$ (right). We discuss only Case 1, the smallest IDP (=38.82) and LID = 54, represented by the blue line in Fig. 3 (on the right). In this case, two vehicles return to HQ at different times, one at 22 (yellow line) and one at 26 (blue line), as shown in Fig. 3 (left), bringing partial information of 0.22 (=19/85) and 0.27 (=0.49−0.22). For the second-round trip, these two vehicles return at time 52 and 54 with 0.23 (=0.72−0.49) and 0.28 (=1−0.72) fraction of information shown by green and red lines in Fig. 3 (left). Similarly, Pattern 2 segmented approach: Fig. 4 shows the optimal routes (left) and good five cases of $u(t)$ (right). Case 1 (IDP = 37.53 and LID = 56) is the best route among the five cases.

4.2 Non-grid Streets

As shown in Fig. 5(a), we consider a non-grid map where the number of links and nodes is 40 and 25. The HQ is at node 7, marked by a red dot, with a degree of 4. We have two vehicles that start and end at HQ, each making four times return to HQ, i.e., R = 4. We compared the effectiveness of the whole area approach and two segmentation approaches (Pattern 1 and 2) based on their IDP and LID values, illustrated in Fig. 5(b) and (c).

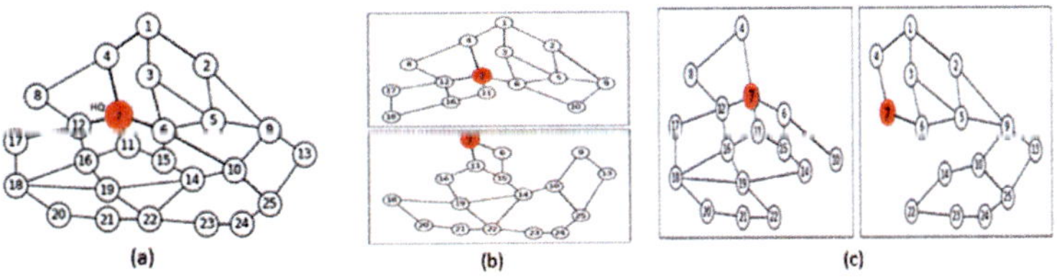

Fig. 5. (a) The whole area, (b) the segmented areas of Pattern 1 and (c) Pattern 2.

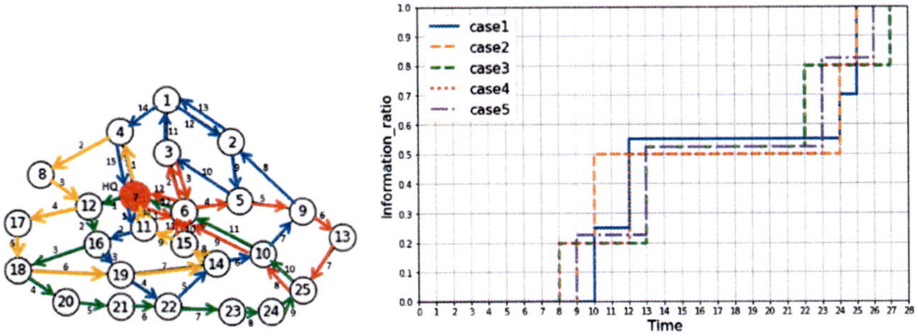

Fig. 6. One of the best routes (left) and the good five cases of $u(t)$ (right) in the whole area approach ($R = 4$, two vehicles).

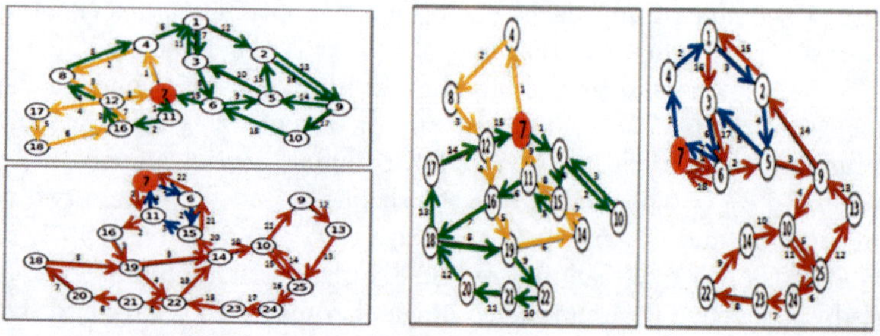

Fig. 7. One of the best routes of the segmented approach Pattern 1 (left) and Pattern 2 (right).

In Fig. 6, the blue line represents Case 1, which indicates the best route with the lowest IDP (=17.2) and LID (=25) for the entire (whole) area approach. On the right side of the figure, the $u(t)$ of the top five cases are shown. Pattern 1 segmentation approach: Fig. 7 (left) displays the best route (Case 1), with the smallest $IDP(= 20.45)$ and $LID = 27$. Figure 7 (right) shows the optimal Pattern 2 segmentation route. In Case 1, $IDP = 17.92$ and $LID = 25$.

4.3 Partially Damaged Condition

We analyzed the decline in information-collecting performance for five successful cases, explicitly examining the resiliency of area segmentation in the presence of damaged streets that caused significant delays. However, vehicles following established routes did not recognize these streets. Seven damaged links are placed in location 1 and location 2 on the grid map, as shown in Fig. 8(a) and 8(d). Figures 9(a) and 9(d) display eight damaged links on location 1 and location 2 of the non-grid. Damages are likely to occur side by side during a disaster due to their proximity. Thus, we assume two groups of damaged links in different parts of the graph.

We also assume two levels of additional delays incurred on damaged links: Level 1 damage and Level 2 damage. Each damaged link increases travel time by 2 and 4 units for Level 1 and Level 2 damage, respectively, for passing vehicles. We analyzed the increase in degradation of IDPs and LIDs for five cases using both the whole area and area segmentation approaches (Patterns 1 and 2) under normal and damaged conditions. Tables 1 and 2 display the results. In each case, the IDP and LID should increase when in a damaged state. The extent of the increase varies based on the situation.

When using the whole area approach, Case 1 IDP increases from 36.59 to 38.8 (Level 1 damage case) and 43.2 (Level 2 damage case) as shown in Table 1, and from 38.82 to 42.92 (Level 1 damage case) and 51.11 (Level 2 damage case), from 37.53 to 40.31 (Level 1 damage case) and 45.86 (Level 2 damage case) in the segmentation approach Patterns 1 and 2 for location 1. Case 1 remains the best among the five cases on both damage cases. Case 2 becomes the best IDP

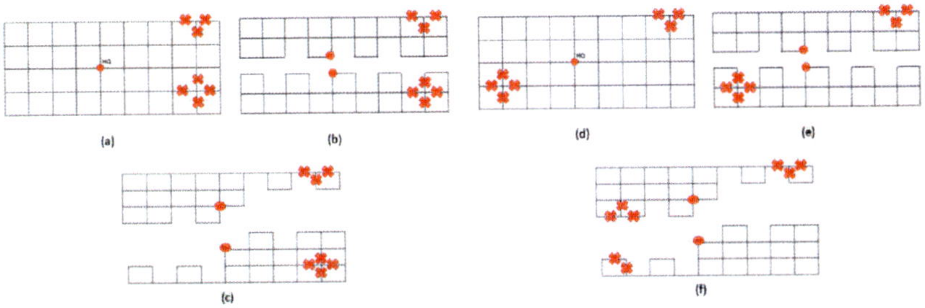

Fig. 8. Two patterns of damaged link locations on the grid map: Location 1 in (a) the whole area, (b) the segmented areas of Pattern 1 and (c) Pattern 2, and Location 2 in (d) the whole area, (e) the segmented areas of Pattern 1 and (f) Pattern 2.

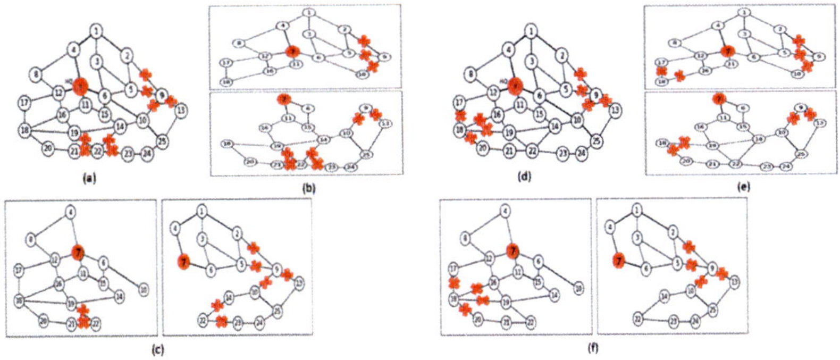

Fig. 9. Two patterns of damaged link locations on the non-grid map: Location 1 in (a) the whole area, (b) the segmented areas of Pattern 1 and (c) Pattern 2, and Location 2 in (d) the whole area, (e) the segmented areas of Pattern 1 and (f) Pattern 2.

in pattern 1 (Level 1 damage case) and pattern 2 (Level 1 and 2 damage cases). Similarly, the best case for location 2 also degraded the IDP's values on both Level 1 and 2 damage cases. In location 2, Case 2 becomes the smallest IDP for the whole area approach (Level 1 and 2 damage cases) and Pattern 1 (Level 2 damage).

In Table 2, the IDP of Case 1 increases from 17.2 to 19.6 (Level 1 damage) and 24.40 (Level 2 damage) for the non-grid whole area approach, from 20.45 to 24.08 (Level 1 damage) and 32.08 (Level 2 damage) for segmented Pattern 1, and from 17.92 to 20.22 (Level 1 damage) and 24.82 (Level 2 damage) for segmented Pattern 2 for damaged location 1. Case 1 remains the best IDP for the whole area approach on both Level 1 and 2 damage cases. Case 2 becomes the smallest IDP for segmented areas of Pattern 1 (Level 2 damage case). Case 3 becomes the smallest IDP for segmented areas of Pattern 2 (Level 2 damage case). Similarly, for the damaged location 2, the IDP of the best case (Case 1) has increased from 17.92 to 20.10 for Level 1 damage and 25.90 for Level 2 damage in the whole area approach. In segmented Pattern 1, it has increased from 20.45 to 24.15 for Level 1 damage and 31.55 for Level 2 damage, and in Pattern 2, it has risen to

Table 1. The impact of damaged links in locations 1 and 2 (with damage levels 1 and 2) on the IDPs and LIDs of the five good cases in each approach for the grid map.

Whole area approach										
	Normal		Damaged Location 1				Damaged Location 2			
			Damaged (level 1)		Damaged (level 2)		Damaged (level 1)		Damaged (level 2)	
Case	IDP	LID	IDP	LID	IDP	LID	IDP	LID	IDP	LID
1	36.59	52	38.8	58	43.2	70	39.86	60	46.40	76
2	36.71	52	40.09	62	46.87	82	39.58	56	45.32	68
3	36.82	54	41.01	64	49.39	84	40.16	58	46.85	68
4	37.11	54	40.26	58	46.56	66	39.69	56	44.80	66
5	37.13	54	40.19	58	46.31	68	40.19	58	46.31	68
Segmented Pattern 1										
	Normal		Damaged Location 1				Damaged Location 2			
			Damaged (level 1)		Damaged (level 2)		Damaged (level 1)		Damaged (level 2)	
Case	IDP	LID	IDP	LID	IDP	LID	IDP	LID	IDP	LID
1	38.82	54	42.92	58	51.11	70	41.29	58	46.24	70
2	38.92	54	42.96	58	51.06	70	41.34	58	46.19	70
3	39.20	54	43.34	58	51.62	70	41.72	58	46.75	70
4	39.46	56	43.51	60	51.60	70	41.88	60	46.73	70
5	39.60	54	43.69	60	51.88	72	42.14	60	47.22	72
Segmented Pattern 2										
	Normal		Damaged Location 1				Damaged Location 2			
			Damaged (level 1)		Damaged (level 2)		Damaged (level 1)		Damaged (level 2)	
Case	IDP	LID	IDP	LID	IDP	LID	IDP	LID	IDP	LID
1	37.53	56	40.31	60	45.86	68	40.05	62	44.99	78
2	37.60	56	39.79	60	44.16	68	40.14	62	45.18	78
3	37.62	56	40.49	60	46.24	68	41.13	66	45.53	78
4	37.69	56	39.98	60	44.54	68	40.4	62	45.72	78
5	37.74	56	40.99	62	47.44	74	41.08	66	45.46	78

20.32 for Level 1 damage and 25.12 for Level 2 damage, from the initial value of 17.92. Case 4 becomes the best IDP for the whole area approach on Level 1 and 2 damage cases. However, Case 1 remains the best IDP for Pattern 1. Case 3 becomes the smallest IDP for Pattern 2 (Level 1 damage case).

About IDP, in general, when a vehicle encounters a damaged road, the impact on IDP increases if it happens earlier in the route and the vehicle is moving slowly. Therefore, the degradation of IDP is highly dependent on the correlation between the challenged areas, vehicle routes, and segmentation patterns. On the other hand, about LID, when a vehicle returns to HQ last in normal condition and meets a damaged link on the way, LID is increased by 3. Therefore, this makes it easier to estimate the degradation of LID. When a vehicle returns to HQ without damage but encounters a damaged road, the LID increases by 3.

Table 2. The impact of damaged links in locations 1 and 2 (with damage levels 1 and 2) on the IDPs and LIDs of the five good cases in each approach for the non-grid.

Whole area approach

	Normal		Damaged Location 1				Damaged Location 2			
			Damaged (level 1)		Damaged (level 2)		Damaged (level 1)		Damaged (level 2)	
Case	IDP	LID	IDP	LID	IDP	LID	IDP	LID	IDP	LID
1	17.20	25	19.60	29	24.40	37	20.10	29	25.90	37
2	17.25	25	19.80	29	24.90	37	20.38	30	26.57	42
3	17.27	27	19.83	31	24.92	39	20.23	31	26.12	39
4	17.30	26	19.90	30	25.10	38	20.30	30	26.30	38
5	17.38	26	19.87	30	24.88	38	19.87	30	24.88	38

Segmented Pattern 1

	Normal		Damaged Location 1				Damaged Location 2			
			Damaged (level 1)		Damaged (level 2)		Damaged (level 1)		Damaged (level 2)	
Case	IDP	LID	IDP	LID	IDP	LID	IDP	LID	IDP	LID
1	20.45	27	24.08	32	32.08	44	24.15	33	31.55	45
2	20.87	27	24.47	32	31.58	44	24.63	33	32.12	45
3	21.27	27	24.88	32	34.23	50	25.02	33	32.52	45
4	21.30	27	25.00	32	32.30	44	25.15	33	32.85	45
5	21.43	27	25.23	32	32.73	44	25.38	33	33.27	45

Segmented Pattern 2

	Normal		Damaged Location 1				Damaged Location 2			
			Damaged (level 1)		Damaged (level 2)		Damaged (level 1)		Damaged (level 2)	
Case	IDP	LID	IDP	LID	IDP	LID	IDP	LID	IDP	LID
1	17.92	25	20.22	31	24.82	43	20.32	29	25.12	37
2	18.00	25	20.75	31	20.25	43	21.70	33	28.10	44
3	18.12	26	20.27	32	24.58	44	20.43	30	25.02	38
4	18.15	25	20.45	31	25.05	43	22.05	30	26.70	37
5	18.20	26	20.8	32	26.00	44	20.85	30	26.15	38

4.4 Discussion

We focus on how the damaged links (Level 2; Locations 1 and 2) affect the performance degradation of the best IDP routes in normal conditions in each of the three routing cases: the whole area approach and the area-segmentation approach (Pattern 1 and Pattern 2). Comparing IDPs and LIDs of those three cases in normal conditions (top) and the damaged states with location 1 (middle) and location 2 (bottom) for both the grid (left) and non-grid (right) maps are shown in Fig. 10.

In normal conditions (top), the whole area approach shows the best performance (IDP and LID), the segmentation Pattern 2 is the next, and Pattern 1 is the worst for both the grid and non-grid. On the other hand, in the damaged cases, the situation changes. In the case of damaged location 1, in terms of LID, using the whole area approach yields results equivalent to Pattern 1 and worse than Pattern 2 for the grid (middle-left). In the case of damaged location 2 for the grid (bottom-left), the whole area approach is almost as good or slightly infe-

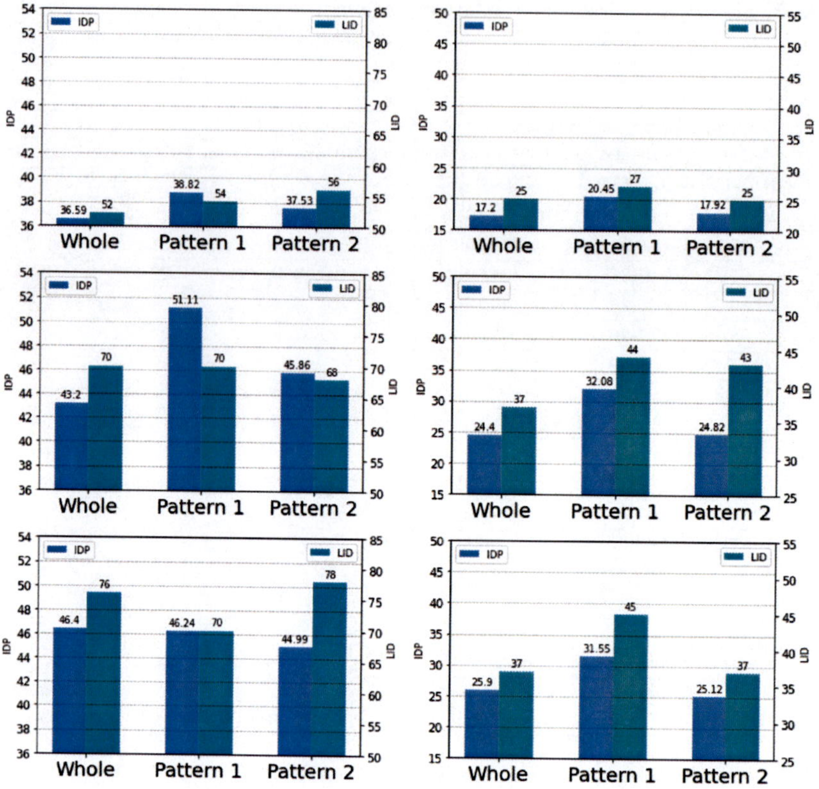

Fig. 10. IDP/LID of good cases by three approaches for grid map (left) and non-grid (right): in the normal condition (top), the damaged conditions of location 1 (middle) and 2 (bottom).

rior to Patterns 1 and 2 in terms of IDP; Pattern 1 can offer the best-balanced performance. In addition, when it comes to the non-grid with the damaged location 2 (bottom-right), the entire (whole) approach is almost as good as Pattern 2 or slightly worse in terms of IDP and LID. Pattern 2 and the entire (whole) approach may yield the same optimum results. It is possible for Pattern 2 and the entire (whole) area approach to have equally optimal performance. In general, such complex situations happen because the degradation of IDP and LID are caused by how the vehicle routes meet the damaged links. Moreover, the number of damaged links to be passed by vehicles may change depending on the segmentation as shown in Figs. 8 and 9. In both cases of damaged locations 1 and 2, the (original) numbers damaged links are 7 and 8 for the grid and non-grid, respectively. However, for the grid, the numbers of damaged links of two sub-areas in Patterns 1 and 2 are the same $(3+4)$ in damaged location 1, while those in Patterns 1 and 2 are $(3+4)$ and $(6+2)$, respectively, in damaged location 2. For the non-grid. those in Patterns 1 and 2 are $(3+6)$ and $(2+6)$, respectively, in damaged location 1, while those are $(5+4)$ and $(4+4)$, respectively, in damaged location 2. Those observations indicate that the area-segmentation approach can have the potential to be more effective and robust than the whole area approach for collect information in disaster situations.

We drill down more details especially in the case of damaged location 2. For the grid, Pattern 2 shows the smallest IDP. This is mainly because, in the first round trip, one vehicle passes the damaged link two times in Pattern 2, whereas, in the whole area approach, a vehicle passes the damaged link four times. Pattern 1 shows a good IDP and the smallest LID. This is mainly because the total damaged link passed by the vehicles in Pattern 1 is less than that in Pattern 2, and also Pattern 1 has no damaged link for the first round trip. For the non-grid, Pattern 2 shows the smallest IDP with a good LID. This is mainly because two vehicles for Pattern 2 do not pass the damaged link in the first round trip and they pass the damaged link 8 times $(4+4)$ in the second round trip. In contrast, for the whole area approach and Pattern 1, both the vehicles pass the damaged links in the first (4 $(2+2)$ times and 2 times) and second round trips (4 $(2+2)$ times and (8 $(2+6)$ times). Passing the damaged links in the first round trip might worse the IDP. However, please note that not passing the damaged links in the first round means the HQ cannot get the information of damaged links in an early stage. Therefore, we should consider that a small IDP is not simply preferable in this case.

Those observations suggest a more resilient segmentation design. If both two vehicles meet the damaged links in their first round trip, no information cannot come to HQ quickly. On the other hand, if multiple damaged links are located in a single sub-area, both LID and IDP are increased significantly. Therefore, if we can estimate the risk of damage of each location with its damage level depending on the disaster situation, such high-risk areas should be evenly distributed among segmented sub-areas.

Finally, we note the effects of the number of patrolling vehicles. In general, it is better to use vehicles more than two to collect information from a large area. While we only showed the results in the case of two vehicles in this paper, our framework and the systematic search scheme for finding a good set of routes of vehicles are applicable to any number of vehicles. In particular, the area-segmentation approach can reduce the computational complexity when the network size and the number of vehicles are large. As the number of vehicles increases, both IDP and LID are decreased (i.e., the performance is improved) because more links can be concurrently monitored by using more vehicles, shortening the time to monitor multiple links. On the other hand, if the number of vehicles is large, the number of duplicated link transits by vehicles is also large (i.e., the efficiency is degraded) because the number of link-disjoint routes on the graph is limited regardless of the number of vehicles which enlarges the time taken to return to HQ. From this trade-off, the performance improvement on IDP and LID cannot be proportional to the number of vehicles and will gradually saturate. Furthermore, maintaining many vehicles will suffer from the costs.

5 Conclusion

We considered a routing problem for vehicles starting from an emergency disaster response headquarters (HQ), monitoring the information along all the streets in a region, and bringing back such information to HQ The study compared the effectiveness of area-segmentation and whole-area approaches in minimizing IDP, representing the average delay time in incrementally collected information. In normal cases, the area-segmentation approach can find only semi-optimal routes because it restricts the search space, assuming no unknown damaged links. In this paper, we have observed that the area-segmentation approach is resilient in cases where there is damage to links causing delays in a vehicle's travel. This approach may reduce the risk of a significant increase in IDP and could be more effective than the whole-area approach.

References

1. Khwairakpam, S., Shibata, M., Tsuru, M.: Vehicle routing in whole and segmented areas to incrementally collect the disaster information. In: Barolli, L. (ed.) BWCCA 2022. LNNS, vol. 570, pp. 236–247. Springer, Cham (2022). https://doi.org/10.1007/978-3-031-20029-8_23
2. Toth, P., Vigo, D. (ed.): The Vehicle Routing Problem. SIAM Discrete Mathematics and Applications, Philadelphia (2002)
3. Hsueh, C.-F., Chen, H.-K., Chou, H.-W.: Dynamic Vehicle Routing for Relief Logisticsin Natural Disasters. In: Vehicle Routing Problem. IntechOpen (2008)
4. Grotschel, M., Yuan, Y.-X,.: Euler-Mei-Ko Kwan, Konigsberg, and a Chinese Postman. Documenta Mathemetica: Optim. Stories, pp. 43–50 (2012)
5. Maki, Y., Mu, W., Shibata, M., Tsuru, M.: Vehicle routing for incremental collection of disaster information along streets. In: Hara, T., Yamaguchi, H. (eds.) MobiQuitous 2021. Lecture Notes of the Institute for Computer Sciences, vol. 419, pp. 487–492. Springer, Cham (2021). https://doi.org/10.1007/978-3-030-94822-1_28
6. Khwairakpam, S., Shibata, M., Tsuru, M.: Vehicle routing to minimize the average delay for collecting the disaster damage information. In: Proceedings of the IEEE WF-IoT 2022 (2022)
7. Anuar, W.K., Moll, M., Lee, L., Pickl, S., Seow, H.: Vehicle Routing optimization for humanitarian logistics in disaster recovery: a survey. In: Proceedings of the 2019 International Conference on Security and Management (SAM 2019), pp. 161–167 (2019)
8. Farahani, R.Z., Lotfi, M., Baghaian, A., Ruiz, R., Rezapour, S.: Mass casualty management in disaster scene: a systematic review of OR&MS research in humanitarian operations. Eur. J. Oper. Res. **287**, 787–819 (2020)
9. Bruni, M.E., et al.: A fast heuristic for routing in post-disaster humanitarian relief logistics. ScienceDirect, Trans. Res. Procedia **30**, 304–313 (2018)

The Probability of Encounters of Nomadic Lévy Walk on Unit Disk Graphs

Kazuma Matsubara[1] and Naohiro Hayashibara[2](✉)

[1] Graduate School of Frontier Informatics, Kyoto Sangyo University, Kyoto, Japan
i2386120@cc.kyoto-su.ac.jp
[2] Faculty of Information Science and Engineering, Kyoto Sangyo University, Kyoto, Japan
naohaya@cc.kyoto-su.ac.jp

Abstract. Random walks, including the Nomadic Lévy Walk, play a crucial role in various computer science domains such as networking, distributed systems, and optimization. The Nomadic Lévy Walk, a variant of the Homesick Lévy Walk, holds promise as a potential candidate for message ferry routing schemes. This approach combines the homing behavior characteristic of the Homesick Lévy Walk with the base relocation strategy employed in Lévy walks. In Delay-Tolerant Network (DTN) routing, encounter probability is a critical factor. In this paper, we present simulation results that highlight the impact of base relocation in the Nomadic Lévy Walk on encounter probability. Additionally, we conduct thorough comparisons of encounter probabilities using unit-disk graphs with varying diameters to further investigate the efficacy of the Nomadic Lévy Walk approach.

1 Introduction

Random walks have been employed as a modeling tool for human behavior in the context of developing protocols and assessing their efficacy within mobile ad-hoc networks, as evidenced by existing studies [5,6,20]. Furthermore, random walks have proven to be valuable in the pursuit of information retrieval and aggregation within peer-to-peer (P2P) networks, as well as in tackling large-scale global optimization problems [22,23].

Recently, the Lévy walk has been gaining attention for its ability to optimize search in animal foraging [7,21] and its statistical similarity to human mobility [13]. This mathematical fractal pattern consists of long segments followed by shorter hops in random directions, based on the Lévy distribution [11]. It has also been observed as a naturally selected strategy for animals searching for sparse targets to survive [7]. In fact, mobile robot simulations have shown the effectiveness of this strategy in target exploration [10].

In the areas of computer science, Lévy walk has been used to improve the spatial coverage of mobile sensor networks [8,19], to analyze the properties of an evolving network topology of a mobile network [3], and to enhance grayscale images combined with other bio-inspired algorithms (i.e., bat and firefly algorithms) [4]. It is also considered to be particularly useful for data collection and finding outliers of sensor values to detect incidents in a large-scale delay-tolerant sensor network.

Despite many articles on Lévy walk, they evaluated it on continuous fields, and only a few results on graphs have been available. Shinki et al. showed Lévy walk pattern is

L. Barolli (Ed.): BWCCA 2023, LNDECT 186, pp. 15–25, 2024.
https://doi.org/10.1007/978-3-031-46784-4_2

an effective way for detecting a target on a unit disk graph and found that its efficiency emphasizes by increasing the distance between the target and the initial position of the searcher [15].

Homesick Lévy Walk (HLW) has been proposed as a variant of Lévy Walk [8]. Each agent starts from the base position and returns to it with the given probability of α in addition to the Lévy Walk movement pattern. This walk model statistically resembles human mobility traces regarding human encounters.

Nomadic Lévy Walk (NLW) [18] that is a variant of HLW, is proposed to improve search efficiency with homing behavior. Each agent changes its base position according to the given strategy with the probability γ. This strategy is called the *base relocation strategy* (it is also called the sink relocation strategy in the context of Wireless Sensor Networks).

The agent based on NLW always starts from the base and goes back to it. Thus, base relocation helps the agent to explore a graph widely. It means that the strategy for base relocation is essential for improving search efficiency. This movement model would be useful for autonomous electric vehicles with mobile charging stations in cyber-physical systems [1].

The probability of encounters is one of the important factor for a routing scheme to improve the message reachability with opportunistic communication. Although we've proven the efficiency of NLW for randomized routing of message ferry in Delay-Tolerant Networks (DTNs) and Wireless Sensor Networks (WSNs) by numerous simulation results, we did not evaluate the base relocation strategy of NLW regarding the probability of encounters. The base relocation of NLW is a notion to spread the area to be explored by each agent. It is obvious that NLW is efficient to explore a field wider than HLW. However, there is no quantitative discussion based on the experimental/simulation result.

Contribution

This paper aims to assess the influence of the base relocation strategy in Nomadic Lévy Walk (NLW) on encounter probabilities within unit-disk graphs. We conduct measurements to evaluate the impact of NLW's base relocation strategy and compare it with both NLW and Homesick Lévy Walk (HLW) in terms of encounter probabilities. Through quantitative analysis of simulation results, we discuss the implications of the base relocation strategy on encounter probabilities and provide insights into its effectiveness.

2 Related Work

We now introduce several research works related to Lévy walk.

Birand et al. proposed the Truncated Levy Walk (TLW) model based on real human traces [3]. The model gives heavy-tailed characteristics of human movement. Authors analyzed the properties of the graph evolution under the TLW mobility model.

Valler et al. analyzed the impact of mobility models, including Lévy walk on the spread of the epidemic in MANET [20]. They adopted the scaling parameter $\lambda = 2.0$ in the Lévy walk mobility model. From the simulation result, they found that the impact of velocity of mobile nodes does not affect the spread of virus infection.

Thejaswini et al. proposed the sampling algorithm for mobile phone sensing based on Lévy walk mobility model [19]. The authors showed that the proposed algorithm performs significantly better than the existing method in terms of energy consumption and spatial coverage.

Fujihara et al. proposed a variant of Lévy flight which is called Homesick Lévy Walk (HLW) [8]. In this mobility model, agents return to the starting point with a homesick probability after arriving at the destination determined by the power-law step length. As a result, the frequency of agent encounter obeys the power-law distribution through random walks, and Lévy walk do not obey it.

Bartumeus et al. and James et al. clarified the probability of encounter of mutual search with multiple searchers and targets by the movement pattern, velocity, and density of targets in a field [2,9]. According to their simulation result, the ballistic behavior is efficient for searching sparse targets.

Mizumoto et al. measured the encounter probability of Lévy walk and Random walk regarding mating of sexual dimorphism (e.g., male and female) on the one-dimensional and the two-dimensional periodic space with limited lifespans [12]. This work clarified that the movement pattern influences the probability of mating success. The mating of males and females is also a kind of mutual search.

The current results of the mutual search by Lévy walk and Random walk show that the ballistic behavior contributes to the improvement of the encounter probability in two-dimensional continuous plane [2,9,12].

Most of the works related to Lévy walk assume a continuous plane, and hardly any results on graphs are available. Shinki et al. defined the algorithm of Lévy walk on unit disk graphs [15]. They also found that the search capability of Lévy walk emphasizes increasing the distance between the target and the initial position of the searcher. It is also efficient if the average degree of a graph is small [14].

3 System Model

We assume a unit disk graph $UDG = (V, E)$ with constant radius r. Each node $v \in V$ is located in the Euclidean plane. An undirected edge (link) $\{v_i, v_j\} \in E$ between two nodes v_i and v_j exists if and only if the distance of these nodes is less than $2r$. The nodes in the network represent landmarks, such as shelters in disasters. Note that r is the Euclidean distance in the plane. We assume that any pair of nodes in the graph has a path (i.e., connected graph). It does not mean network connectivity. It just means that there is a geometric route to any node in the field. UDGs are often used for modeling road networks. We assume the field restricts the movement of mobile entities by links.

We also assume a mobile entity called an *agent*. Each of them belongs to one of two groups. This assumption is similar to the existing work [12]. Agents can move to the neighboring node from the current node on a graph. Each agent knows its own position (e.g., obtained by GPS), and it has a compass to get the direction. On the other hand, agents have no prior knowledge of the environment (e.g., the link density, etc.) and the positions of other agents.

4 Nomadic Lévy Walk on Unit Disk Graphs

The Lévy walk has garnered significant recognition as a highly effective technique for exploring sparsely distributed targets and facilitating message dissemination on unit disk graphs, as evidenced by our previous research [14,15]. Empirical evidence presented by Sugihara and Hayashibara [17] further supports the notion that the Homesick L'evy walk is capable of exploring unit disk graphs, albeit at the expense of the wide-area search characteristic inherent to the Lévy walk. Nevertheless, the homing behavior exhibited by the Lévy walk continues to hold value and serves as a valuable model for understanding animal and human behavior in domains such as bio-logging and Mobile Ad hoc Networks (MANETs). Moreover, its applicability extends to exploration scenarios in disaster-stricken areas, where cautious navigation and prioritizing safety are of paramount importance.

4.1 Definition

We now define a random walk called Nomadic Lévy walk, which is a variant of Homesick Lévy walk [8], to improve the ability of the broad area search while preserving the homing behavior. It holds the following properties in addition to Homesick Lévy walk.

Nomadicity: Each agent moves the position of its base (e.g., sink) with the given probability γ.

Base movement strategy: The trajectory of base positions is like Lévy walk. The next position of the base is decided by a particular strategy, for example, to cover the untracked area based on the information exchanged with other agents.

Figure 1 illustrates the trajectory of two types of random walks, namely the Nomadic Lévy walk and the Homesick Lévy walk, within a two-dimensional plane. The key distinction between these two walks lies in their relationship to a fixed base. In the case of the Homesick Lévy walk, the agent moves radially away from the base, adhering to a certain level of attachment. Conversely, the agent following the Nomadic Lévy walk explores a wider expanse of the field in comparison to the Homesick Lévy walk, indicating a lesser degree of attachment to a specific location.

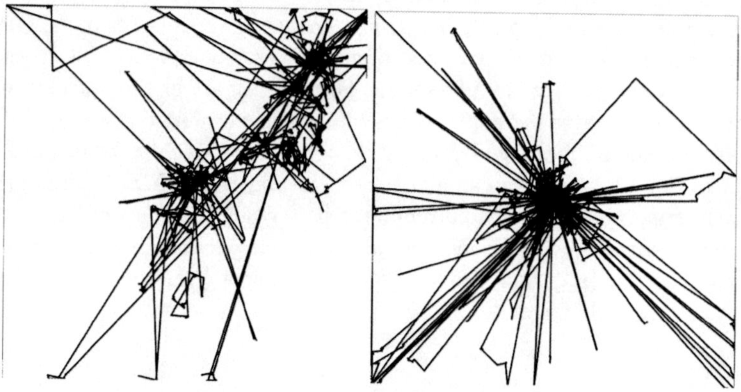

Fig. 1. The trajectory of NLW (left) and HLW (right) in a two-dimensional plane.

All agents move along Homesick Lévy walk. Thus, their trajectory is radially from their base. Moreover, they move their base at the probability γ in Nomadic Lévy walk. In fact, the fixed base restricts the area that each agent explores. The property of Nomadicity is expected to improve coverage by each agent.

4.2 The Base Relocation Strategy

NLW has two different trajectories on agents and their base. First, each agent set the base position as an initial position and then it changes the base position with the given probability γ. Thus, the path of an agent includes that of its base.

The coverage of a graph by agents depends on the base relocation strategy of the base position. We focus on the following two base relocation strategy of NLW in this paper and clarify the impact of those strategies on the efficiency of message dissemination.

4.2.1 Lévy Walk (LW) Strategy The next position of the base is changed obeying the Lévy Walk movement pattern. The orientation o of the next base is selected at random from $[0,2\pi)$ and the distance d of it from the current position is decided by the power-law distribution.

4.2.2 Reverse Prevention (RP) Strategy In this strategy, each agent assumes a set of the history of the base positions B_{hist} with the size h to store the coordinates of the nodes' position at which its base has been located in the past. It calculates the reverse orientation of each position in the set. For each position $x \in B_{hist}$, the line $L_{\perp}$ that is orthogonal to the line between x and the current base position b is computed. It also indicates the range of the orientation of the next base. The intersection of the ranges $\bigcap_{x \in B_{hist}} L_x$ is defined as θ_{cand}. The orientation of the next base is determined randomly from θ_{cand}. The procedure of the RP and LW strategies is shown in Algorithm 1.

4.2.3 Clustering-Based Reverse Prevention Strategy (CRPS) The sink positions could be biased at a particular area of the graph while relocating. As a result, the cover ratio could not be improved with MFs. For this reason, we use the unweighted pair group me thod with arithmetic mean (UPGMA) clustering method [16] to detect a bias of sink positions.

UPGMA method is an agglomerative hierarchical clustering method based on the average distance of members in two clusters. It calculates a pairwise distance $pdist(A, B)$ between two clusters A and B. $pdist(A, B)$ is computed as follows.

$$pdist(A, B) = \frac{1}{|A| \cdot |B|} \sum_{a \in A} \sum_{b \in B} d(a, b) \tag{1}$$

$d(a, b)$ is the distance between the element a and b. In each step, a cluster A merges another cluster B so that $pdist(A, B)$ is the minimum value. Finally, all elements are in one cluster, and the merging process can be represented as a binary tree.

First, each sink constructs a matrix of $pdist(\{x\}, \{y\})$, where $x, y \in H$. $pdist(\{x\}, \{y\})$ is calculated by the Eq. 1. Note that the distance $d(x, y)$ is Euclidean distance. Then, two clusters $\{x\}$ and $\{y\}$ merge into one if $pdist(\{x\}, \{y\})$ is the minimum value and $pdist(\{x\}, \{y\}) \leq T$. This merging phase is repeated until there is no candidate to merge. As a result, the sink determines the past sink positions are biased if the size of the biggest cluster is more significant than $\lceil \frac{|H|}{2} \rceil$.

In the case of detecting a bias of sink positions, the next sink s will be relocated with the following condition, where $C \subseteq H$ is the biggest cluster.

$$pdist(C, \{s\}) > T$$

Algorithm 1. baseRelocation($strat$, B_{hist})

Initialize:
$b \leftarrow$ the position of the base
$B'_{hist} \leftarrow B_{hist} \setminus \{b\}$
$\theta_{cand} \leftarrow [0, 2\pi)$
if $strat$ = 'LW' **then** ▷ For the LW strategy.
  $return$ the orientation selected randomly from $[0, 2\pi)$
else if $strat$ = 'RP' **then** ▷ For the RP strategy.
  for $x \in B'_{hist}$ **do**
    $\theta_{cand} \leftarrow \theta_{cand} \cap L_x$ ▷ L_x is the angle π formed by b and x.
  end for
  if $\theta_{cand} \neq \emptyset$ **then**
    $return$ the orientation selected randomly from θ_{cand}
  else
    $return$ the orientation selected randomly from $[0, 2\pi)$
  end if
end if

4.3 Algorithm on Unit Disk Graphs

We now explain the detail of the algorithm of Nomadic Lévy walk on unit disk graphs in Algorithm 2.

At the begging of the algorithm, each agent holds the initial position as the base position b. In every walk, each agent determines the step length d by the power-law distribution and selects the orientation o of a walk randomly from $[0, 2\pi)$. It can obtain a set of neighbors $N(c)$ and a set of possible neighbors $PN(c) \subseteq N(c)$, to which agents can move, from the current node c. In other words, a node $x \in PN(c)$ has a link with c that the angle θ_{ox} between o and the link is smaller than $\frac{\pi}{2}$.

In unit disk graphs, it is not always true that there exist links to the designated orientation. We introduce δ that is a permissible error against the orientation. In this paper, we set $\delta = 90$. It means that agents can select links to move in the range ± 90 with the orientation o as a center.

Algorithm 2. A walk using Nomadic Lévy Walk on Unit Disk Graphs

```
1: Initialize:
       b ← the position of the base
       c ← the current position
       o ← 0                                    ▷ orientation for a walk.
       B_hist.initialize(h)   ▷ The sliding window on base positions is initialized
       with the size i.
       strat ∈ {LW, RP}                         ▷ The given strategy of NLW.
       PN(c) ← the possible neighbors to move.
2: if Probability: α then
3:     d ← the distance to b
4:     o ← the orientation of b
5: else
6:     d is determined by the power-law distribution
7:     o is randomly chosen from [0, 2π)
8: end if
9: if Probability: γ then
10:    move to b                                ▷ going back to the base position.
11:    B_hist.enqueue(b)                        ▷ add the base position into the sliding window.
12:    d is determined by the power-law distribution
13:    o ← baseReposition(strat, B_hist)        ▷ change the base position.
14: end if
15: while d > 0 do
16:    PN(c) ← {x|abs(θ_ox) < δ, x ∈ N(c)}
17:    if PN(c) ≠ ∅ then
18:        d ← d − 1
19:        move to v ∈ PN(c) where v has the minimum abs(θ_ov)
20:        c ← v
21:    else
22:        break                                ▷ no possible node to move.
23:    end if
24: end while
```

In a given probability α, the agent goes back to the base node (line 2 in Algorithm 2). In this case, it sets the orientation to the base node as o and the distance to the base as d.

Each agent changes its base position with the given probability γ (line 10 in Algorithm 2). The next position of the base will be determined by the given strategy stated in Sect. 4.2.

5 Simulation

This simulation aims to elucidate the effects of the base relocating of NLW on the likelihood of encounters. Additionally, we provide a quantitative analysis of the impact using simulation results. We employ HLW as a reference point for comparison with NLW and investigate the base relocation's effects on encounters. The simulation in this work is performed using a discrete event simulator developed in Java. This simulator is specifically designed to execute simulations of Lévy walks on unit-disk graphs, utilizing the GraphStream library.

5.1 Simulation Setup

We establish the scaling parameter λ as 1.2 for both NLW and HLW since it has been demonstrated to be effective in facilitating extensive graph-based searches [14]. Likewise, the homesick probability α is uniformly applied to both NLW and HLW, with a fixed value of $\alpha = 0.2$. Furthermore, we specify the configuration of γ, which represents the probability of NLW base relocation, as $\gamma = 0.2$. We use the Lévy walk (LW) strategy for the base relocation of NLW. It means that each agent moves based on HLW and its base moves based on Lévy walk.

For both NLW and HLW, we configure the number of agents $n \in \{10, 25\}$ and the simulation time $t \in \{200, 800\}$.

We created several unit-disk graphs as the environment of the simulation. We show the average degree of nodes, diameter and density of links of unit disk graphs in the Table 1.

Table 1. Properties of the unit-disk graphs that are used in the simulation.

Avg. diamater	40.4	30.5	24.7	20.8	18.1	16.1
Avg. degree	7.4	10.7	14.5	18.7	23.5	28.7
Avg. density	0.007	0.010	0.014	0.018	0.023	0.028

The mobility of individual agents within the system is constrained by the connections established by the underlying graph structure. As illustrated in Table 1, there exists a correlation between the average diameter of the graph and the degree of freedom in agent movement. Specifically, an increase in the average diameter leads to a decrease in the available degrees of freedom. This decrease in degrees of freedom can be attributed to the inverse relationship between the graph's diameter and its link density. Consequently, a reduction in link density imposes greater restrictions on the movement capabilities of each agent.

5.2 Simulation Result

We show the simulation results of NLW and HLW regarding the probability of encounters in unit-disk graphs (see Fig. 2 to 3). The probability of encounters is proportional to the average degree of the graph. It means that agents can meet or pass by another one easily if the links in a graph is dense.

According to the results, we have confirmed that NLW is 10 to 30% higher than HLW regarding the probability of encounters with $t = 800$ though they don't have much difference at the beginning of the simulation. The difference between them increases by time elapsed. This is because the base relocation of NLW gradually effects the encounters of agents.

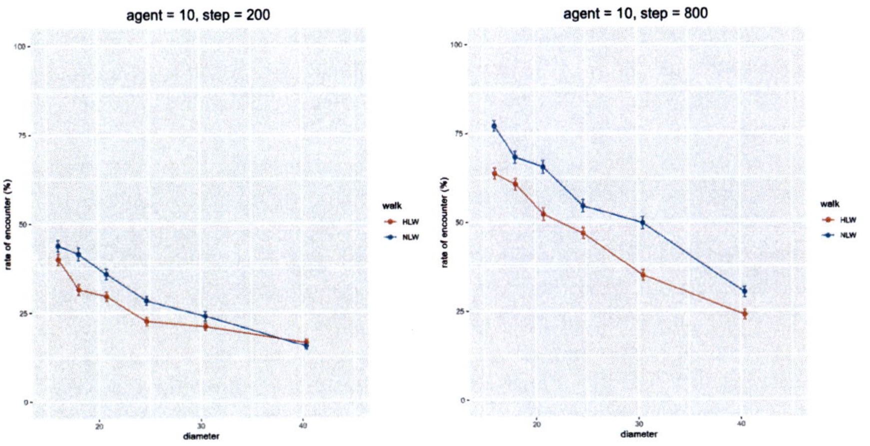

Fig. 2. The result of NLW and HLW regarding the probability of encounters and the diameter of the graph with $n = 10$ and $t \in \{200, 800\}$.

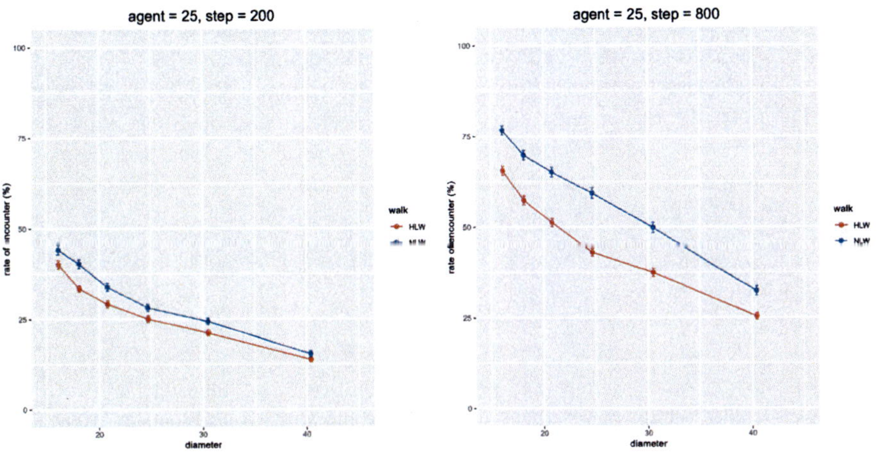

Fig. 3. The result of NLW and HLW regarding the probability of encounters the diameter of the graph with $n = 25$ and $t \in \{200, 800\}$.

To determine whether the result of NLW is statistically significant compared to HLW, we performed a statistical test. Due to the simulation results not following a normal distribution, we utilized the Wilcoxon rank-sum test with a significance level of $p = 0.01$. Our findings indicated that, the probability of encounters was significantly better for NLW than HLW except the result with the diameter = 40.4.

6 Conclusion

To investigate the impact of base relocation in Nomadic Lévy walk (NLW) on encounter probabilities, we conducted simulations and performed a quantitative comparison between the encounter probabilities of NLW and a baseline scenario represented by Homesick Lévy walk (HLW). The purpose of this examination was to assess and measure the extent to which NLW's base relocation affects the likelihood of encounters among agents.

The effect of base relocation on the encounters of agents is twofold. First, by relocating the base, the reachable area for agents expands, allowing them to explore a broader range of the field. As a result, agents following NLW have a higher probability of encountering each other compared to those following Homesick L'evy walk (HLW), where the base remains fixed. This increased encounter probability is particularly evident at a specific time point, such as $t = 800$.

Second, as time elapses and the base continues to relocate, the encounters between agents become increasingly influenced by the changing positions of the base. This means that the encounters are not only dependent on the agents' individual movement patterns but also on the evolving position of the base. Consequently, the difference in encounter probabilities between NLW and HLW tends to grow over time due to the cumulative impact of base relocation on the agents' movement trajectories.

In our simulation, we elucidated the influence of base relocation in Nomadic L'evy walk (NLW) on unit-disk graphs with varying diameters (representing different link densities). Our objective was to gain a deeper understanding of how NLW's base relocation affects the encounter probabilities within these graphs.

Furthermore, we are currently engaged in the exploration of base relocation strategies specifically designed to optimize encounter probabilities. Our focus lies in devising effective strategies that can maximize encounter probabilities, even in scenarios where agents or links are sparsely distributed.

References

1. Afshar, S., Macedo, P., Mohamed, F., Disfani, V.: Mobile charging stations for electric vehicles - a review. Renew. Sustain. Energy Rev. **152**, 111654 (2021). https://doi.org/10.1016/j.rser.2021.111654. https://www.sciencedirect.com/science/article/pii/S1364032121009291
2. Bartumeus, F., Catalan, J., Fulco, U.L., Lyra, M.L.: Optimizing the encounter rate in biological interactions: Lévy versus Brownian strategies. Phys. Rev. Lett. **88**(9), 097901 (2002)
3. Birand, B., Zafer, M., Zussman, G., Lee, K.W.: Dynamic graph properties of mobile networks under levy walk mobility. In: Proceedings of the 2011 IEEE Eighth International Conference on Mobile Ad-Hoc and Sensor Systems, MASS 2011, pp. 292–301. IEEE Computer Society (2011)
4. Dhal, K.G., Quraishi, M.I., Das, S.: A chaotic Lévy flight approach in bat and firefly algorithm for gray level image. Int. J. Image Graph. Sig. Process. **7**, 69–76 (2015)
5. Dolev, S., Schiller, E., Welch, J.L.: Random walk for self-stabilizing group communication in ad hoc networks. IEEE Trans. Mob. Comput. **5**(7), 893–905 (2006). https://doi.org/10.1109/TMC.2006.104
6. Draief, M., Ganesh, A.: A random walk model for infection on graphs: spread of epidemics & rumours with mobile agents. Discrete Event Dyn. Syst. **21**(1), 41–61 (2011)
7. Edwards, A.M., et al.: Revisiting Lévy flight search patterns of wandering albatrosses, bumblebees and deer. Nature **449**, 1044–1048 (2007)

8. Fujihara, A., Miwa, H.: Homesick Lévy walk and optimal forwarding criterion of utility-based routing under sequential encounters. In: Bessis, N., Xhafa, F., Varvarigou, D., Hill, R., Li, M. (eds.) Internet of Things and Inter-cooperative Computational Technologies for Collective Intelligence. Studies in Computational Intelligence, vol. 460, pp. 207–231. Springer, Heidelberg (2013). https://doi.org/10.1007/978-3-642-34952-2_9
9. James, A., Plank, M.J., Brown, R.: Optimizing the encounter rate in biological interactions: Ballistic versus Lévy versus Brownian strategies. Phys. Rev. **78**(5), 051128 (2008)
10. Katada, Y., Nishiguchi, A., Moriwaki, K., Watanabe, R.: Swarm robotic network using levy flight in target detection problem. In: Proceedings of The First International Symposium on Swarm Behavior and Bio-Inspired Robotics (SWARM 2015), pp. 310–315 (2015)
11. Lévy, P.: Théorie de L'addition des Variables Aléatoires. Gauthier-Villars (1937)
12. Mizumoto, N., Abe, M.S., Dobata, S.: Optimizing mating encounters by sexually dimorphic movements. J. R. Soc. Interface **14**(130), 1044–1048 (2017)
13. Rhee, I., Shin, M., Hong, S., Lee, K., Kim, S.J., Chong, S.: On the levy-walk nature of human mobility. IEEE/ACM Trans. Netw. **19**(3), 630–643 (2011)
14. Shinki, K., Hayashibara, N.: Resource exploration using Lévy walk on unit disk graphs. In: the 32nd IEEE International Conference on Advanced Information Networking and Applications (AINA-2018) (2018)
15. Shinki, K., Nishida, M., Hayashibara, N.: Message dissemination using Lévy flight on unit disk graphs. In: The 31st IEEE International Conference on Advanced Information Networking and Applications (AINA 2017) (2017)
16. Sokal, R.R., Michener, C.D.: A statistical method for evaluating systematic relationships. Univ. Kansas Sci. Bull. **38**, 1409–1438 (1958)
17. Sugihara, K., Hayashibara, N.: Collecting data in sensor networks using homesick Lévy walk. In: The 11th International Workshop on Advanced Distributed and Parallel Network Applications (ADPNA-2017), pp. 779–786 (2017)
18. Sugihara, K., Hayashibara, N.: Performance evaluation of nomadic Lévy walk on unit disk graphs using hierarchical clustering. In: Barolli, L., Amato, F., Moscato, F., Enokido, T., Takizawa, M. (eds.) AINA 2020. AISC, vol. 1151, pp. 512–522. Springer, Cham (2020). https://doi.org/10.1007/978-3-030-44041-1_46
19. Thejaswini, M., Rajalakshmi, P., Desai, U.B.: Novel sampling algorithm for human mobility-based mobile phone sensing. IEEE Internet Things J. **2**(3), 210–220 (2015)
20. Valler, N.C., Prakash, B.A., Tong, H., Faloutsos, M., Faloutsos, C.: Epidemic spread in mobile ad hoc networks: determining the tipping point. In: Domingo-Pascual, J., Manzoni, P., Palazzo, S., Pont, A., Scoglio, C. (eds.) NETWORKING 2011. LNCS, vol. 6640, pp. 266–280. Springer, Heidelberg (2011). https://doi.org/10.1007/978-3-642-20757-0_21
21. Viswanathan, G.M., Afanasyev, V., Buldyrev, S.V., Murphy, E.J., Prince, P.A., Stanley, H.E.: Lévy flight search patterns of wandering albatrosses. Nature **381**, 413–415 (1996)
22. Yang, X.S.: Cuckoo search via lévy flights. In: Proceedings of World Congress on Nature & Biologically Inspired Computing (NaBIC 2009), pp. 210–214 (2009)
23. Yang, X.S.: Firefly algorithm, Lévy flights and global optimization. In: Bramer, M., Ellis, R., Petridis, M. (eds.) Research and Development in Intelligent Systems XXVI, pp. 209–218. Springer, London (2010). https://doi.org/10.1007/978-1-84882-983-1_15

Investigating SIC Performance Using Sequential Power Allocation for Downlink NOMA

Hurianti Vidyaningtyas[1,2](✉), Iskandar[1], Hendrawan[1], Aloysius Adya Pramudita[2], and Desti Madya Saputri[1,2]

[1] School of Electrical Engineering and Informatics, ITB, Bandung, Indonesia
{33220014,33220020}@mahasiswa.itb.ac.id, {iskandar,hend}@itb.ac.id

[2] School of Electrical Engineering, Telkom University, Bandung, Indonesia
pramuditaadya@telkomuniversity.ac.id

Abstract. The performance of successive interference cancellation (SIC) employing Sequential Power Allocation (SePA) on the NOMA system downlink is discussed in this study. The simulation scenario is based on the variation of the SePA constant value and the number of users ranging from users who are far from the BS to users who are close to the BS. Bit error rate (BER) performance will be investigated using SePA constants and different number of users. Users who are farthest from the BS can directly decode their own signals, while users who are the second farthest to the closest to the BS must perform SIC first. In addition, the performance of the SePA algorithm is compared to Static Power Allocation (SPA) using QPSK modulation and Rayleigh fading channels. The simulation results show that the greater the SePA constant used, the greater the power transmit required to reach BER 10^{-3} will be even greater.

1 Introduction

NOMA is one of the solutions proposed in the framework of 5G and beyond as a solution for multiple users sharing the same time-frequency resources in a wireless communication system. As 5G network needs increase, NOMA offers a viable answer to the issues of increasing data rates, huge connection, ultra-low latency, and dependable communication [1–3].

One of the important parameters in NOMA is the division of Power allocation for each user. Channel conditions are assumed to be perfect so that the user's distance to the BS is also interpreted as the channel gain condition of the user itself. NOMA with 2 users has better total throughput performance than OMA [4]. In research [5], the performance of NOMA with more than two users was investigated. Based on the simulation results, NOMA sumrate is higher than OMA for a certain number of users. SIC performance on the receiver in the form of a comparison of the bit error rate (BER) to Eb/No for the fading channel is also investigated [6]. However, the power allocation used in this study has a comparison of values that are close to each other.

However, most of the studies on power allocation at NOMA have not analyzed the performance of the SIC against more than 2 users. The performance of the SIC will be

L. Barolli (Ed.): BWCCA 2023, LNDECT 186, pp. 26–34, 2024.
https://doi.org/10.1007/978-3-031-46784-4_3

seen through the results of the bit error rate (BER) on the transmit power of the BS. This paper investigates the performance of SIC using Sequential Power Allocation on downlink direction NOMA with Rayleigh channel. The contributions of this paper are as follows.

- Investigation of SIC performance on NOMA was carried out on a single cell using Rayleigh fading channels and QPSK modulation. The power allocation ratio uses Sequential Power allocation (SePA) which divides the power ratio based on the user's distance from the BS.
- SIC performance simulation on NOMA in the receiver is carried out to analyze Bit error rate (BER) performance compared to several SePA constants with the same number of users. In addition, BER analysis was also carried out for more than 2 users.
- Comparing SIC performance using SePA and SPA for the same number of users.

This paper consists of 4 sections. The first section is an introduction related to previous NOMA research. Sect. 2 focuses on discussing the system model of downlink Non-Orthogonal Multiple Access (NOMA), superposition coding (SC), Rayleigh fading channel, successive interference cancellation (SIC), and the decoder. The purpose of this section is to provide a comprehensive understanding of the proposed system's theoretical framework and key components. Section 3, SIC performance on NOMA using sePA on Rayleigh channel will be compared for different constants. In Sect. 4 contains the conclusions of the research that has been done.

2 System Model

As demonstrated in (Fig. 1), the NOMA system downlinks a single cell with one user BS and M. The user who is farthest away from the BS is referred to as UE_1, the user who is second farthest is referred to as UE_2, and the user who is closest to the BS is referred to as UE_M. The power allocation delivered to the user is determined on the user's distance from the BS under ideal channel conditions. The user who is the furthest away is given the most authority, while the user who is closest is given the least.

In Fig. 1, the power allocation for each user given by the BS is given a lighter color. The distribution of user power allocation uses Sequential power allocation (SePA) [5] based on distance and channel gain conditions $|h_1|^2 < |h_2|^2 < |h_3|^2 < \ldots < |h_M|^2$. Parameter γ is the initial SePA constant and $\beta = (1\text{-}\gamma)$. The SePA equation is as follows

$$P_m = \begin{cases} \gamma(\beta)^{m-1} P_t, m < M \\ P_{m-1}(\beta/\gamma) P_t, m = M \end{cases} \tag{1}$$

where M is the number of users. P_t is the total transmit power of the BS and P_m is the ratio of power allocation for the m-*th* user. There are 2 conditions for SePA to work optimally with the following conditions [7]

$$C1 : 1 = \sum_{m=1}^{M} P_m \tag{2}$$

$$C2 : P_m \geq P_{m+1} + P_{m+2} + \cdots + P_M \tag{3}$$

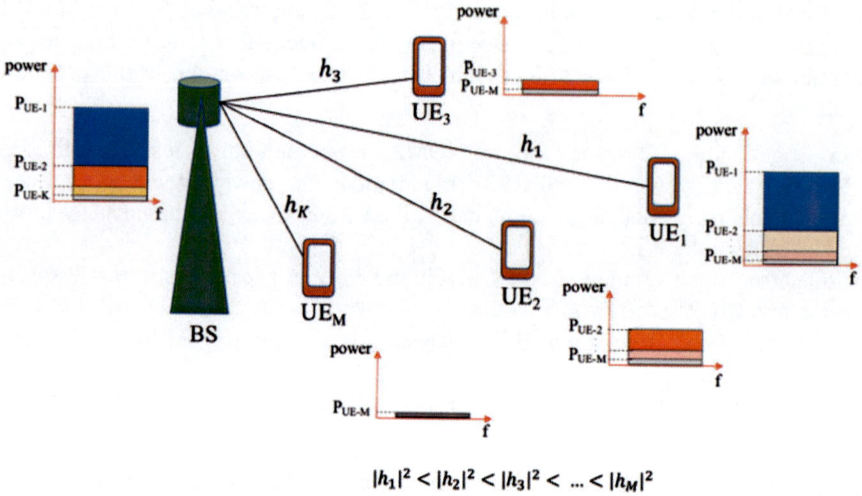

Fig. 1. NOMA Downlink system with M users

The process of merging user signals on the transmitter uses superposition coding (SC). The superposition technique allows the signals of different users to be stacked on top of each other, forming a composite signal. The power levels assigned to each user determine the signal's amplitude within the composite signal. Signal for each user can be written as follows

$$x_m = \sqrt{P_m} S_m \tag{4}$$

where P_m is the power allocated to each UE_M and S_m is the signal from the m-th user. The total signal after Superposition coding can be written

$$x = \sqrt{P_1} S_1 + \sqrt{P_2} S_2 + \ldots + \sqrt{P_M} S_M \tag{5}$$

$$x = \sum\nolimits_{m=1}^{M} \sqrt{P_m} S_m \tag{6}$$

The signal amplitude that the receiver receives has a Rayleigh distribution in a Rayleigh channel. The probability distribution of the signal amplitudes in this channel is known as the Rayleigh distribution. The following is how the Rayleigh distribution equation is written:

$$f(x) = \begin{cases} \frac{x}{\sigma^2} e^{-\frac{x^2}{2x^2}} \; x \geq 0 \\ 0 \; otherwise \end{cases} \tag{7}$$

At the receiver, each user will receive the same signal. However, channel gain parameters vary depending on the user's distance from the BS. Received signal for each user can be written as follows

$$y_m = h_m \sum\nolimits_{m=1}^{M} \left(\sqrt{P_m} S_m \right) + n_m \tag{8}$$

where n_m represents Additive White Gaussian Noise (AWGN) and h_m is Rayleigh fading channel. Rayleigh fading is a model for a non line-of-sight path where the signal undergoes local scattering and has no dominant component [8]. The PDF of a Rayleigh distribution is given by the following equation:

$$P(s) = \begin{cases} \frac{s}{\sigma^2} exp\left(-\frac{s^2}{2\sigma^2}\right) fors \geq 0 \\ 0\, otherwise \end{cases} \tag{9}$$

where s is the amplitude of the fading signal and σ is the scale parameter that determines the spread of the distribution.

The UE_1 that is farthest from the BS can directly decode its own signal. Whereas UE_2, which is the second furthest from the BS to the closest user from the BS, will go through the SIC process first. The following is a picture of the SIC work process at NOMA.

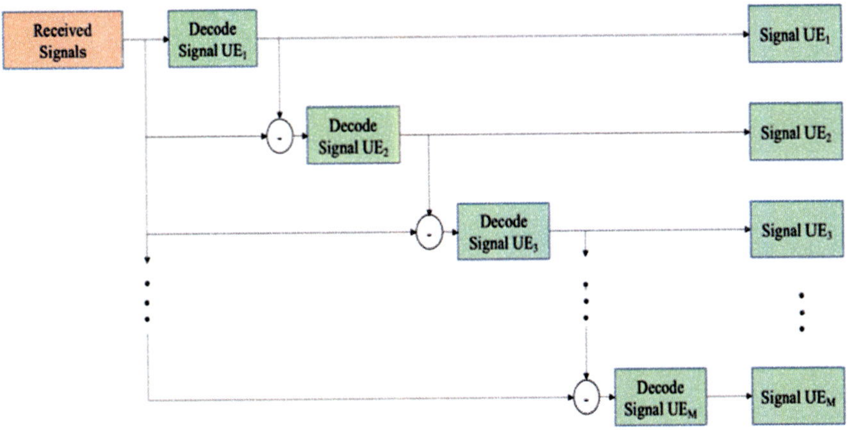

Fig. 2. SIC work process at NOMA

Figure 2 illustrates the operation of the SIC on a NOMA receiver. The UE_1 with the highest power allocation ratio can decode its own signal directly. The interference from the UE_1 signal must be cancelled by UE_2. UE_2 can decode its own signal after deducting the UE_1 signal. After SIC on UE_1 and UE_2, UE_3 can decode its own signal. And so on, all the way up to UE_M.

3 Result and Analysis

This section shows the number of bit error rates (BER) using sequential power allocation (SePA) with different constant values. The parameters used in the simulation are as follows (Table 1):

Static Power Allocation (SPA) is a division of power allocation that is determined only based on rules (2) and (3). SPA does not have a specific pattern like the Sequential

Table 1. Simulation parameters

Bandwidth		10^6 Hz
Number of bits		5 x 10^5 bit
Number of users (M)		3, 5, 7 user
User distance (d)	3 user	1000, 500, 100 m
	5 user	1000, 750, 500, 250, 100 m
	7 user	1000, 850,700, 550, 400, 250, 100 m
Modulation		QPSK
Channel		Rayleigh

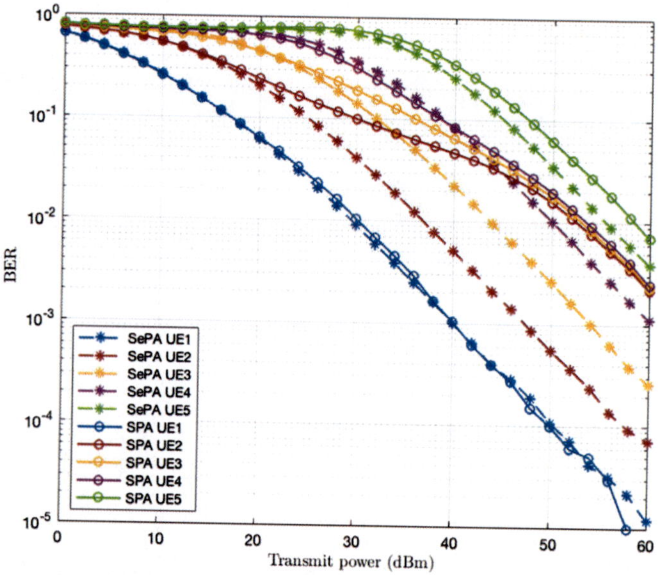

Fig. 3. BER performance using SePA and SPA

Power Allocation (SePA) used in this study. The BER performance simulation results using Sepa and SPA for 5 users are shown in Fig. 3. The SePA constant used is $\gamma = 0.8$ and the power allocation used in SPA for 5 users is [0.8; 0.14; 0.044; 0.015; 0.001]. on SPA only UE_1 achieves BER 10^{-3} with transmit power of 40 dBm. UE_2 to UE_5 require more than 60 dBm transmit power to reach BER 10^{-3}. Whereas in SePA, UE_1 to UE_4 can reach BER 10^{-3} with transmit power of 40–60 dBm. T he difference in transmit power required between users to achieve BER 10^{-3} is 6–8 dBm.

Figure 4 shows BER performance at SePA constant 0.8 for different number of users. The number of users 3 on the target BER 10^{-3} requires a transmit power of 39.3dBm for UE_1, 45.9 dBm for UE_2, and 51.7dBm for UE_3. Based on the three graphs, it can be seen that the required transmit power is close to each other for the same user. This

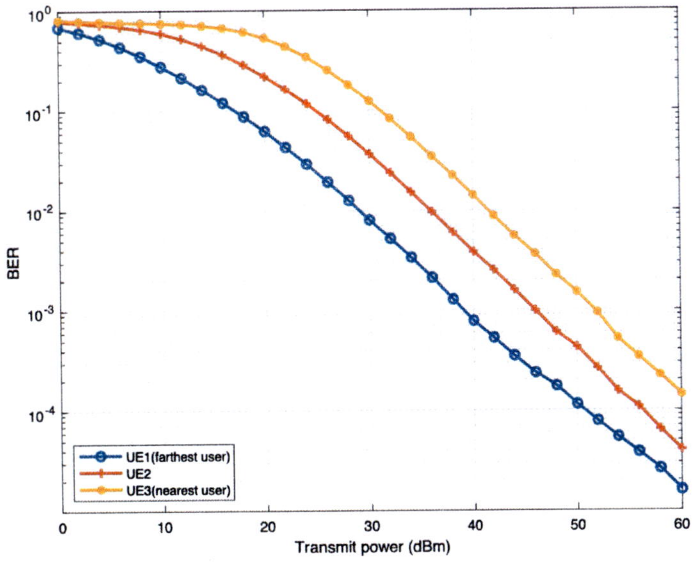

(a) SePA [$\gamma = 0.8$] for 3 users

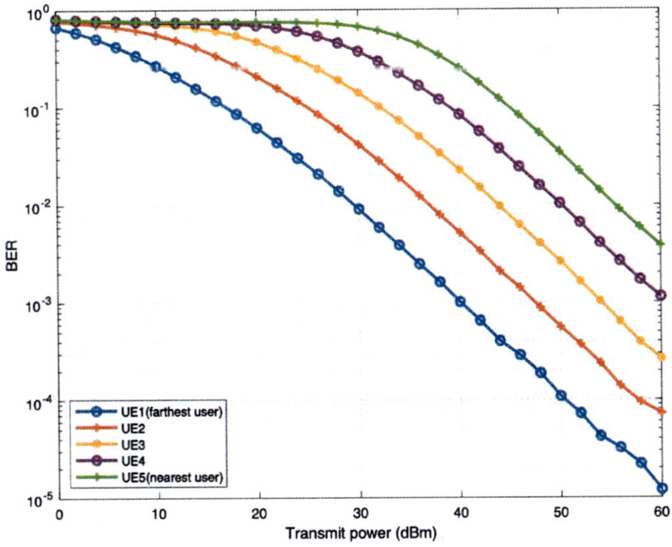

(b) SePA [$\gamma = 0.8$] for 5 users

Fig. 4. BER performance for different number of users

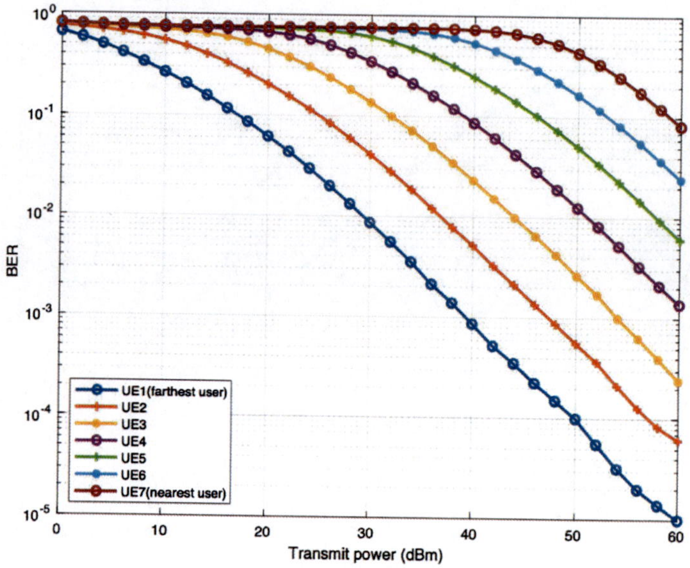

(c) SePA [$\gamma = 0.8$] for 7 users

Fig. 4. (*continued*)

is because in SePA the distribution of power allocation is done sequentially so that the value of the power allocation ratio is the same except for the 2 closest users. In addition, the amount of transmit power required to achieve BER 10^{-3} is directly proportional to the number of users on the NOMA downlink system. This is due to the way the SIC works on the receiver which cancels signals one by one so that if there is an error in the first process, there will be a propagation error.

BER performance for different SePA constants is shown in Fig. 5. The transmit power value needed to achieve a BER of 10^{-3} on UE_1 with $\gamma = 0.9$ is 37.1 dBm, $\gamma = 0.8$ is 39.3 dBm, and $\gamma = 0.7$ is 49.2 dBm. The smaller constant will require greater transmit power. This is because the received signal at the receiver does not have a significant difference so the decoding process will be more difficult in separating the signals.

Transmit power required from UE_1 to UE_2 and UE_2 to UE_3 has a power that is close to each SePA constant. At $\gamma = 0.9$ it takes about 20 dBm, $\gamma = 0.8$ it takes about 6 dBm, and $\gamma = 0.7$ it takes about 2.7 dBm. The greater the constant value, the greater the difference in power between users, so that the transmit power required will be greater so that the receiver side can decode the signal properly.

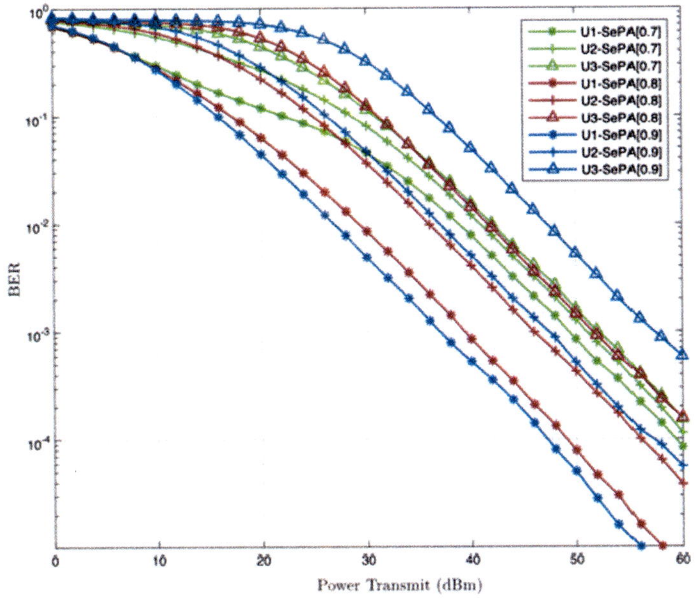

Fig. 5. BER performance for 3 users with different SePA constants

4 Conclusion

This paper investigates SIC performance using SePA on Rayleigh fading channel for more than 2 users. Based on the simulation results, a large SePA constant can achieve BER with a smaller transmit power for the first user. But to achieve the same BER target for other users, a large SePA constant will require a greater transmit power than a smaller SePA constant. For future work, SePA algorithm can be investigated for other channels.

Acknowledgement. Thank you for providing funding in this research, Indonesia Endowment Fund for Education (LPDP), Ministry of Finance Republic Indonesia.

References

1. Islam, S.R., Avazov, N., Dobre, O.A., Kwak, K.S.: Power-domain in non orthogonal multiple access (NOMA) in 5G systems: potentials and challenges. IEEE Commun. Surv. Tutor. **19**(2), 721–742 (2017)
2. "5G radio access: Requirements, concepts and technologies. NTT DOCOMO, Inc., Tokyo, Japan, 5G Whitepaper (2014)
3. Ding, Z., Lei, X., Karagiannidis, G.K., Schober, R., Yuan, J., Bhargava, V.K.: A survey on non-orthogonal multiple access for 5G networks: research challenges and future trends. IEEE J. Sel. Areas Commun. **35**(10), 2181–2195 (2017). https://doi.org/10.1109/JSAC.2017.2725519
4. Higuchi, K., Benjebbour, A.: Non-orthogonal Multiple Access (NOMA) with successive interference cancellation for future radio access. IEICE Trans. **98-B**, 403–414 (2015)

5. Vidyaningtyas, H., Kurniawan, A., Pramudita, A.A., Saputri, D.M.: The optimum number of users using sequential power allocation on PD-NOMA. In: 2023 International Conference on Artificial Intelligence in Information and Communication (ICAIIC), Bali, Indonesia, pp. 158–162 (2023). https://doi.org/10.1109/ICAIIC57133.2023.10066962
6. El-Mokadem, E.S., El-Kassas, A.M., Elgarf, T.A., El-Hennawy, H.:BER performance evaluation for the downlink NOMA system over different fading channels with different modulation schemes. In: 2019 5th International Conference on Science and Technology (ICST), Yogyakarta, Indonesia, pp. 1–6 (2019). https://doi.org/10.1109/ICST47872.2019.9166288
7. Ali, M.S., Hossain, E., Al-Dweik, A., Kim, D.I.: Downlink power allocation for CoMP-NOMA in multi-cell networks. IEEE Trans. Commun. **66**(9), 3982–3998 (2018)
8. Mohammed, A.G.A.: Performance evaluation of BPSK modulation based spectrum sensing over wireless fading channels in cognitive radio. IOSR J. Electron. Commun. Eng. **9**, 24–28 (2014)

Information Flow Control in the Fog Computing Model Based on a Component Degree Concept

Shigenari Nakamura[1(✉)], Tomoya Enokido[2], and Makoto Takizawa[3]

[1] Tokyo Denki University, Tokyo, Japan
s.nakamura@mail.dendai.ac.jp
[2] Rissho University, Tokyo, Japan
eno@ris.ac.jp
[3] Hosei University, Tokyo, Japan
makoto.takizawa@computer.org

Abstract. The FC (Fog Computing) model is discussed to implement the IoT (Internet of Things). Here, it is significant to protect fog nodes and devices from malicious accesses since the both are only supported with poor computation resources. The CBAC (Capability-Based Access Control) model is useful because each node can just check a capability token carried with an access request. Sensor data is obtained from a device and stored in another device by an authorized subjects who are granted tokens. However, even unauthorized subjects get the data from another device, i.e. illegal information flow. In the FC model, nodes receive sensor data from devices and send processed data to another nodes. In addition to devices, we have to prevent illegal information flow among nodes. Since sensor data for devices is processed and new data is generated by nodes, it is critical to discuss how much sensor data is included in data sent by nodes. We discuss a concept of component degree of each data which nodes exchange. By taking advantage of the concept, we newly propose an RR (Request Rejection) protocol for preventing illegal information flow among nodes. In the evaluation, we show the ratio of requests rejected to all the requests of nodes in the RR protocol.

Keywords: IoT (Internet of Things) · FC (Fog Computing) model · CBAC (Capability-Based Access Control) model · Information flow among fog nodes · Component degree

1 Introduction

In order to implement the efficient IoT (Internet of Things) [2,5,9,18], the FC (Fog Computing) model [1,7] is proposed. Here, application processes are executed to process sensor data in not only servers but also fog nodes. Here, devices send sensor data to nodes. The nodes process the data and send the processed data to other nodes.

L. Barolli (Ed.): BWCCA 2023, LNDECT 186, pp. 35–44, 2024.
https://doi.org/10.1007/978-3-031-46784-4_4

Access control models [4,19,20] are widely used in information systems to protect from malicious accesses. However, it is difficult to prevent illegal information flow [3]. In our previous studies [10,12], types of protocols to prevent illegal information flow in database and P2PPSO (Peer-to-Peer Publish/Subscribe with Object concept) systems are proposed and discussed. In papers [11,13–16], types of the CBAC (Capability-Based Access Control) model [6,8] are proposed. Furthermore, protocols to prevent types of illegal information flow among devices by taking advantage of the CBAC model are proposed, implemented, and evaluated [11,13–16].

A node receives and processes sensor data from other nodes and obtain a new data. The new data is generated from the collection of the sensor data. Thus, nodes exchange the data. In our previous studies [17], a concept of *component degree* of data which nodes and devices exchange is proposed. The component degree of data shows how much the sensor data is included in the data. Here, the smaller the component degree of data is, the more difficult to get the sensor data from the data. In this paper, we newly propose an RR (Request Rejection) protocol to preventing illegal information flow among nodes by taking advantage of the concept of the component degree. In the evaluation, we show the ratio of requests rejected to all the requests of nodes in the RR protocol.

In Sect. 2, we discuss a system model and the concept of component degree. In Sect. 3, we present the information flow relations among nodes by taking advantage of the concept of component degree. In Sect. 4, we newly propose the RR protocol. In Sect. 5, we evaluate the RR protocol.

2 System Model

2.1 CBAC Model

In the FC model, data is exchanged in networks through communications among nodes. We consider the *request-response* model.

Let OP_j be a set of operations supported by a node f_j. Let CAP_i be a set of capability tokens granted to a node f_i in the CBAC model [6,8]. A token is a pair $\langle op, f_j \rangle$ of an operation *op* and a node f_j. In this paper, each node f_i is assumed to support operations *get* and *put* for devices. A node f_i sends a request message *req* with a token $cap = \langle op, f_j \rangle$ to another node f_j. If the token *cap* is valid, the node f_j executes the operation *op* and then sends a reply to the node f_i. Otherwise, the operation *op* is not executed.

Let id_{ij} denote input data which a node f_i receives from a node f_{ij} ($j = 1, \ldots, b_i$ ($b_i \geq 0$)). Let ID_i be a set $\{id_{i1}, \ldots, id_{ib_i}\}$ of input data from the nodes $f_{i1}, \ldots, f_{ib_i}$. Output data od_i is made by calculating on the input data ID_i at a node f_i. We assume a node f_i supports processes *min*, *max*, *any*, *count*, *sum*, and *avg* which output one value od_i by using ID_i in this paper. In the *min*, *max*, and *any* processes, one output data od_i is selected in the input data set ID_i. Here, one in the input data ID_i is output data od_i ($\in ID_i$). In the *count*, *sum*, and *avg* processes, one output data od_i is calculated on ID_i. Here, the output data od_i is different from the input data set ID_i.

Example 1. Suppose a pair of tokens $cap_1 = \langle get, d_1\rangle$ and $cap_2 = \langle get, d_2\rangle$ are granted to a node f_i as shown in Fig. 1. Suppose a node f_i sends a *get* request with the tokens cap_1 and cap_2 to the devices d_1 and d_2. Then, the devices d_1 and d_2 check the tokens of the request and send sensor data to the node f_i. The node f_i gets sensor data id_{i1} and id_{i2} from the devices d_1 and d_2, respectively. Nest, the node f_j sends a request with a token $cap_j = \langle sum, f_i\rangle$ to the node f_i. If the token cap_j is valid, a summation value od_i of the input data id_{i1} and id_{i2}, i.e. $od_i = sum(\{id_{i1}, id_{i2}\}) = id_{i1} + id_{i2}$, is calculated by the node f_i and od_i is sent to f_j. Here, some information on the sensor data id_{i1} and id_{i2} are carried by the output data od_i to f_j, i.e. information flow occurs from f_i to f_j.

2.2 Component Degree

Sensor data is processed and exchanged by nodes in the FC model. This means, data in a node is sent to another node, i.e. the information flow among nodes. It is critical to discuss how much sensor data of a device is included in the data received from another node. In order to show that, a concept of *component degree* is proposed in our previous studies [17]. Let D be a set $\{d_1, \ldots, d_m\}$ of devices in the FC model. Each data du sent by a device and a node carries a vector **cd** $= [cd_1 \ldots cd_m]$ where each cd_k is a component degree on sensor data of a device d_k ($k = 1, \ldots, m$). If a device d_k sends its sensor data du, $du.cd_k$ is 1. $du.cd_h$ ($h \neq k$) is 0 because no sensor data of d_h is included in the data du.

Suppose a node f_i receives input data id_{ij} from every node f_{ij} ($j = 1, \ldots, b_i$), and obtains the output data od_i by a process $proc_i$. The *component degree* $od_i.cd_k$ of the output data od_i is defined as follows:

[component degree rule]

1. If a token $\langle get, d_k\rangle$ is granted to the node f_i or a device d_k sends some input data id_{ij}, $od_i.cd_k = 1$ and $od_i.cd_h = 0$ ($h \neq k$).
2. If $proc_i$ is *min*, *max*, or *any*, $od_i.cd_k = \{1 - \prod_{j=1}^{b_i}(1 - id_{ij}.cd_k)\}/|ID_i|$.
3. If $proc_i$ is *count*, *sum*, or *avg*, $od_i.cd_k = \{1 - \prod_{j=1}^{b_i}(1 - id_{ij}.cd_k)\}/|ID_i|^2$.

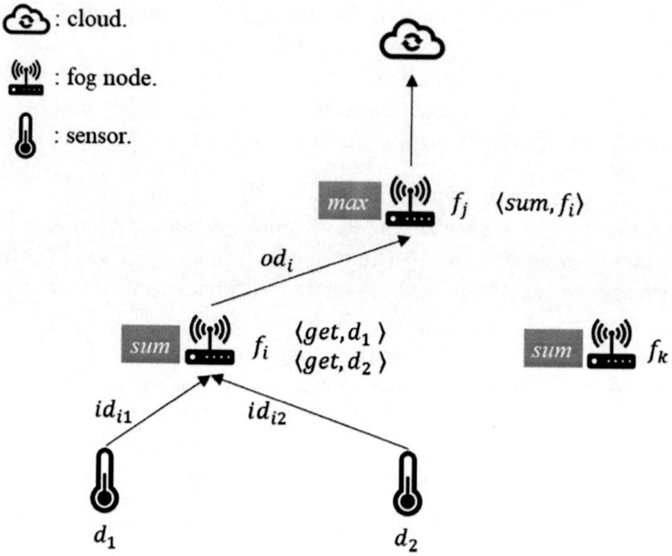

Fig. 1. Information flow among fog nodes.

Example 2. Vectors $\boldsymbol{cd}$ of component degrees are manipulated as shown in Fig. 2. First, a node f_i receives sensor data id_{i1} and id_{i2}. Here, $id_{i1}.\boldsymbol{cd} = [1\ 0\ 0]$ and $id_{i2}.\boldsymbol{cd} = [0\ 1\ 0]$. Next, a maximum value od_i of the input data id_{i1} and id_{i2}, i.e. $od_i = max(\{id_{i1}, id_{i2}\})$, is taken by the node f_i. Here, $od_i.cd_1 = 0.5$ $(= (1 - (1 - id_{i1}.cd_1)(1 - id_{i2}.cd_1))/|ID_i| = 1/2)$. $od_i.cd_2 = 0.5$ $(= (1 - (1 - id_{i1}.cd_2)(1 - id_{i2}.cd_2))/|ID_i| = 1/2)$. $od_i.cd_3 = 0$ $(= (1 - (1 - id_{i1}.cd_3)(1 - id_{i2}.cd_3))/|ID_i| = 0/2)$. Hence, $od_i.\boldsymbol{cd} = [0.5\ 0.5\ 0]$.

Another node f_k also receives sensor data id_{k1} and id_{k2} from the devices d_2 and d_3, respectively. Here, $id_{k1}.\boldsymbol{cd} = [0\ 1\ 0]$ and $id_{k2}.\boldsymbol{cd} = [0\ 0\ 1]$. A minimum value od_k of the input data id_{k1} and id_{k2}, i.e. $od_k = min(\{id_{k1}, id_{k2}\})$ is taken by the node f_k. Here, $od_k.cd_1 = 0$ $(= (1 - (1 - id_{k1}.cd_1)(1 - id_{k2}.cd_1))/|ID_i| = 0/2)$. $od_k.cd_2 = 0.5$ $(= (1 - (1 - id_{k1}.cd_2)(1 - id_{k2}.cd_2))/|ID_i| = 1/2)$. $od_k.cd_3 = 0.5$ $(= (1 - (1 - id_{k1}.cd_3)(1 - id_{k2}.cd_3))/|ID_i| = 1/2)$. Hence, $od_i.\boldsymbol{cd} = [0\ 0.5\ 0.5]$.

Finally, a node f_j receives the output data od_i and od_k. The summation value od_j of the output data od_i and od_k, i.e. $od_j = sum(\{od_i, od_k\})$ $= od_i + od_k$, is calculated by the node f_j. Here, $od_j.cd_1 = 0.125$ $(= (1 - (1 - od_i.cd_1)(1 - od_k.cd_1))/|ID_i|^2 = 0.5/4)$. $od_j.cd_2 = 0.186$ $(= (1 - (1 - od_i.cd_2)(1 - od_k.cd_2))/|ID_i|^2 = 0.75/4)$. $od_j.cd_3 = 0.125$ $(= (1 - (1 - od_i.cd_3)(1 - od_k.cd_3))/|ID_i|^2 = 0.5/4)$. Hence, $od_j.\boldsymbol{cd} = [0.125\ 0.186\ 0.125]$.

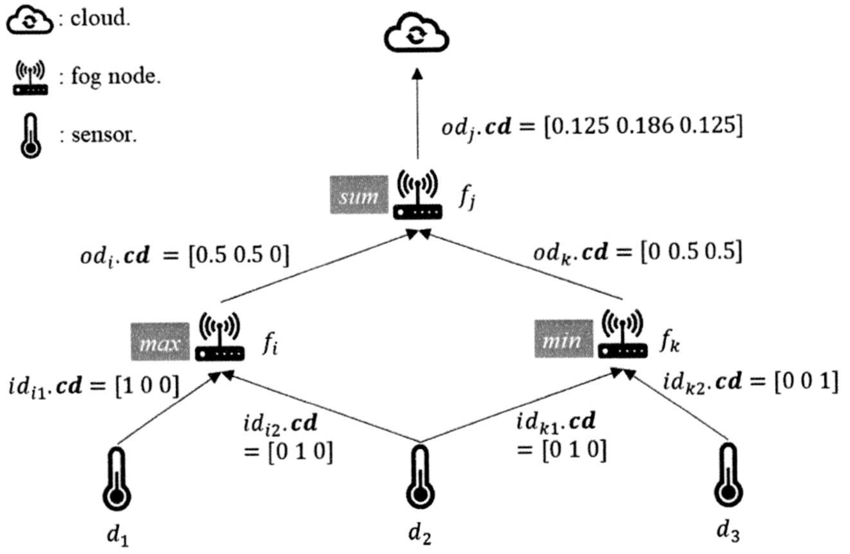

Fig. 2. Manipulation of ***cd*** vectors.

3 Information Flow Relations

In Example 1, the node f_i gets the summation (*sum*) value od_i of the input data id_{i1} from the device d_1 and id_{i2} from d_2. If the node f_j is granted a pair of the tokens $cap_1 = \langle get, d_1 \rangle$ and $cap_2 = \langle get, d_2 \rangle$, sensor data of the devices d_1 and d_2 in the summation value od_i from the node f_i is allowed to be got by f_j. Otherwise, the sensor data of the devices d_1 and d_2 is not allowed to be got by the node f_j. However, some data on the devices d_1 and d_2 can be got by the node f_j by getting the summation data $od_i = id_{i1} + id_{i2}$, i.e. *illegal information flow* occurs. Types of information flow relations are defined as follows:

Definition 1. Information *legally* flows from a node f_i to a node f_j ($f_i \rightarrow f_j$) iff (if and only if) $\forall cd_k \in id_{ji}.\boldsymbol{cd}$ ($cd_k = 0 \vee \langle get, d_k \rangle \in CAP_j$).

Here, since a node f_i obtains its output data od_i and sends od_i to another node f_j, od_i is the input data id_{ji} for f_j. If the node f_j receives only data which is allowed to be got by f_j from the node f_i, $f_i \rightarrow f_j$ holds. If no sensor data of a device d_k flow to the node f_j, f_j does not have to be granted a token $\langle get, d_k \rangle$ to get the sensor data.

In Example 2, the node f_j receives the output data od_i. Here, the vector $od_i.\boldsymbol{cd}$ of component degrees of the output data od_i is [0.5 0.5 0] as shown in Fig. 2. This means, sensor data of a pair of devices d_1 and d_2 is included in the output data od_i. Hence, if f_j is granted the tokens $\langle get, d_1 \rangle$ and $\langle get, d_2 \rangle$, the legal information flow relation among f_i and f_j holds, i.e. $f_i \rightarrow f_j$. Since sensor data of a device d_3 is not included in the output data od_i, the node f_j does not have to be granted a token $\langle get, d_3 \rangle$.

Definition 2. Information *illegally* flows from a node f_i to a node f_j iff $f_i \not\rightarrow f_j$.

In Example 2, if the node f_j is not granted the tokens $\langle get, d_1 \rangle$ and $\langle get, d_2 \rangle$, the illegal information flow relation among f_i and f_j holds, i.e. $f_i \not\rightarrow f_j$. In addition, if the node f_j is not granted a token $\langle get, d_3 \rangle$ and sensor data of a device d_3 is included in the output data od_i, $f_i \not\rightarrow f_j$ holds.

Definition 3. Information *partially legally* flows from a node f_i to a node f_j ($f_i \overset{\alpha}{\dashrightarrow} f_j$) iff $\forall cd_k \in id_{ji}.\mathbf{cd}$ ($cd_k \leq \alpha \vee \langle get, d_k \rangle \in CAP_j$).

In Example 2, although the node f_i obtains its output data od_i form the sensor data of the devices d_1 and d_2, od_i includes either sensor data because f_i executes a process max. Here, it is not sure f_j gets both sensor data. Hence, it is critical to discuss how much sensor data is included in the output data od_i. In Definition 3, we consider the concept of component degree with respect to a parameter α (≤ 1). In Fig. 2, if $\alpha \geq 0.5$, the partially legal information flow relation among f_i and f_j holds, i.e. $f_i \overset{\alpha}{\dashrightarrow} f_j$.

4 Information Flow Control

If the illegal information flow occurs, a node can get some information on devices even if the node is not allowed to get the sensor data from the devices. In this paper, we newly propose an RR (Request Rejection) protocol to prevent the illegal information flow at nodes by taking advantage of the concept of component degree. Here, requests to occur illegal information flow are rejected at nodes. Suppose a node f_j sends a request message to another node f_i to get output data of f_i.

1. The node f_i processes its input data according to the request of the node f_j and makes output data od_i.
2. The node f_i makes the vector $od_i.\mathbf{cd}$ of component degrees of the output data od_i by the component degree rule.
3. The node f_i rejects the request of the node f_j if $f_i \overset{\alpha}{\not\dashrightarrow} f_j$.

Suppose the RR protocol is performed in Example 2. We suppose the nodes f_i and f_j are granted the tokens $\langle get, d_1 \rangle$ and $\langle get, d_2 \rangle$ and a token $\langle get, d_1 \rangle$, respectively. Since the node f_j requests a maximum value, the node f_i takes the maximum value od_i of the set of input data ID_i. The vector $od_i.\mathbf{cd}$ of component degrees of the output data od_i is made ($= [0.5\ 0.5\ 0]$). If the parameter α in

Definition 3 is 0.3, $f_i \overset{0.3}{\not\dashrightarrow} f_j$ holds because $od_i.cd_2 > 0.3$. Hence, the node f_i rejects the request of the node f_j. However, if the parameter α is 0.6, $f_i \overset{0.6}{\dashrightarrow} f_j$ holds. Here, the node f_j can get the maximum value od_i even if the node f_j is not granted the token $\langle get, d_2 \rangle$.

5 Evaluation

In this paper, we evaluate the RR protocol in terms of the ratio of requests rejected to all the requests in a simulation. We consider a binary tree of nodes. OP is $\{min, max, any, count, sum, avg\}$. Each node executes one process and the process is randomly decided. Every node f_i is granted a token to get output data from each child node f_j. Tokens to get sensor data of devices granted to nodes except for leaf nodes are randomly decided. The number of devices is twenty. Each device sends sensor data to some of the leaf nodes which are randomly decided. Every leaf node is granted the token to get sensor data of devices which send their sensor data to the leaf node.

Let st be the simulation steps. In one simulation step, devices and nodes behave as follows:

1. Some of the devices send their sensor data to leaf nodes. Here, the devices are randomly decided.
2. Each leaf node which receives sensor data from devices processes the sensor data and sends the data to its parent node.
3. Each node processes its input data and sends the output data to its parent node.

The sets D and F of devices and nodes are generated fifty times. For each pair of given the sets D and F, the above procedures for st are iterated twenty five times. The total number of requests sent by nodes and the number of requests rejected to prevent illegal information flow are calculated. Finally, the average ratio of requests rejected to all the requests is calculated.

Figure 3 shows the ratio of requests rejected to prevent illegal information flow to all the requests of nodes in the RR protocol for the simulation steps st. The height of tree is five. Here, the tree has thirty one nodes and sixteen leaf nodes. As the parameter α increases, the ratio of requests rejected to all the requests decreases. For example, the average ratios are about 0.029 for $\alpha = 0.2$ and about 0.007 for $\alpha = 0.8$, respectively.

Figure 4 shows the ratio of requests rejected to prevent illegal information flow to all the requests of nodes in the RR protocol for the height of tree. The simulation steps st is 300 and the parameter α is 0.4. As the height of tree increases, the ratio of requests rejected to all the requests increases. For example, the average ratios are about 0.0065 for the height of tree is 2 and about 0.0135 for the height of tree is 5, respectively.

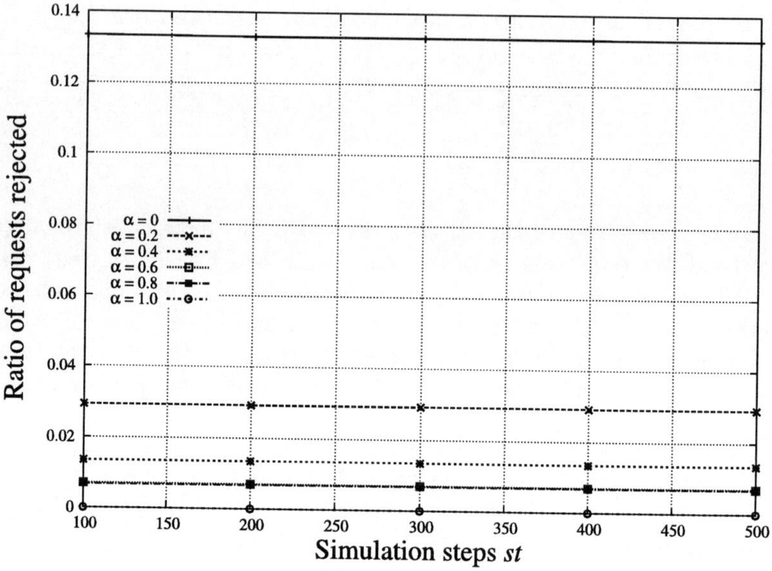

Fig. 3. Ratio of requests rejected to all the requests for the simulation steps *st*.

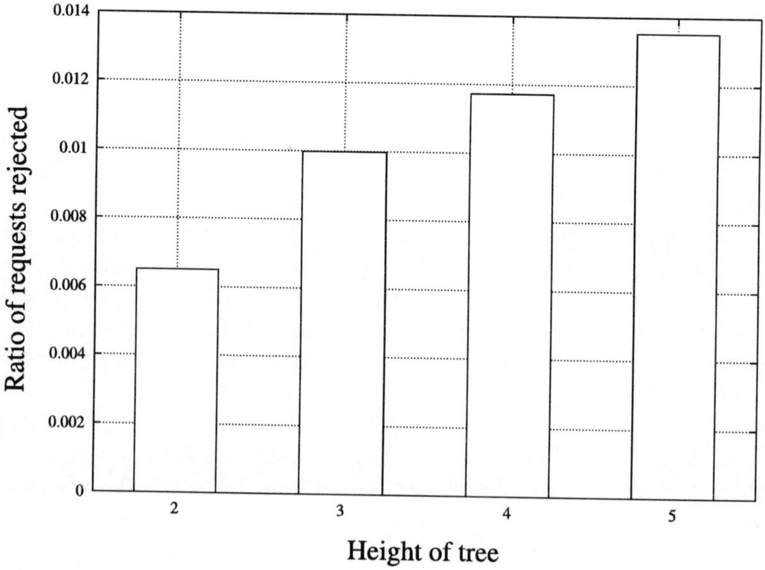

Fig. 4. Ratio of requests rejected to all the requests for the height of tree.

6 Concluding Remarks

In the FC model, each node processes sensor data and sends the processed data to other nodes, i.e. nodes exchange the data. Here, data exchanged among nodes includes data obtained from sensor data of devices. It is critical to make clear how much sensor data is included in data exchanged among nodes. We discussed information flow relations among nodes by taking advantage of the concept of component degree. Based on the relations, we newly propose the RR protocol to prevent the illegal information flow in this paper. In the evaluation, we show the ratio of requests rejected to all the requests of nodes in the RR protocol.

Acknowledgements. This work is partially supported by Japan Society for the Promotion of Science (JSPS) KAKENHI Grant Numbers JP22K12018 and JP23K16887.

References

1. Bonomi, F., Milito, R., Zhu, J., Addepall, S.: Fog computing and its role in the internet of things. In: Proceedings of the First Edition of the MCC Workshop on Mobile Cloud Computing, pp. 13–16 (2012)
2. Chavhan, S., Gupta, D., Chandana, B.N., Khanna, A., Rodrigues, J.J.P.C.: Iot-based context-aware intelligent public transport system in a metropolitan area. IEEE Internet Things J. **7**(7), 6023–6034 (2020)
3. Denning, D.E.R.: Cryptography and Data Security. Addison Wesley, Boston (1982)
4. Fernandez, E.B., Summers, R.C., Wood, C.: Database Security and Integrity. Adison Wesley, Boston (1980)
5. Gungor, V.C., et al.: Smart grid and smart homes: key players and pilot projects. IEEE Ind. Electron. Mag. **6**(4), 18–34 (2012)
6. Gusmeroli, S., Piccione, S., Rotondi, D.: A capability-based security approach to manage access control in the internet of things. Math. Comput. Model. **58**(5–6), 1189–1205 (2013)
7. Hanes, D., Salgueiro, G., Grossetete, P., Barton, R., Henry, J.: IoT Fundamentals: Networking Technologies, Protocols, and Use Cases for the Internet of Things. Cisco Press, Indianapolis (2018)
8. Hernández-Ramos, J.L., Jara, A.J., Marín, L., Skarmeta, A.F.: Distributed capability-based access control for the internet of things. J. Internet Serv. Inf. Secur. **3**(3/4), 1–16 (2013)
9. Islam, S.M.R., Kwak, D., Kabir, M.H., Hossain, M., Kwak, K.S.: The internet of things for health care: a comprehensive survey. IEEE Access **3**, 678–708 (2015)
10. Nakamura, S., Duolikun, D., Enokido, T., Takizawa, M.: A read-write abortion protocol to prevent illegal information flow in role-based access control systems. Int. J. Space-Based Situated Comput. **6**(1), 43–53 (2016)
11. Nakamura, S., Enokido, T., Takizawa, M.: Information flow control based on the CapBAC (capability-based access control) model in the IoT. Int. J. Mobile Comput. Multimed. Commun. **10**(4), 13–25 (2019)
12. Nakamura, S., Enokido, T., Takizawa, M.: Information flow control in object-based peer-to-peer publish/subscribe systems. Concurrency Comput.: Pract. Exp. **32**(8), e5118 (2020)

13. Nakamura, S., Enokido, T., Takizawa, M.: Implementation and evaluation of the information flow control for the internet of things. Concurrency Comput.: Pract. Exp. **33**(19), e6311 (2021)
14. Nakamura, S., Enokido, T., Takizawa, M.: Information flow control based on capability token validity for secure IoT: implementation and evaluation. Internet Things **15**, 100423 (2021)
15. Nakamura, S., Enokido, T., Takizawa, M.: Traffic reduction for information flow control in the IoT. In: Barolli, L. (ed.) BWCCA 2021. LNNS, vol. 346, pp. 67–77. Springer, Cham (2022). https://doi.org/10.1007/978-3-030-90072-4_7
16. Nakamura, S., Enokido, T., Takizawa, M.: Capability token selection algorithms to implement lightweight protocols. Internet Things **19**, 100542 (2022)
17. Nakamura, S., Enokido, T., Takizawa, M.: Capability-based access control model for fog computing model. In: Barolli, L. (ed.) NBiS 2023, vol. 183, pp. 79–85. Springer, Cham (2023). https://doi.org/10.1007/978-3-031-40978-3_9
18. Oma, R., Nakamura, S., Duolikun, D., Enokido, T., Takizawa, M.: An energy-efficient model for fog computing in the internet of things (IoT). Internet Things **1–2**, 14–26 (2018)
19. Sandhu, R.S., Coyne, E.J., Feinstein, H.L., Youman, C.E.: Role-based access control models. IEEE Comput. **29**(2), 38–47 (1996)
20. Yuan, E., Tong, J.: Attributed based access control (ABAC) for web services. In: Proceedings of the IEEE International Conference on Web Services (ICWS 2005), p. 569 (2005)

Implementation and Benchmarking of Kubernetes Horizontal Pod Autoscaling Method to Event-Driven Messaging System

Xavier Pilyai[1], Rafsanjani Nurul Irsyad[1], Ikhwan Nashir Zaini[1], Ridha Muldina Negara[1(✉)], Sofia Naning Hertiana[1], and Rohmat Tulloh[2]

[1] School of Electrical Engineering, Telkom University, Bandung, Indonesia
xavier@student.telkomuniversity.ac.id, ridhanegara@telkomuniversity.ac.id

[2] School of Applied Science, Telkom University, Bandung, Indonesia

Abstract. The rapid development of technology is accompanied by a large number of demands and users. One of the solutions is to create reliable infrastructure that can support application performance. Kubernetes is a platform which offers one of the solutions by providing container application orchestration and enabling high availability and scalability through various automatic scaling mechanisms such as Horizontal Pod Autoscaler (HPA), which dynamically scales the number of pod resources without restarting the entire system. By default, Kubernetes only monitors built-in resource metrics such as CPU and memory usage for each host machine and pod. To use Custom Metrics, external software such as Prometheus can be employed to monitor metrics as needed. On the other hand, Kubernetes Event-driven Autoscaling (KEDA) software offers a solution to simplify and facilitate autoscaling by efficiently implementing event-based automatic scaling and achieving the scale-to-zero capability. In this paper, we investigate the performance of autoscaling through various experiments to understand the behavior of each custom metrics autoscaling solution.

1 Introduction

In recent years, both large-scale and medium-small companies have largely transitioned to Cloud Computing. This transformation has brought about new challenges, such as optimal resource scheduling, efficient power usage, network latency, and others [1, 2]. These challenges arise because this technology is a new breakthrough with the hope of creating efficiency and reducing operational costs [3]. The development of Cloud Computing continues and requires solutions to address all the issues that arise, such as power efficiency and handling an increasing number of tasks [4, 5].

To address the challenges resulting from the rapid development of Cloud Computing technology, various approaches have emerged. One example is the emergence of microservices technology, which divides each part of a service into smaller components [6]. Then, to manage microservices, orchestration technology for microservices has emerged, which is used to manage and ensure the smooth and error-free operation of

L. Barolli (Ed.): BWCCA 2023, LNDECT 186, pp. 45–56, 2024.
https://doi.org/10.1007/978-3-031-46784-4_5

microservices. Within this orchestration technology, there is a feature called autoscaling, which is useful to prevent a service from being overwhelmed when experiencing excessive requests or traffic [4, 7].

Several papers have studied the autoscaling feature in microservice orchestration. We found a paper that compares the implementation of one type of autoscaling, namely Horizontal Pod Autoscaling (HPA). The paper compares the performance of HPA when using Kubernetes Resource Metrics and Prometheus Custom Metrics [8]. Their research successfully evaluated HPA and provided insights gained from the experiments they conducted.

We proposed Horizontal Pod Autoscaling Method for Event-Driven Messaging System using autoscale. we compare the performance of Horizontal Pod Autoscaling using metrics from KEDA, Prometheus Custom Metrics, and the collaboration of KEDA and Prometheus Custom Metrics. KEDA is a new autoscaler breakthrough that leverages event-driven technology [9]. Since this is a new breakthrough, we attempt to compare the performance of the proposed autoscale with existing auto scalers. The parameters used for comparison are CPU usage, RAM usage, and replica count.

2 Literature Research

2.1 Microservice

A system that adopts microservices is a variation of the architecture that focuses on separating services. This architecture usually consists of several small services. Each service runs for its own purpose and communicates with other services through a protocol [10].

2.2 Kubernetes

Kubernetes is an open-source orchestrator platform for deploying containerized applications. Originally developed by Google, inspired by a decade of experience deploying scalable container systems. Kubernetes as known as k8s has necessary feature to make our service high available like Horizontal Pod Autoscaler (HPA) [11].

2.3 Event-Driven

An event-driven architecture is a software architecture that captures events that occur within an application. These events are then used as a way to integrate an application. In this architecture, there are three main components: the event producer, the event consumer, and the message broker [12].

2.4 RabbitMQ

RabbitMQ is an open-source message broker platform. RabbitMQ has a function as a broker or intermediary for independent services to be able to communicate with each other. RabbitMQ implements the Advanced Message Queuing Protocol (AMQP), which supports complex message grouping and message routing features [13].

2.5 Prometheus

Prometheus is an open-source monitoring tool originally developed at SoundCloud. Since 2012, Prometheus has been used by many companies and organisations around the world, and the project has a very active community of developers and users. Prometheus joined the Cloud Native Computing Foundation in 2016 as the second project hosted after Kubernetes [14].

2.6 Horizontal Pod Autoscaler

Horizontal Pod Autoscaler (HPA) provides support for ensuring high availability by adjusting the number of pods and resources used based on varying demands. When there is an increase in workload, HPA will create new pods to distribute the workload without affecting the pods already running in the current cluster [15].

2.7 Keda

KEDA (Kubernetes Event-Driven Autoscaler) is a Kubernetes-based event-driven autoscaler. KEDA works with other Kubernetes components, such as HPA, and can be used without initiating existing components. KEDA has the advantage of being flexible and secure to run with several applications or other Kubernetes components. [9]. With KEDA we can scale applications based on event while other applications are running.

2.8 Grafana

Grafana is an open-source data visualization and analysis tool. Grafana can be useful for querying, visualizing, alerting, and analysing various metrics, logs, and traces stored in different locations. Grafana has a feature to convert a time series database (TSDB) into an attractive and insightful graph and visualization [16].

2.9 Siege

Siege is an open-source benchmarking and testing tool. Siege can track the total number of clicks, bytes transferred, response time, concurrency, and return status. The features in the siege can be adjusted according to user preferences and needs [17].

3 Proposed Architecture and Autoscaling Method

3.1 System Architecture Design and System

The producer application generates messages that are sent using the AMQP protocol and processed by the RabbitMQ Cluster message broker. Subsequently, the consumer application consumes these messages. The scraping server periodically retrieves metrics, and the HPA reads the exposed metrics as custom metrics. Scaling occurs based on the comparison between the number of metrics at retrieval time and the total metrics, which serves as the comparison threshold. Both applications, the consumer and producer, were developed using the Go programming language with the AMQP library.

The following pseudocode outlines the essential workflow of these interconnected components (Fig. 1).

```
// Producer App
while true:
    message = generate_message()
    send_message_via_amqp(message)
    wait_for_next_iteration()

// Consumer App
while true:
    message = receive_message_via_amqp()
    process_message(message)
    wait_for_next_iteration()

// Scraping Server
while true:
    metrics = get_metrics()
    expose_metrics(metrics)
    wait_for_next_iteration()

// HPA
while true:
    custom_metrics = read_custom_metrics()
    total_metrics = read_total_metrics()

    if custom_metrics > total_metrics:
        scale_up()
    else if custom_metrics < total_metrics:
        scale_down()

    wait_for_next_iteration()
```

Our Kubernetes cluster is running on 3 main compute machines that act as nodes for running applications, using the Ubuntu Server 20.04 operating system. We have ensured that all nodes have identical specifications to achieve balanced distribution of container resources. The specifications can be seen in Table 1.

The system we have designed requires a monitoring system to monitor the resources of virtual machines, as well as the performance and behavior of KEDA and Prometheus. Kubernetes by default provides the metrics server application to expose resource metrics for each object. The built-in metrics server and kube state metrics application generate metrics that are then captured and stored by the time series database, Prometheus [18]. The stored data is then displayed in the Grafana dashboard application, allowing real-time monitoring of resource usage, performance, and scaler behavior. Figure 3 illustrates the design of the monitoring system for this experiment (Fig. 2).

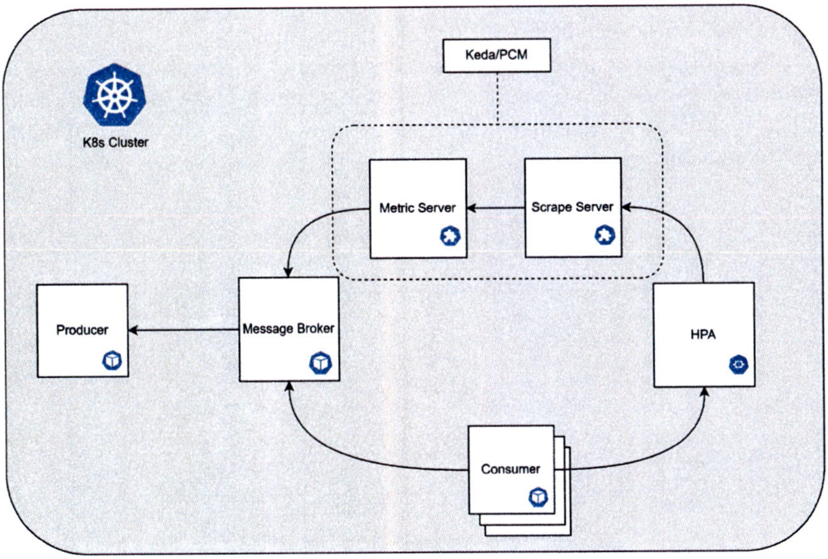

Fig. 1. Experimental System Methodology

Table 1. Cluster Specification

Node	CPU	RAM	DISK
Node 1	4 Core	4 GB	50 GB
Node 2	4 Core	4 GB	50 GB
Node 3	4 Core	4 GB	50 GB

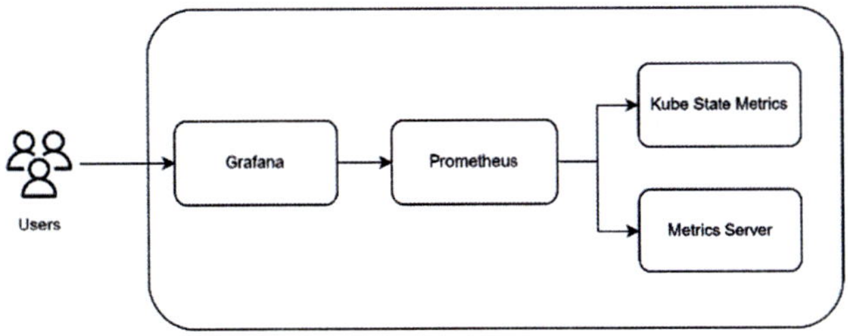

Fig. 2. Monitoring System

3.2 Autoscaling Method

In this study, three autoscaling methods will be used to monitor the performance and behavior of the scaler when using KEDA and Prometheus. The three scenarios, in sequence, are scenarios using KEDA.

The method can be seen in Fig. 3. Utilizes KEDA, which already has built-in scalers, metrics adapters, and controllers. Therefore, when creating objects from KEDA used for autoscaling, namely ScaledObject, HPA will be automatically configured to use metrics according to the built-in scaler we use, which is the RabbitMQ Queue Scaler with the QueueLength metric.

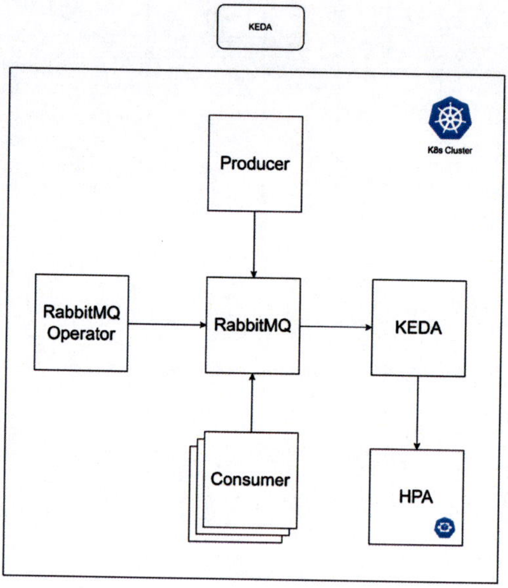

Fig. 3. Autoscaling method using KEDA

The second method utilizes Prometheus, which exposes custom metrics as an API that can be read by HPA [15]. Prior to that, HPA needs to be configured to read the desired custom metrics, such as RabbitMQ message ready queue, and set the target deployment to be scaled and the metric threshold value before HPA can perform autoscaling. Illustration can be seen in Fig. 4.

The last method utilizes the built-in Prometheus scaler in KEDA to read metrics exposed by Prometheus. For this, we create a KEDA object called ScaledObject and set the trigger type to Prometheus, with the metricName being the same as in the second scenario, which is the RabbitMQ message ready queue metric. The following autoscaling method can be seen in Fig. 5.

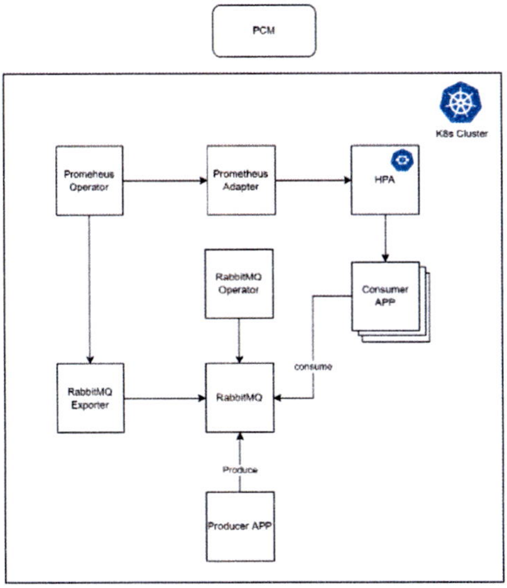

Fig. 4. Autoscaling method using PCM (Prometheus Custom Metric)

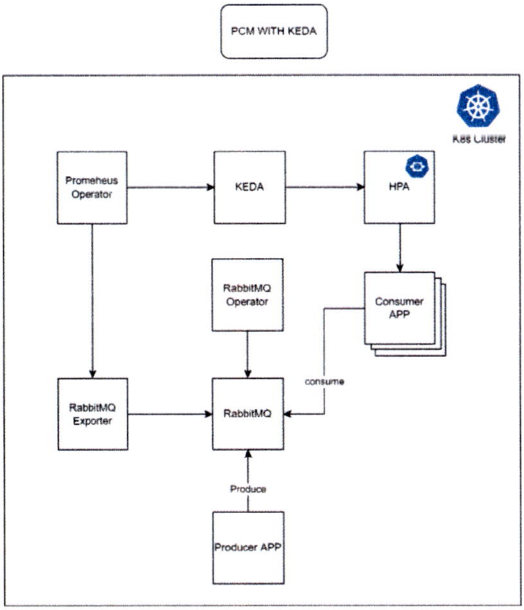

Fig. 5. Autoscaling method using PCM + KEDA

4 Experimental Result

Based on the discussed architecture and methods above, all components or applications are deployed on the Kubernetes cluster. The parameters used in this experiment are CPU Usage of the RabbitMQ server, Memory Allocation of the RabbitMQ server, and Number of Replicas in the HPA (Horizontal Pod Autoscaler).

These three parameters will be tested in the three proposed architectures and evaluated based on a specific number of concurrent users, ranging from 25, 50, 75, 100, 125, up to 150, 175, 200. These concurrent users will send messages to the RabbitMQ server, and the messages stored in the RabbitMQ queue will be consumed by the consumer application, which will scale automatically with the HPA (Horizontal Pod Autoscaler) with a maximum of 40 pods.

The simulation time used in this experiment is 1 min per concurrent user, with a scrape interval of 15 s. The request limit used in this experiment can be seen in Table 2.

Table 2. Request Limit RabbitMQ Server

	CPU		Memory	
RabbitMQ Server	request	limit	request	limit
	500 m	1	500 Mi	1Gi

4.1 Comparison of Number of Replicas on KEDA, PCM, and KEDA + PCM

The graph in Fig. 6 represents the average of 5 experimental runs, with each run lasting one minute, for each concurrent user group in the three different autoscaling methods: PCM, KEDA, and KEDA+PCM. The graph in Fig. 6 shows the X-axis as the concurrent user count and the Y-axis as the number of replicas.

In concurrent users above 100 autoscaling methods using PCM scale more pods with an average of 32 pods compared to the other two methods, namely KEDA with an average of 16 pods and KEDA PCM at 19 pods. The method using KEDA is arguably more efficient than other methods because the cluster does not accommodate as much load due to fewer pods. With KEDA, pods are very well managed so that no unneeded pod allocation.

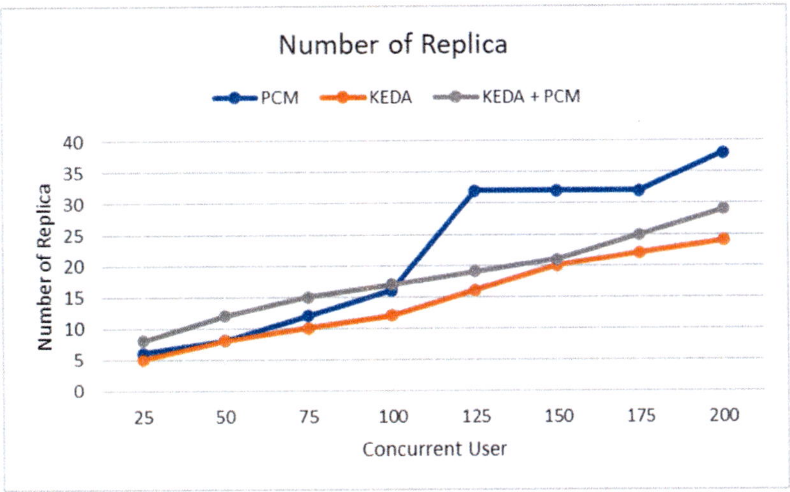

Fig. 6. Number of Replica

4.2 Comparison Against CPU Usage RabbitMQ Server

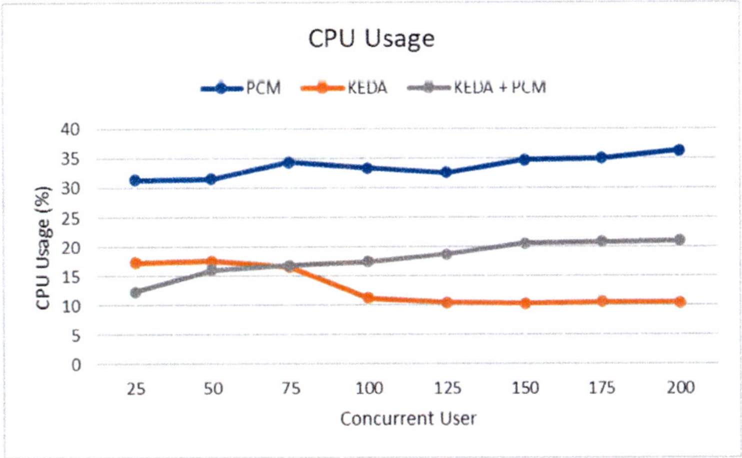

Fig. 7. Comparison against CPU Usage RabbitMQ Server

In the next scenario, we analyze the comparison of CPU usage in the RabbitMQ server among the different methods. The PCM method tends to consume more CPU, with an average usage of 32.75%. The KEDA+PCM method shows a lower CPU consumption compared to PCM alone, with an average usage of 16.96%. The KEDA-only method performs even better compared to the other methods mentioned, with an average CPU consumption of 13.84% (Fig. 7).

From the obtained results, it can be concluded that KEDA has the most positive impact on the RabbitMQ server in terms of CPU usage. As more pods are scaled by KEDA, the workload on the RabbitMQ server becomes lighter. This suggests that KEDA efficiently distributes the workload and optimizes resource utilization, resulting in lower CPU usage for the RabbitMQ server.

4.3 Comparison Against Memory Usage RabbitMQ Server

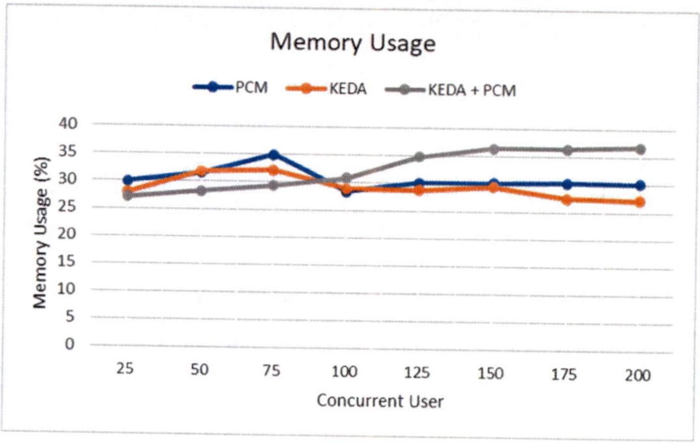

Fig. 8. Comparison against Memory Usage RabbitMQ server

The last scenario compares the three methods against the memory usage consumption of the RabbitMQ server. On Fig. 8 You can see a graph showing the trend of memory usage in a particular concurrent user. The chart does not show a significant difference in each scaling method, but in concurrent user 75 there is a decreasing trend in the KEDA and PCM methods, while in PCM+KEDA the trend on the chart continues to rise which matches the increasing scaling will also increase consumption in memory. PCM has an average memory usage of 30.85%, KEDA 29.9%, and KEDA+PCM has a memory usage of 31.15%. In the comparative analysis, KEDA again excelled in memory efficiency.

5 Conclusion

It can be concluded that the KEDA method has better efficiency in managing workloads and has a positive impact on CPU and memory usage on the RabbitMQ server. This method can optimize the use of resources efficiently, maintain server performance, and avoid overloading the CPU and memory. In addition, the number of pods scaled by KEDA is more efficient than other methods, reducing the overall Kubernetes cluster workload.

In the future, it is expected that further analysis of other scaling methods and event-driven implementations will also need to be applied not only to Horizontal Pod

Autoscaler (HPA) in Kubernetes but also to Vertical Pod Autoscaler (VPA) that scales the resources in single pod. According to this paper the message broker that been used is RabbitMQ, it also can be opportunity in the future research to compare the Autoscaling method in several message broker platform like Apache Kafka or cloud-based message broker like Amazon SQS.

References

1. Qi, W., Sun, M., Hosseini, S.R.A.: Facilitating big-data management in modern business and organizations using cloud computing: a comprehensive study. J. Manag. Org. 1–27 (2022)
2. Ma, Z., et al.: Virtual machine migration techniques for optimizing energy consumption in cloud data centers. IEEE Access (2023)
3. Kuryazov, D., Jabborov, D., Khujamuratov, B.: Towards decomposing monolithic applications into microservices. In: 2020 IEEE 14th International Conference on Application of Information and Communication Technologies (AICT). IEEE (2020)
4. Tulloh, R., Ginting, J.G.A., Mulyana, A., Lutfi, M.: Performance comparison of file transfer protocol service between link state and distance vector routing protocol in software defined network. In: IOP Conference Series: Materials Science and Engineering, vol. 982, no. 1, pp. 012026, December 2020. https://doi.org/10.1088/1757-899X/982/1/012026
5. Saboor, A., et al.: Containerized microservices orchestration and provisioning in cloud computing: a conceptual framework and future perspectives. Appl. Sci. **12**(12), 5793 (2022). https://doi.org/10.3390/app12125793
6. Nathaniel, L., Perdana, G.V., Hadiana, M.R., Negara, R.M., Hertiana, S.N.: Istio API gateway impact to reduce microservice latency and resource usage on kubernetes. In: 2023 International Seminar on Intelligent Technology and Its Applications (ISITIA), Surabaya, Indonesia, pp. 43–47 (2023). https://doi.org/10.1109/ISITIA59021.2023.10221035
7. Adinegoro, F., Rahmania, C., Zaini, I.N., Negara, R.M., Hertiana, S.N.: Latency and RAM usage comparison of advanced and lightweight service mesh. In: 2022 5th International Seminar on Research of Information Technology and Intelligent Systems (ISRITI), Yogyakarta, Indonesia, pp. 369-372 (2022). https://doi.org/10.1109/ISRITI56927.2022.10052905
8. Donca, I.C., Corches, C., Stan, O., Miclea, L.: Autoscaled RabbitMQ Kubernetes Cluster on single-board computers. In: 2020 IEEE International Conference on Automation, Quality and Testing, Robotics (AQTR), Cluj-Napoca, Romania, pp. 1–6 (2020). https://doi.org/10.1109/AQTR49680.2020.9129886
9. Pereira-Vale, A., Fernandez, E.B., Monge, R., Astudillo, H., Márquez, G.: Security in microservice-based systems: a multivocal literature review. Comput. Secur. **103**, 102200 (2021). https://doi.org/10.1016/j.cose.2021.102200
10. Burns, B., Beda, J., Hightower, K., Evenson, L.: Kubernetes: Up and Running. Google Books (2022)
11. Nguyen, T.-T., Yeom, Y.-J., Kim, T., Park, D.-H., Kim, S.: Horizontal pod autoscaling in kubernetes for elastic container orchestration. Sensors **20**(16), 4621 (2020). https://doi.org/10.3390/s20164621
12. Medium: "Event-Driven Architecture" (2021). https://medium.com/tlabcircle/event-driven-architecture-5e78d4e96508. Accessed 23 June 2023
13. Johansson, L., Dossot, D.:RabbitMQ Essentials: Build distributed and scalable applications with message queuing using RabbitMQ, 2nd edn. Google Books (2020)
14. Prometheus: Overview What is Prometheus? (2014). https://prometheus.io/docs/introduction/overview/. Accessed 23 June 2023
15. Keda: "Feature" (2014). https://keda.sh/. Accessed 23 June 2023

16. Grafana Labs: "Introduction to Grafana" (2014). https://grafana.com/docs/grafana/latest/introduction/. Accessed 23 June 2023
17. Github JoeDog: "Siege What is it?" (2008). https://github.com/JoeDog/siege. Accessed 23 June 2023
18. Marraskuu: Opinnäytetyö. MONITORING A KUBERNETES APPLICATION (2019)

Application of Convolutional Neural Network Method with MobileNet V1 and ResNet-152 V2 Architecture in Batik Motif Classification

Aulia Chusnyriani Sani Zulkarnaen[1], I Gusti Ngurah Rejski Ariantara Putra[1], Nada Fauzia Reviana[1], Rahmawati Hidayah[1], Nur Ibrahim[1,2](✉), Nor Kumalasari Caecar Pratiwi[1], and Yunendah Nur Fuadah[1]

[1] Faculty of Electrical Engineering, Telkom University, Bandung, Indonesia
{auliachusnyriani,gustiari,nadafauzia, rahmawatihidayah}@student.telkomuniversity.ac.id, {nuribrahim, caecarnkcp,yunendah}@telkomuniversity.ac.id

[2] Department of Electrical Engineering, Universitas Indonesia, Depok City, Indonesia

Abstract. Indonesia is a country that has diverse natural resources, cultures, and languages. One of the cultural diversities in Indonesia is Batik, which is an Indonesian cultural heritage consisting of cloth drawn by hand using traditional techniques. To assist the public in recognizing various batik motifs, a classification method was developed to identify the types of batik through input images. The classification method uses Convolutional Neural Network (CNN) based on MobileNet V1 and ResNet-152 V2 architecture. This research uses a dataset consisting of 3300 batik images from six different batik motifs, namely ceplok, parang, nitik, megamendung, kawung, and tambal. The optimal classification model was obtained using ResNet-152 V2 architecture with shear pre-processing method and RMSprop optimizer with test accuracy value of 89.67% and validation loss of 0.44.

1 Introduction

Indonesia is a country rich in natural resources, diverse ethnicities, cultures, and languages. Indonesia has one of the cultures known as Batik. Batik is a craft and one of the typical cultural heritages of the archipelago that has become a high artistic and cultural heritage since ancient times [1]. Batik has several definitions, one of the definitions of Batik is a visual art form from Indonesia that is produced using traditional drawing techniques on materials [2]. Each region in Indonesia has a distinctive batik that characterizes the diversity of types and motifs. In every motif that exists in this type of batik has its own philosophical meaning where each motif has a long historical value [3]. The types and motifs of batik cannot be separated from the elements inherent in each region where it is made. Batik has received recognition by the world, namely the award from the United Nations Educational, Scientific and Cultural Organization (UNESCO) on October 2, 2009 by stating that batik is the intellectual cultural rights of the Indonesian nation [4]. The number of batik motifs circulating in the community requires the classification of batik motifs so that they can be identified based on the name of the batik.

L. Barolli (Ed.): BWCCA 2023, LNDECT 186, pp. 57–68, 2024.
https://doi.org/10.1007/978-3-031-46784-4_6

One way to do the classification is to use the deep learning method implemented in the mobile application. This application aims to demonstrate the classification results using deep learning. Deep learning is one part of artificial intelligence. The purpose of deep learning is to develop algorithms and computational models that are used to enable a system to learn independently and improve its performance based on the data provided. According to the findings, this method is widely applied in batik motif classification.

Based on previous research, batik classification using the Naïve Bayes method based on texture feature extraction can produce accuracy of up to 97.22% using 420 training data and 180 test data covering three different types of batik [2]. Then, in another study conducted in 2020, it was found that batik classification using a combination of Artificial Neural Network (ANN), K-Nearest Neighbor (KNN), Decision Tree, and Support Vector Deep (SVM) methods produced an accuracy rate of up to 96.43% in classifying two types of batik motifs, with a dataset consisting of 76 batik images [3]. However, both studies still have shortcomings in terms of a separate feature extraction process, a simple model, and a limited number of batik motifs classified. Therefore, a new classification method was developed with the Convolutional Neural Network (CNN) model which is a development of JST. As in research in 2020, the classification of 6 types of batik from a total dataset of 944 images with the CNN method based on the 201 denset model can produce an accuracy rate of up to 99% [5]. This research also compares the accuracy value with other models such as Alexnet which is able to produce up to 98% accuracy, Resnet-152 with 100% accuracy, and Squeezenet 1 which produces an accuracy rate of 99%. CNN has advantages and disadvantages in the computational process. The advantages of CNN include its ability to select features without modifying or developing feature extraction in image data [6]. However, CNN also has a drawback, namely the need for data augmentation, so it requires data augmentation at the pre-processing stage. Based on the background description that has been presented, researchers are interested in conducting research on "Application of Convolutional Neural Network Method with MobileNet V1 and ResNet-152 V2 Architecture on Batik Classification". The purpose of this research is to determine the accuracy and loss results of the input image data. In this research, two models will be made with different architectures, namely MobileNetV1 and ResNet-152 V2. Modeling with these two architectures aims to see a comparison of the performance generated from the two models. This research is intended to fulfill the topic of Multimedia Communication and Applications at the Broadband and Wireless Computing, Communication, and Applications conference. It is hoped that with this research design, the general public can distinguish or identify batik patterns through research designed by researchers.

2 Convolutional Neural Network

Convolutional Neural Network (CNN) is a development of the Multi-Layer Perceptron (MLP) deep learning method designed to process two-dimensional data [7]. As seen from the name, the main process that occurs in CNN is convolution. Convolution is the repeated application of a function to the output of another function. CNN is a popular deep learning method that has been widely applied in previous research and has been proven to achieve good accuracy rates. For example, in a study titled "Wood Species Identification using Convolutional Neural Network with Mobilenet Architecture," a classification system with an accuracy rate of up to 95% was achieved [8]. Currently, various CNN architectures have been developed, such as AlexNet, ResNet, VGG Net, GoogLeNet, NASNet, and MobileNet [9].

2.1 Mobilenet Architecture

MobileNet is one of the CNN architectures specifically developed for mobile computing purposes [10]. The most fundamental difference between the MobileNet architecture and other CNN architectures lies in the use of depthwise separable convolutions, where the thickness of the filters is adjusted according to the thickness of the input image [11]. The operation of the MobileNet architecture is based on the concept of depthwise separable convolutions [12]. Depthwise separable convolutions consist of two main layers: depthwise convolution layers and pointwise convolution layers [13]. The depthwise convolution layer filters the input image without creating new features, while the pointwise convolution layer is responsible for creating new features [14].

2.2 ResNet Architecture

ResNet or Residual Network is one of the popular artificial neural networks used in the CNN method and was proposed by a group of researchers from Microsoft, namely Kaiming He et al. [15, 16]. In 2015, the ResNet architecture won the ILSVRC-2015. From this research, they have solved the problem of gradient loss or explosion by introducing the concept of shortcut connections that are more flexible and capable of training deep neural networks [17]. The input layer of the ResNet architecture consists of several residual blocks and each building block plays an important role in improving accuracy [18]. It is also accompanied by the number of building blocks that determines the depth of the network [18]. The number of layers in this architecture has several variations ranging from 18, 34, 50, 101, and 152.

3 Methodology

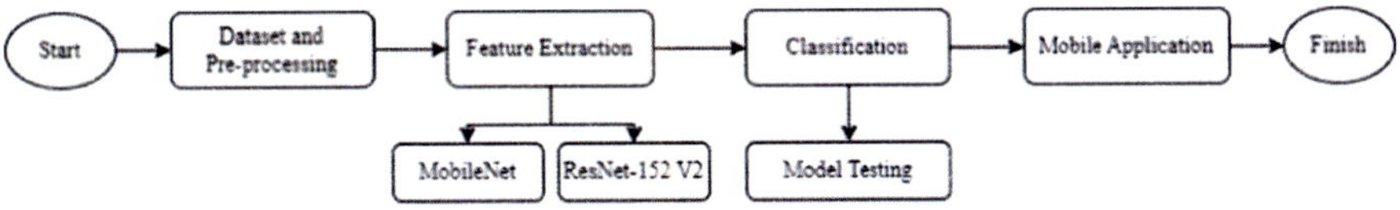

Fig. 1. Flowchart of the system design

Figure 1 is an overview of the flow of this research. Starting from the deep learning process such as pre-processing, feature extraction and classification stages. After the process, the results of deep learning are inputted into the mobile application. The pre-processing stage is carried out by augmenting the shear type, brightness, rotation, width, horizontal flip using the batik dataset, then from this process the best results will be obtained to proceed to the next stage. Next is the feature extraction stage by performing weight, top layer and layer configuration using the best results from the pre-processing process. The last stage of the deep learning process is classification. This stage is a continuation of the feature extraction process so that the data used is the data used by the previous stage. In this stage optimizer, learning rate, batch size, and epoch are performed. The result that will be put into mobile application is the best result from the whole deep learning process. The mobile application process is used to produce accuracy that is equivalent to the deep learning process.

3.1 Dataset and Pre-processing

In this research, the dataset used consists of images of Indonesian batik images with six different motifs namely ceplok, parang, nitik, kawung, megamendung, and tambal. This dataset consists of a total of 660 batik images, the existing data is continued in the process, namely the pre-processing technique. This technique has a type of work method including the one used is the type of data augmentation. When data enters data augmentation, the data will be processed by means of shear, brigthness, width and rotation so as to produce new data. The images resulting from data augmentation amount to 3.300 images which are divided into 2400 images for training data and 600 images for validation data, with 300 testing images. All images used in this research dataset were sourced from platforms such as Kaggle and through independent searches on photo sharing platforms such as Pinterest, Freepik, and Shutterstock. The following are examples of images used in this research dataset. Figure 2 shows an image of a batik motif and Table 1 shows the details of each batik image dataset class used in this research dataset.

The data listed in Table 1. then underwent pre-processing. Pre-processing begins with the use of 'ImageDataGenerator' to perform image augmentation using various methods such as brigthness range, width range, shear range, rotation range, and horizontal shift. Furthermore, batik image data will be entered into stages so that it can produce the expected classification of the designed research.

3.2 Implementation of MobileNet V1 and ResNet-152 V2 Architecture

In this research, the architectures implemented are MobileNet V1 and ResNet-152 V2. The two architectures used in this research have two purposes, namely to extract features from the dataset and classify batik motifs. Figure 3 shows an overview of the MobileNetV1 architecture used in this research.

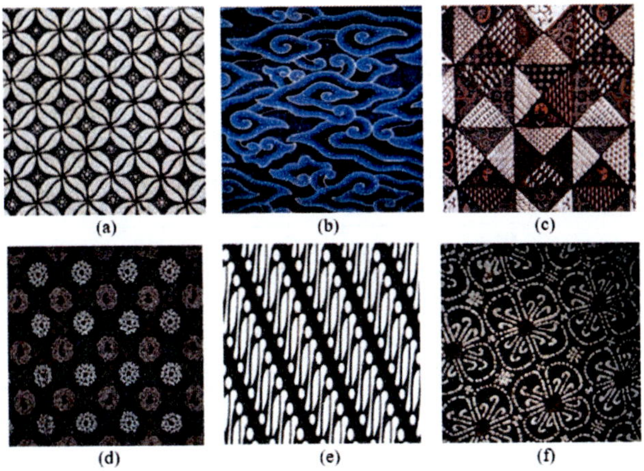

Fig. 2. The batik motif images with details are as follows: (a) Kawung Motif, (b) Megamendung Motif, (c) Tambal Motif, (d) Ceplok Motif, (e) Parang Motif, (f) Nitik Motif

Table 1. Split data from each type of batik

Motifs / Classes of Batik	Data Categories			Total per Class
	Training	Validation	Testing	
Ceplok	400	100	50	550
Parang	400	100	50	550
Nitik	400	100	50	550
Kawung	400	100	50	550
Megamendung	400	100	50	550
Tambal	400	100	50	550
Total per Categories	2.400	600	300	3.300

From Fig. 3, it can be seen that the model used has seven layers. The layers used include:

1. One layer of MobileNet is placed at the beginning to invoke the MobileNet architecture. The invoked MobileNet architecture has been previously trained using the ImageNet dataset, to extract features from the input images
2. One Global Average Pooling 2D layer is used to take the average value of the features generated in the form of a 2D matrix
3. One Batch Normalization layer is used to improve the stability of the classification model's performance by normalizing input values, providing regularization, and reducing dependency on weight initialization
4. Three Dense layers with ReLU activation are used to connect the input layer with the output layer

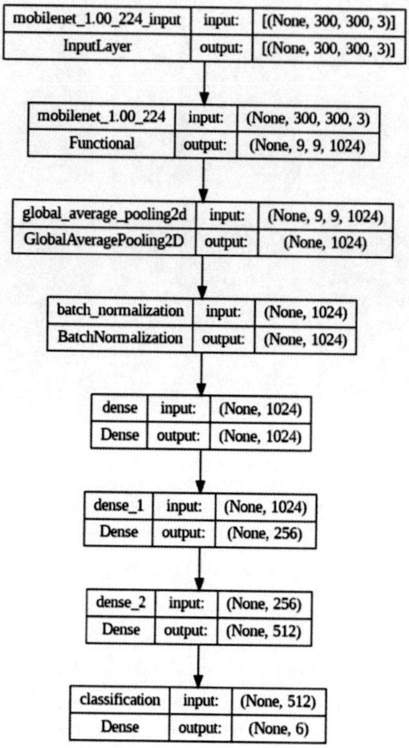

Fig. 3. Structure of MobileNet V1 architecture

5. One Dense Output layer with Softmax activation is used to generate the output along with the predictions for each class

Furthermore, Fig. 4 shows an overview of the ResNet-152 V2 architecture used in this study.

From Fig. 4, it can be seen that the model arrangement has almost the same layer arrangement as the previous architecture, MobileNet V1. The layers used in this ResNet-152 V2 architecture include:

1. One ResNet-152 V2 layer is placed at the beginning to invoke the ResNet-152 V2 architecture. The called ResNet-152 V2 architecture has been previously trained using the imageNet dataset, to extract features from the input images.
2. One layer of global average pooling 2D is used to take the average value of the generated features in the form of a 2D matrix.
3. One flatten layer is used to recreate the feature map from the originally multidimensional image matrix into a one-dimensional vectors. This will make it easier for the subsequent layers to process the extracted features.
4. One dense layer with ReLU activation is used to connect the input layer with output layer.

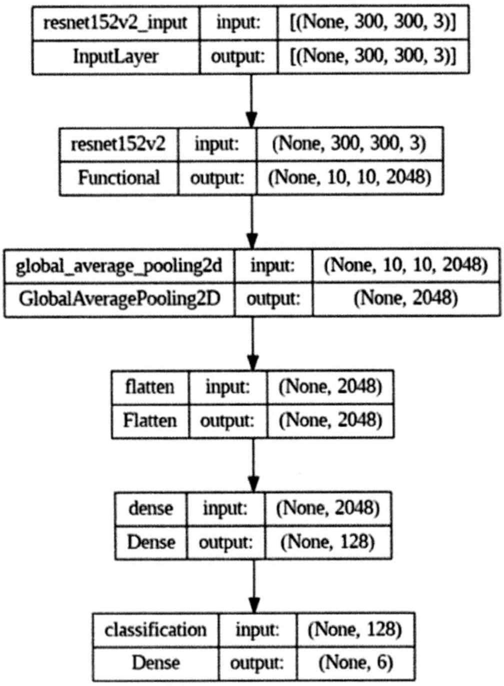

Fig. 4. Structure of the ResNet-152 V2 architecture

5. Dense output layer with Softmax activation is used to generate outputs along with predictions for each class.

3.3 Model Testing

In this research, the classification model created will be tested through training, validation, and testing stages to see its performance. Testing will be carried out in several experimental scenarios related to changes in several parameters. The experiments that will be carried out are:

1. The first experiment was conducted by comparing the values of the data augmentation type parameters on the two architectures. The augmentation types used were brigthness, rotation, shear and width. The comparison architectures are MobileNet V1 and ResNet-152 V2.
2. The last experiment was conducted by comparing the type of optimizer used. The optimizer types to be tested are Adam, Adamax, and RMSprop. All experiments will be conducted using an epoch value of 100 and a batch size value of 64 on both architectures.

4 Experimental Result

This stage contains the experimental results of testing the system designed to classify batik image using MobileNet V1 architecture and ResNet-152 V2. The testing stages carried out are by comparing image results using different pre-processing and optimization techniques. The test is carried out on the type of augmentation as a test on pre-processing, and optimizer as a classification on two architectures as a comparison. This test will use parameters such as target size 300, batch size 64, learning rate 0.0001 and epoch 100 using early stopping. Table 2 parts a and b show the first test results.

Table 2. (a) Test results of pre-processing techniques on MobileNet V1 architecture (b) Test results of pre-processing techniques on ResNet-152 V2 architecture.

(a)

MobileNet V1	Train Accuracy	Train Loss	Validation Accuracy	Validation Loss	Test Accuracy
Brigthness	95,79%	0,20	70%	1,60	56,67%
Rotation	94%	0,22	66,83%	1,63	57,86%
Shear	95,83%	0,11	73%	1,06	**65,67%**
Width	98%	0,05	64%	1,68	57,67%

(b)

ResNet-152 V2	Train Accuracy	Train Loss	Validation Accuracy	Validation Loss	Test Accuracy
Brigthness	99,96%	0,02	87%	0,55	80%
Rotation	100%	0,01	87%	0,45	79%
Shear	100%	0,01	87,83%	0,41	**88%**
Width	100%	0,01	85,33%	0,52	83%

Based on Table 2, the results of pre-processing testing are carried out by comparing data augmentation in the form of brightness, rotation, shear and width. In the first test, the MobileNet V1 architecture showed that the best test results were obtained with the shear augmentation type which resulted in a train accuracy of 95.83% and a test accuracy of 65.67%. The second test was conducted on the ResNet-152 V2 architecture which resulted in a train accuracy of 100% and a test accuracy of 88%. After knowing the results of the pre-processing performance, the last test was continued in the form of testing the system optimizer. The parameters used for optimizer testing are Adam, Adamax, and RMSProp. This test will use the same architecture as the previous test, with a target size of 300, batch size of 64, and epoch 100 using early stopping. Table 3 shows the results of the last test.

Table 3. Optimizer test results on MobileNet V1 architecture, (b) Optimizer test results on ResNet- 152 V2 architecture.

(a)

MobileNet V1 Optimizer	Train Accuracy	Train Loss	Validation Accuracy	Validation Loss	Test Accuracy
Adam	94.37%	0.18	66.66%	1.05	64.99%
Adamax	93.95%	0.22	65.83%	0.95	68.33%
RMSProp	96.04%	0.14	67.5%	1.96	**69.5%**

(b)

ResNet-152V2 Optimizer	Train Accuracy	Train Loss	Validation Accuracy	Validation Loss	Test Accuracy
Adam	99,96%	0,02	87,17%	0,40	86,33%
Adamax	99,96%	0,03	87,67%	0,41	86,67%
RMSProp	99,87%	0,01	89,17%	0,44	**89,67%**

Table 3 describes the performance results of the designed system. From Table 3, it can be seen that testing on the MobileNet V1 optimizer resulted in a train accuracy of 96.04% and a test accuracy of 69.5%. The last optimizer test was carried out on the ResNet-152 V2 architecture where the test results obtained a train accuracy of 99.87% with a test accuracy of 89.67%. Thus, t can be concluded in the four tests that have been carried out that the test uses the ResNet-152 V2 architecture. The following is a summary of the results described in Fig. 5, where the results show the accuracy graph as well as the loss and confusion matrix of the test results.

The results on the accuracy and loss graphs in Fig. 5 parts a and b show a fairly stable performance. This can be seen from the graph line which does not fluctuate and shows that model learning tends not to experience large changes. Furthermore, Fig. 5 parts c shows the confusion matrix results of the model made in classifying six types of batik motifs. This confusion matrix is made based on the testing results. Overall, the model's performance is quite good in predicting the type of batik motif, but the model still makes many mistakes in recognizing the ceplok batik motif. This can be caused because the motifs in ceplok batik have many similarities with other types of batik motifs and the majority have been modified. In Nitik batik and Parang batik motifs, there are still errors although not as many as Ceplok batik motifs. From the model created, there are performance results in the form of accuracy, precision, recall, and F1- Score values which can be seen in Table 4.

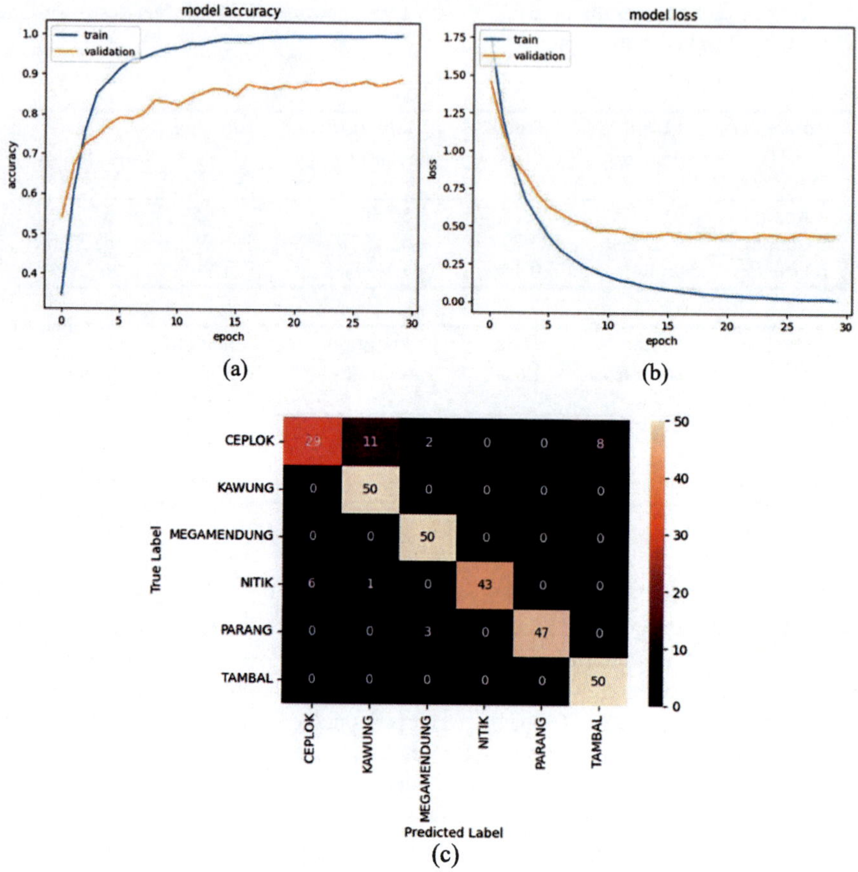

Fig. 5. (a) Accuracy graph of ResNet152 V2, (b) Loss graph of ResNet152 V2, (c) Confusion Matrix of ResNet-152 V2 architecture model

Table 4 Classification Report of ResNet-152 V2 architecture model

Batik	Precision	Recall	F1-score
Ceplok	0.80	0.79	0.79
Kawung	0.81	1.00	0.89
Megamendung	1.00	1.00	1.00
Nitik	0.92	0.84	0.88
Parang	1.00	0.88	0.93
Tambal	0.91	0.90	0.91

5 Conclusion

Batik is a traditional craft from Indonesia that has been recognized worldwide for its diverse types and motifs. With the many types and motifs of batik spread in various regions, each has historical value, characteristics, and certain rules that must be considered. This requires innovation in developing a system that can classify batik motifs using deep learning. Based on testing the Convolutional Neural Network method with MobileNet V1 and ResNet-152 V2 architectures for batik motif classification, the best results were obtained using the ResNet-152 V2 architecture with the shear pre-processing method and the RMSprop optimizer. From this configuration, the most optimal results were obtained with a test accuracy value of 89.67% and a validation loss of 0.44. Overall, the performance of the ResNet-152 V2 architecture is better than the MobileNet V1 architecture as shown by the significant performance difference. The test results presented show that this research can run well and ensure that the system operates according to the designed framework and is aligned with related topics, namely Communication and Multimedia Applications.

References

1. Damayanti, N.Y., Pandanwangi, A., Sukapura, B., Dewi, A.M.A.: Innovation of Gutha Tamarin's Batik Colet technique and visual language on Batik Bercerita ornaments by female artists. In: Damayanti, N.Y., Sayahdikumullah, D., Maulina, R. (eds.) ARTESH: 2nd International Conference on Art for Technology, Science, and Humanities, Bandung: Visual Art Study Program, Bandung Institute of Technology, pp. 67–76 (2020)
2. Rasyidi, M.A., Bariyah, T.: Batik pattern recognition using convolutional neural network. Bull. Electr. Eng. Inform. **9**(4), 1430–1437 (2020). https://doi.org/10.11591/eei.v9i4.2385.
3. Qiram, I., Buhani, Rubiono, G.: Batik Banyuwangi: aesthetic and technical comparison of Coastal Batik. LEKESAN: Interdiscip. J. Asia Pac. Arts **1**(2), 79–85 (2018)
4. Dewi Girsang, N., Muhathir: Classification of batik images using multilayer perceptron with histogram of oriented gradient feature extraction. In: Lin Liu, H., Shafiq, M., Wen Kin, H. (eds.) 4th Proceeding International Conference on Science and Engineering (ICSE), Medan: Faculty of Science and Technology Sunan Kalijaga State Islamic University Yogyakarta, pp. 197–204 (2021)
5. Sartana, Saptandari, E.W., Helmi, A.F.: Youth imagination about being an Indonesian. In: Hidayati, I., Kurniawan, R., Aulia, R. (eds.) Proceedings of the 1st International Conference on Psychology and Health Issues (ICOPHI 2022). Atlantis Press, Yogyakarta pp. 65–81 (2023). https://doi.org/10.2991/978-94-6463-212-5_8.
6. Bhatt, D., et al.: CNN variants for computer vision: history, architecture, application, challenges and future scope. Electronics **10**(20), 1–28 (2021). https://doi.org/10.3390/electronics10202470.
7. Purwaningsih, T., Ayu Anjani, I., Bekti Utami, P.: Convolutional neural networks implementation for chili classification. In: International Symposium on Advanced Intelligent Informatics (SAIN), pp. 190–194 (2018)
8. Hendriyana, H., Maulana, Y.H.: Identification of types of wood using convolutional neural network with MobileNet architecture. Jurnal RESTI (Rekayasa Sistem dan Teknologi Informasi) **4**(1), 70–76 (2020). https://doi.org/10.29207/resti.v4i1.1445.

9. Sae-Lim, W., Wettayaprasit, W., Aiyarak, P.: Convolutional Neural Networks Using MobileNet for skin lesion classification. In: 2019 16th International Joint Conference on Computer Science and Software Engineering (JCSSE) , pp. 242–247. IEEE, Songkhla (2019). https://doi.org/10.1109/JCSSE.2019.8864155.
10. Mohapatra, S., Abhishek, N.V.S., Bardhan, D., Ghosh, A.A., Mohanty, S.: Comparison of MobileNet and ResNet CNN architectures in the CNN-based skin cancer classifier model. Bhubaneswar (2021). https://doi.org/10.1002/9781119792611
11. Tseng, F.H., Yeh, K.H., Kao, F.Y., Chen, C.Y.: Mininet: dense squeeze with depthwise separable convolutions for image classification in resource-constrained autonomous systems. ISA Trans. **132**, 120–130 (2023). https://doi.org/10.1016/j.isatra.2022.07.030.
12. Prasetyo, E., Purbaningtyas, R., Adityo, R.D., Suciati, N., Fatichah, C.: Combining MobileNetV1 and depthwise separable convolution bottleneck with expansion for classifying the freshness of Fish Eyes. Inf. Process. Agric. **9**(4), 485–496 (2022). https://doi.org/10.1016/j.inpa.2022.01.002.
13. Koluguri, N.R., Park, T., Ginsburg, B.: TitaNet: neural model for speaker representation with 1D depth-wise separable convolutions and global context. In: IEEE International Conference on Acoustics, Speech and Signal Processing (ICASSP) (2022). https://doi.org/10.48550/arXiv.2110.04410.
14. Wang, W., Hu, Y., Zou, T., Liu, H., Wang, J., Wang, X.: A new image classification approach via improved MobileNet models with local receptive field expansion in shallow layers. Comput. Intell. Neurosci. **2020**, 1–10 (2020). https://doi.org/10.1155/2020/8817849
15. Rachburee, N., Punlumjeak, W.: Lotus species classification using transfer learning based on VGG16, ResNet152V2, and MobileNetV2. IAES Int. J. Artif. Intell. (IJ-AI) **11**(4), 1344–1352 (2022). https://doi.org/10.11591/ijai.v11.i4.pp1344-1352.
16. Meranggi, D.G.T., Yudistira, N., Sari, Y.A.: Batik classification using convolutional neural network with data improvements. Int. J. Inform. Vis. **6**(1), 6–11 (2022). https://doi.org/10.30630/joiv.6.1.716.
17. Hussain, M.A.I., Khan, B., Wang, Z., Ding, S.: Woven fabric pattern recognition and classification based on deep convolutional neural networks. Electronics (Basel) **9**(6), 1–12 (2020). https://doi.org/10.3390/electronics9061048.
18. Zhang, X., Gao, W., Pan, R.: Deep neural network with strip pooling for image classification of Yarn-Dyed plaid fabrics. CMES – Comput. Model. Eng. Sci. **130**(3), 1533–1546 (2022). https://doi.org/10.32604/cmes.2022.018763

Web Viewer for Educational VR Contents of 3D Scene Models Supporting VR Goggles

Yoshihiro Okada[1](✉), Kosuke Kaneko[2], and Wei Shi[3]

[1] ICER (Innovation Center for Educational Resources) of University Library, Kyushu University, Fukuoka, Japan
okada@inf.kyushu-u.ac.jp
[2] AiRIMaQ, Kyushu University, Fukuoka, Japan
[3] Research Institute for Information Technology, Kyushu University, Fukuoka, Japan

Abstract. In this paper, the authors introduce web viewers for educational VR contents. Especially, they propose new web viewer for educational VR contents of 3D scene models. Although there have been many types of VR goggles and a lot of entertainment VR contents, there have been a few educational VR contents. To enhance educational efficiency using VR technology, such educational VR contents should be created more and more. On the other hand, many types of scanning devices have been researched and developed, e.g., Lidar scanners, 360VR cameras and so on. Using these scanning devices, 3D data like Point Cloud Data (PCD), 3D model data and 360VR images/videos, can be obtained easily. To use these 3D data for educational VR contents, the authors have already developed viewer applications as web services. These web viewers have functionality to upload 3D data from users' web-browser to the web-server on that the web viewers work. However, 3D scene models are not supported. Standard videos are still important media for the on-demand online educations. Therefore, the authors propose new web viewer for 3D scene models supporting standard videos and VR goggles in this paper.

Keywords: Web viewers · 3D models · 360VR · Point Cloud Data · 3D Scenes

1 Introduction

We belong to ICER (Innovation Center for Educational Resources) of Kyushu University Library [1]. The mission of ICER is to provide e-learning materials using recent ICT and supports teachers for them. Also, our university established new initiative called Data-Driven Innovation Initiative last year. It consists of five divisions. We also belong to one of the divisions called NOE (Division of Next generation Open Education promotion) [2]. Its mission is to promote Education DX (Digital Transformation), and currently, we focus on the development of e-learning materials using VR technology. We have developed web viewers for educational VR contents as the activities of NOE and ICER, and then, we introduce them in this paper. In addition, we propose new web viewer for educational VR contents of 3D scene models supporting standard videos and VR goggles.

L. Barolli (Ed.): BWCCA 2023, LNDECT 186, pp. 69–80, 2024.
https://doi.org/10.1007/978-3-031-46784-4_7

Recently, many types of VR goggles have been released from many companies and many entertainment VR applications for them have been developed. However, there have been a few educational VR contents so far because the development of VR applications needs much time and human resources. However, our division, NOE wants to develop many educational VR contents for attracting students' interests and enhancing educational efficiency using VR technology. On the other hand, there have been many types of scanning devices so far, e.g., Lidar scanners, 360VR cameras and so on. Our NOE and ICER have these scanning devices for developing educational VR contents. By these scanning devices, we can obtain 3D data like Point Cloud Data (PCD), 3D model data, 360VR images/videos. To allow teachers to use these obtained 3D data as educational contents, we have already developed web viewer applications for them. In this paper, we introduce such web viewers [3]. Although our web viewers support 3D models, they do not support 3D scene models. Also, standard videos are still important media for the on-demand online educations. Therefore, we propose new web viewer application for 3D scene models and standard videos in this paper. Our web viewer applications also have functionality to upload 3D data from users' browser to the web servers that the web viewers work. By this functionality, even standard users like teachers can provide users like students with educational VR contents as web services. Also, our web viewers support VR goggles, they can provide students with immersive learning environments.

The remainder of this paper is organized as follows: Next Sect. 2 describes related work. In Sect. 3, we introduce current web viewers for 3D data like 3D model data, Point Cloud Data(PCD) and 360VR images/videos. And then, Sect. 4 introduce new web viewer for 3D scene models. Finally, we conclude the paper in Sect. 5.

2 Related Work

For developing interactive 3D graphics applications like VR contents, any toolkit systems should be used. As a research system of our laboratory, we use IntelligentBox [4] that is a development system for interactive 3D graphics desktop applications. We have already developed many 3D desktop applications including educational materials [5–8] actually using IntelligentBox. As shown in the left part of Fig. 1, there are web viewer applications developed using the web-version of IntelligentBox [9]. However, IntelligentBox and its web-version do not support VR goggles such as Meta Quest 2. Other toolkit systems should be used for developing educational VR contents. Unity [10] is one of the most popular game engines in the world. We have developed a couple of serious games [11, 12] for education purposes using Unity. Although Unity is very powerful, it requests developers programming knowledge and skills. So, Unity is not easy for common end-users like teachers to use it. Our web viewers [3] and VR tour development systems [13] of PCD, 3D model data and 360VR images/videos are developed using Three.js [14], one of the popular WebGL-based 3D graphics JavaScript library systems. We have developed several web-based 3D educational materials using our development frameworks [15–20] based on Three.js.

There are many commercial services for creating interactive web contents using 360VR camera images. The service of Ricoh Theta360.zi [21] does support 360VR images but not 360VR videos nor PCD. The service of Matterport [22] does support 360VR images and PCD but not 360VR videos.

Our web viewers [3] are supposed to be released as open web services. Then, it is thought that the number of educational VR contents will grow quickly, and as a result, the education using VR technology will be promoted. The right part of Fig. 1 shows one of the screen images of our current web viewer for 3D models. It does not support 3D scene models like our web-based 3D content for Japanese History study [17] shown in Fig. 2. Therefore, we developed new web viewer that supports 3D scene models.

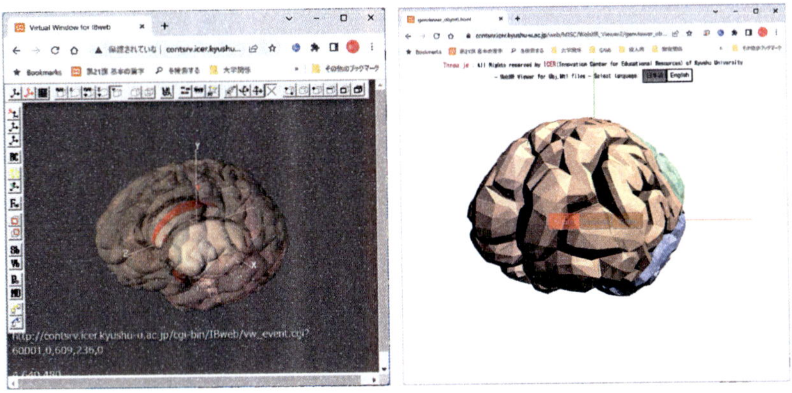

Fig. 1. Web viewer of IntelligentBox and current web viewer for 3D models (left and right).

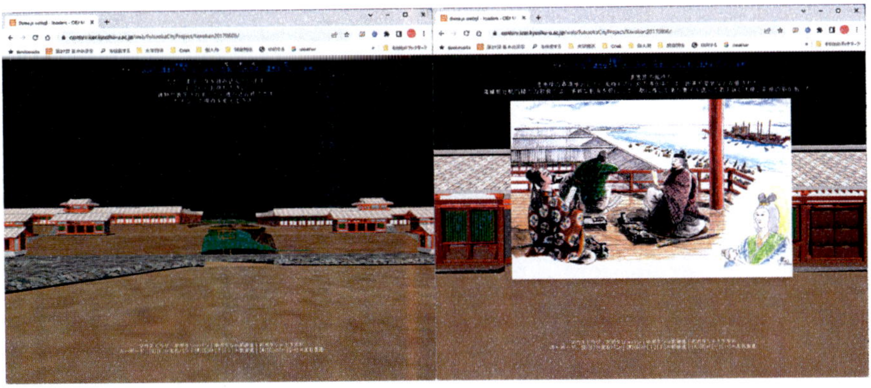

Fig. 2. Web 3D educational content of Korokan(Historical Building) for Japanese History study.

The web-based 3D content of Fig. 2 needs texture mapping images and you need to upload them together with 3D scene models to the server. This content also supports Image Animation functions as shown in the middle image of the right part of Fig. 2. When using VR goggles i.e. WebXR mode, the user cannot access any DOM elements of texts, images and videos like text tags, image tags and video tags. If you want to access these media in WebXR mode, you have to display them as 3D objects existing in a 3D space. For instance, by using the texture mapping technique, texts data, image data and video data can be displayed on the surface of a 3D plane model. Our web viewer for

3D scene models proposed in this paper employs this technique for Image Animation function, i.e., supports standard videos.

3 Current Web Viewers

3.1 Web Viewer for 3D Models

Users can upload their 3D model data to a target web server using its data upload web service as shown in Fig. 3. Its upper part is the screen image of the data upload web service. The texts are displayed in both Japanese and English. Once you choose 3D model file and two subsidiary files in the three boxes and push the send button, the local files will be uploaded to the web server. If the file upload is completed successfully, the message including two URLs (WebGL Ver. and WebXR Ver.) will appear in the browser as shown in the lower part of Fig. 3.

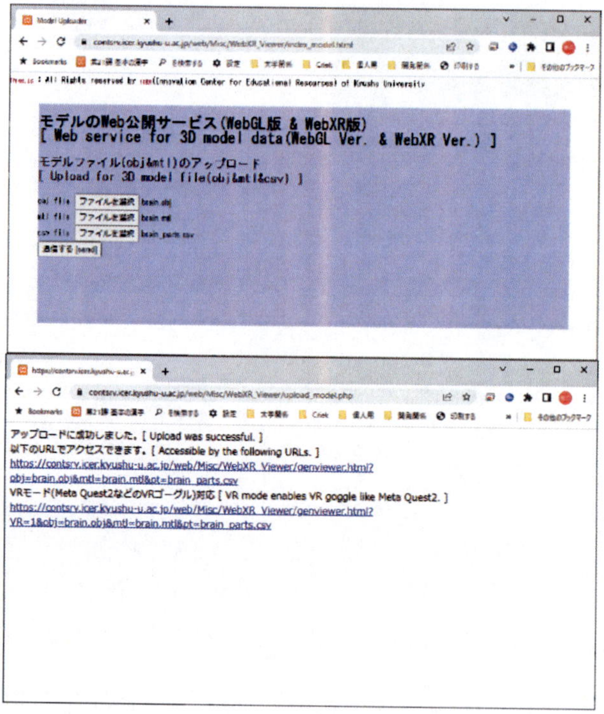

Fig. 3. Screen images of the 3D model data uploader before and after uploading the data files.

If you click on the upper URL(WebGL Ver.) or lower URL(WebXR Ver.), the uploaded model will appear in the browser as shown in Fig. 4. In the middle part of the browser, there is the menu to specify a manipulation type such as rotate, translate and zoom. You can rotate, translate and zoom the 3D model by the mouse drag operation. If you move the mouse cursor on any part of the 3D model, its corresponding name appears those names are defined in one of the subsidiary files. Then, students can learn the details of the 3D model such as a brain interactively. If you choose the lower URL(WebXR Ver.), you can see 'ENTER VR' button in the right lower part of the browser. By clicking on the button, the web viewer will become VR mode as shown in Fig. 5. This means that the user can look the 3D model using a VR goggle like Meta Quest 2.

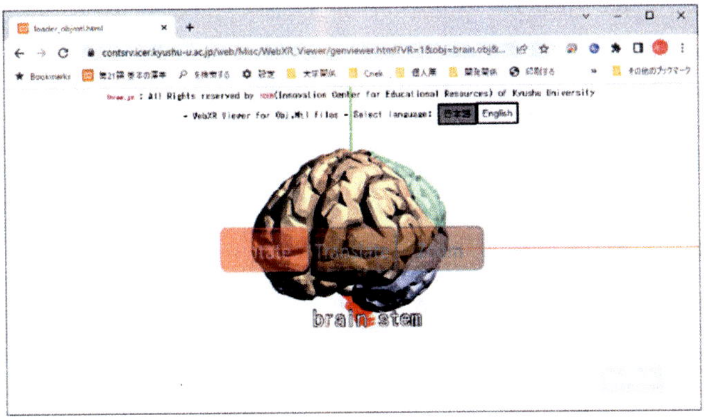

Fig. 4. Screen image of web viewer for 3D model data of WebXR mode.

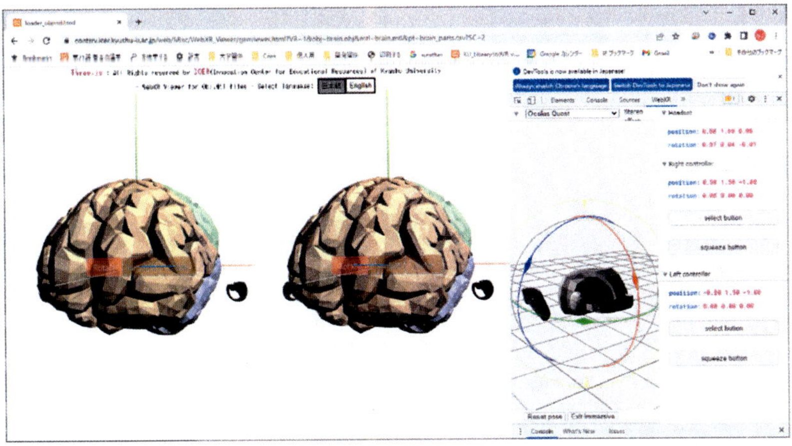

Fig. 5. Screen image of web viewer for 3D model data of WebXR mode (stereo view).

3.2 Current Web Viewer for Point Cloud Data(PCD)

Similarly to the above web viewer, users can upload their PCD using a data upload web service to a target web server. Once you choose PCD file in the input box of the data uploader and push the send button, the PCD file will be uploaded to the web server. If the file upload is completed successfully, the message including two URLs (WebGL Ver. and WebXR Ver.) will appear in the browser like the lower part of Fig. 3. If you click on the upper URL(WebGL Ver.) or the lower URL(WebXR Ver.), the uploaded PCD will appear in the browser as shown in Fig. 6. This case is the PCD of CT training room. You can change your eye position and direction in the room by the mouse drag operation for walk-through.

Fig. 6. Screen image of web viewer for PCD of WebXR mode.

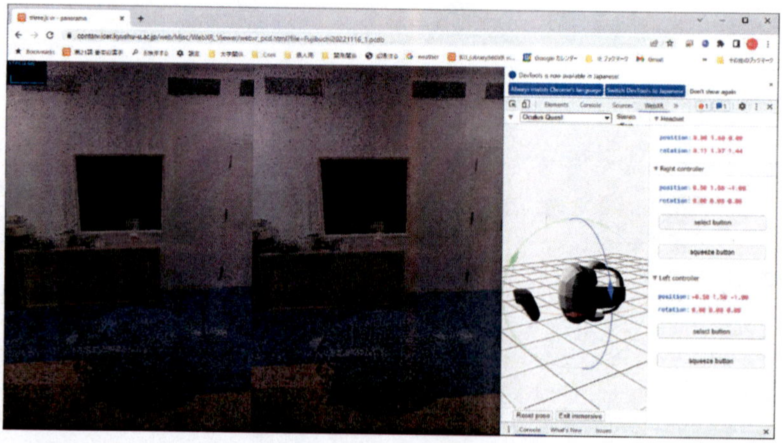

Fig. 7. Screen image of web viewer for PCD of WebXR mode (stereo view).

If you choose the lower URL(WebXR Ver.), you can see 'ENTER VR' button in the right lower part of the browser. By clicking on the button, the web viewer will become VR mode as shown in Fig. 7. This means that the user can see the CT training room using a VR goggle.

3.3 Current Web Viewer for 360VR Images/videos

Similarly to the above two viewers, users can upload their 360VR images/videos using a file upload web service to a target web server. In the uploader, there are two file selection input boxes for 360VR images and for 360VR videos. Once you choose 360VR image file or 360VR video file, and push the send button, the file will be uploaded to the web server. If the file upload is completed successfully, the message including two URLs (WebGL Ver. and WebXR Ver.) will appear in the browser. If you click on the upper URL(WebGL Ver.) or lower URL(WebXR Ver.), the uploaded 360VR image or 360VR video will appear in the browser as shown in Fig. 8 or in Fig. 9. These cases are also the CT training room. You can change your eye direction in the room by the mouse drag operation. In the case of 360VR video of Fig. 9, you can control the video by the buttons located in the middle of the browser.

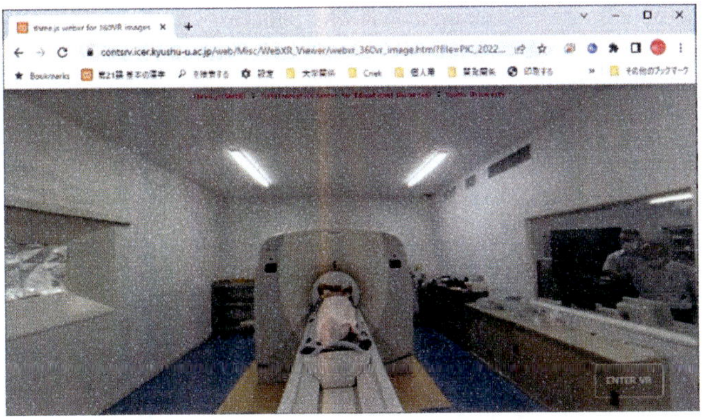

Fig. 8. Screen image of web viewer for 360VR image of WebXR mode.

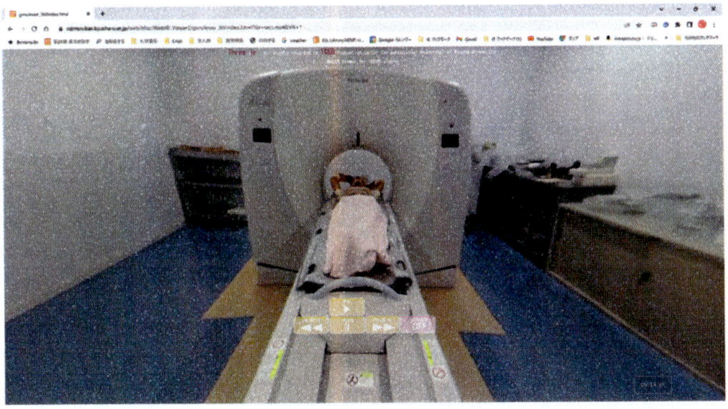

Fig. 9. Screen image of web viewer for 360VR video of WebXR mode.

If you click lower URL(WebXR Ver.), you can see 'ENTER VR' button in the right bottom of the browser. By clicking on the button, the web viewer will become VR mode like the above two web viewers. This means that the user can see the CT training room using a VR goggle. Medical students can learn how the CT therapy device works.

4 Web Viewer for 3D Scene Models

Similarly to the current web viewers, we developed a data upload web service by 'index_scene.html' and 'upload_scene.php'. Users can upload their 3D scene models using this service from their local PC to a target web server. The upper part of Fig. 10 shows the screen images of the data upload web service for 3D scene models. The texts are represented in both Japanese and English. In the upper part of the figure, there is one file selection box, this is for specifying one zip file. In the figure, KorokanAll.zip file is chosen. This zip file contains 'KorokanAll.obj', 'KorokanAll.mtl', 'KorokanAll_files.csv', 'KorokanAll_parts.csv', 'maps' directory and 'ImgAnimMovies' directory. Our web viewer for 3D models usually accepts 3D model data by '*.obj' file and its material data by '*.mtl' file. If you want to specify the names of 3D models individually, the viewer accepts '*_parts.csv' file. Indeed, the viewer also accepts '*_files.csv' file for specifying the details of 3D models such as their size and location in the 3D space.

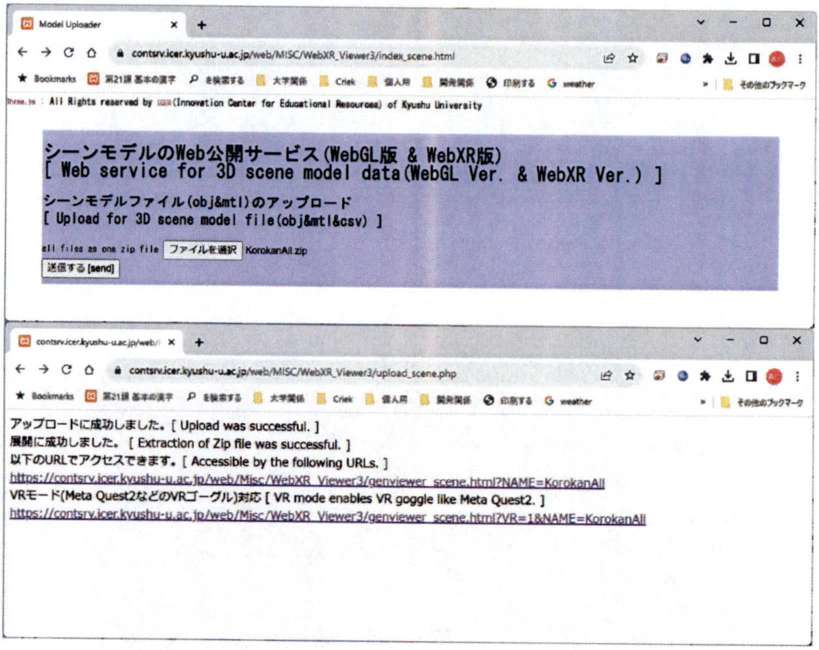

Fig. 10. Screen images of the 3D scene model data uploader before and after uploading the data files (upper and lower).

Similarly, the web viewer for 3D scene models accepts same kinds of files, i.e., 'KorokanAll.obj', 'KorokanAll.mtl', 'KorokanAll_parts.csv' and

'KorokanAll_files.csv' those are included in the zip file. Although the uploader of the web viewer for 3D models cannot upload texture mapping images, the uploader of the web viewer for 3D scene models can upload texture mapping images as one directory like the above 'maps' directory. As explained in Sect. 2, the web content shown in Fig. 2 has Image Animation function. 'ImgAnimMovies' directory includes several movie files for providing Image Animation function as similar movie files. These movies are displayed in a 3D space by the texture mapping technique. Finally, 'movies.js' file defines the details of these movies such as their file name, display size and location in a 3D space.

If the file upload is completed successfully, the message including two URLs (WebGL Ver. and WebXR Ver.) will appear in the browser as shown in the lower part of Fig. 10. This case, Korokan building model was uploaded on the server. Furthermore, if you click on the upper URL(WebGL Ver.) or lower URL(WebXR Ver.), the uploaded 3D scene models will appear in the browser as shown in Fig. 11.

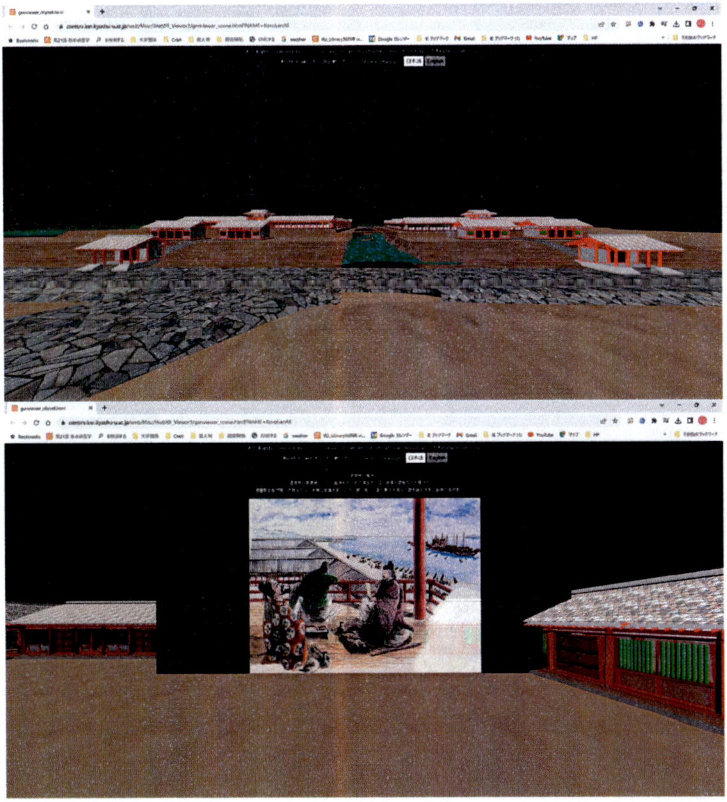

Fig. 11. Screen image of web viewer for 3D scene models, Korokan buildings.

You cannot see the menu to specify a manipulation type of rotate, translate and zoom such as the menu of web viewer for 3D models. Instead, you can change your eye position and direction in the scene by the mouse drag operation for walk-through like the case of web viewer for PCD. If you move the mouse cursor on any part of the 3D models, its corresponding name appears those are defined in the 'KorokanAll_parts.csv' file. Then, students can learn the details of the scene interactively. In the middle of the lower part of Fig. 11, you can see one video that works as Image Animation function. Therefore, students also can learn more from the video. If you click on the lower URL(WebXR Ver.), you can see 'ENTER VR' button in the right lower part of the browser. By clicking on the button, the web viewer will become WebXR mode supporting VR goggle as shown in Fig. 12. This means that the user can walk-through the scene using a VR goggle like Meta Quest 2.

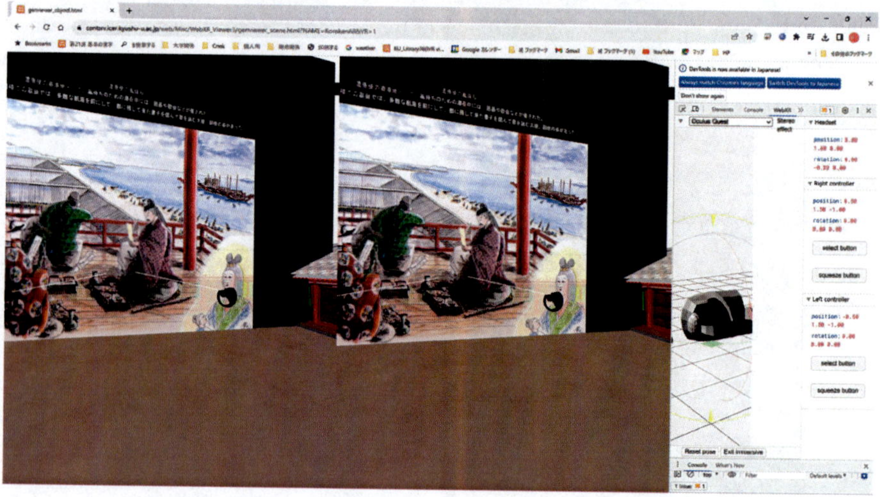

Fig. 12. Screen image of web viewer for 3D scene models of WebXR mode (VR goggle support).

5 Conclusions

Recently, many types of scanning devices have been researched and developed, e.g., Lidar scanners, 360VR cameras and so on. Using these scanning devices, we can obtain 3D data like Point Cloud Data (PCD), 3D model data, 360VR images/videos more easily rather than ever. We have already developed web viewer applications that allow common users like teaches to provide them as educational VR contents to students. These web viewers have functionality to upload 3D data from users' web-browser to the web-server on that the web viewers work. However, these web viewers do not support 3D scene models. Some cases, we need to use 3D scene models for education purposes. Therefore, in this paper, we propose new web viewer for 3D scene models supporting standard videos and VR goggles.

In the near future, it will become possible for common users like teachers to provide other users like students with educational VR contents as web contents easily. In addition, as our web viewers support VR goggles, teachers can provide students with immersive learning environments that are regarded to have higher educational efficiency. As future work, we will try to ask teachers to use our web viewer services and try to collect their comments for improving the services according to the comment. Also, we will consult students for evaluating educational efficiency of some educational VR contents.

Acknowledgements. This research was partially supported by JSPS KAKENHI Grant Number JP22H03705.

References

1. ICER: https://www.icer.kyushu-u.ac.jp/en. 05 Aug 2023
2. NEO: https://dx.kyushu-u.ac.jp/_cms_dir/uploads/2022/12/unit_oe.pdf. 05 Aug 2023
3. Okada, Y., Oki, K.: Web viewers for educational VR contents. In: IADIS International Conferences e-Society and Mobile Learning 2023 Proceedings, pp. 359–366 (2023)
4. Okada, Y., Tanaka, Y.: IntelligentBox: a constructive visual software development system for interactive 3D graphic applications. In: Proceedings of Computer Animation 1995, IEEE CS Press, pp. 114–125 (1995)
5. Okada, Y., Ogata, T. el al, Component-based approach for prototyping of Tai Chi-based physical therapy game and its performance evaluations. Comput. Entertain. **14**(1) (2017). https://doi.org/10.1145/2735383
6. Kosuki, Y., Okada, Y.: 3D visual component based development system for medical training systems supporting haptic devices and their collaborative environments. In: Proceedings of 6th International Conference on Complex, Intelligent, and Software Intensive Systems, CISIS 2012, pp. 687–692 (2012). https://doi.org/10.1109/CISIS.2012.131
7. Nomi, M., Okada, Y.: Improvement of dental treatment training system using a haptic device. In: Barolli, L., Takizawa, M., Enokido, T., Chen, HC., Matsuo, K. (eds.) Advances on Broad-Band Wireless Computing, Communication and Applications. BWCCA 2020. Lecture Notes in Networks and Systems, vol. 159, pp. 143–153. Springer, Cham. https://doi.org/10.1007/978-3-030-61108-8_14
8. Nomi, M., Okada, Y.: Dental Treatment Training System Using Haptic Device and Its User Evaluations. In: Barolli, L. (ed.) Complex, Intelligent and Software Intensive Systems. CISIS 2022. Lecture Notes in Networks and Systems, vol. 497, pp. 569–580. Springer, Cham. https://doi.org/10.1007/978-3-031-08812-4_55
9. Okada, Y.: Web Version of IntelligentBox (WebIB) and Its Integration with Webble World. In: Arnold, O., Spickermann, W., Spyratos, N., Tanaka, Y. (eds.) Webble Technology. WWS 2013. Communications in Computer and Information Science, vol. 372, pp. 11–20. Springer, Heidelberg. https://doi.org/10.1007/978-3-642-38836-1_2
10. Unity: https://unity.com/jp. 05 Aug 2023
11. Sugimura, R., et al.: Mobile Game for Learning Bacteriology. In: Proceedings of IADIS 10th International Conference on Mobile Learning 2014, pp. 285–289 (2014)
12. Sugimura, R., et al.: Serious games for education and their effectiveness for higher education medical students and for junior high school students. In: Proceedings of 4th International Conference on Advanced in Information System, E-Education and Development(ICAISEED 2015), pp. 36–45 (2015)
13. Three.js : https://threejs.org/. 05 Aug 2023

14. Okada, Y., Kaneko, K., Shi, W.: Development framework Using 360VR Cameras and Lidar scanners for web-based XR educational materials supporting VR Goggles. In: Barolli, L. (ed.) Advances in Internet, Data & Web Technologies. EIDWT 2023. Lecture Notes on Data Engineering and Communications Technologies, vol. 161, pp. 401–412. Springer, Cham (2023). https://doi.org/10.1007/978-3-031-26281-4_42
15. Okada, Y., Nakazono, S., Kaneko, K.: Framework for development of web-based interactive 3D educational contents. In: 10th International Technology, Education and Development Conference, pp. 2656–2663 (2016)
16. Okada, Y., Kaneko, K., Tanizawa, A.: Interactive educational contents development framework based on linked open data technology. In: 9th annual International Conference of Education, Research and Innovation, pp. 5066–5075 (2016)
17. Okada, Y., Kaneko, K., Tanizawa, A.: Interactive educational contents development framework and its extension for web-based VR/AR applications. In: Proceedings of the GameOn 2017, Eurosis, pp. 75–79 (2017). ISBN 978-90-77381-99-1
18. Ma, C., et al.: E-learning material development framework supporting VR/AR based on linked data for IoT security education. In: Proceedings of 6th International Conference on Emerging Internet, Data & Web Technologies (EIDWT 2018), pp. 479–491 (2018). ISBN 978-3-319-75928-9
19. Hirayama, D., et al.: Web-based interactive 3D educational material development framework and its authoring functionalities. In: Proceedings of the 22nd International Conference on Network-Based Information Systems (NBiS-2019), pp. 258–269 (2019). ISBN 978-3-030-29028-3
20. Yamamura, H., et al.: A development framework for RP-type serious games in a 3D virtual environment, VENOA 2020. In: Proceedings of the 14th International Conference on Complex, Intelligent, and Software Intensive Systems (CISIS-2020), pp. 166–176 (2020)
21. Ricoh Theta360.zi: https://www.theta360.biz/. 05 Aug 2023
22. Matterport: https://matterport.com/ja. 05 Aug 2023

Human Factors Impacting the Security Actions of Help Recipients

Ayane Sano[1,2](✉), Yukiko Sawaya[1], Takamasa Isohara[1], and Masakatsu Nishigaki[2]

[1] KDDI Research, Inc., Fujimino, Saitama 356-8502, Japan
ay-sano@kddi.com

[2] Graduate School of Science and Technology, Shizuoka University, Hamamatsu 432-8011, Shizuoka, Japan

Abstract. Some users ("Help recipients") delegate necessary security actions to their family, friends, or others close to them. It is important to be able to take appropriate defensive actions against security threats by themselves when help is not available from neighbors ("Helpers"). In this paper, we interviewed 9 users who used to be Help recipients, but who have now started to take security actions by themselves. We investigated the reason why Help recipients delegated their security actions to Helpers and the human factors that have an impact when one takes security actions by oneself. As a result, Help recipients take their own security actions when they try new hobbies or feel a sense of ownership. Based on these findings, we classify Help recipients into four groups and propose an optimized system that shows security action lists according to user situation. These findings are useful when providing appropriate intervention for Help recipients.

1 Introduction

Recently, there have been many cyberattacks such as those exploiting vulnerabilities, and phishing attacks. Some personal computer and smartphone users are required to take their own security measures to prevent such cyberattacks. However, it is reported that more than half of users do not continually take security action [1], because the same approaches are provided for all users despite difference in the level of security literacy, and skill of each user. To improve the rate of security action implementation, it is important to provide personalized interventions that focus on the different levels of security literacy, and skill of users that receive the intervention.

Some personal computer and smartphone users delegate necessary security actions to their family, friends, or others close to them [1]. We define such users as "Help recipients", and their family, friends, or others close to them as "Helpers". The Helpers do not always have close connections with Help recipients and the Help recipients cannot receive help from Helpers every time. For example, we assume the following situation. The Help recipient lives alone and uses an old software version and the Helper is mostly unable to support the Help recipient to take appropriate security action. The Help recipient may have been damaged by attacks that exploit vulnerabilities because the Help recipient could not update to the new software version. Therefore, there is a

L. Barolli (Ed.): BWCCA 2023, LNDECT 186, pp. 81–93, 2024.
https://doi.org/10.1007/978-3-031-46784-4_8

risk that the Help recipient will incur security damage when the Helper does not have a close connection with the Help recipient. In addition, it is reported that a bridging of the digital divide is required [2] and there are various approaches to support digital utilization [3–10]. We consider that it is important to support the Help recipient to take appropriate defensive actions against security threats by oneself.

Our aim is to enable the Help recipients take security action by themselves, and we need to support appropriate approaches for them. Therefore, to understand user characteristics of the Help recipients, we conducted user interviews as a first step, considered appropriate approaches as a second step, and verified the effectiveness of these approaches as a final step. In this paper, we investigated user characteristics of the Help recipients and human factors that impact the Help recipients when taking security actions. We interviewed 9 users who used to be Help recipients, but who now take their own security action. We investigated why the Help recipients delegated their security actions to Helpers and what led them to take security action by themselves. The result revealed that the lacks of literacy, skill, times and security awareness are reasons why the Help recipients delegated their security action to Helpers. We found that the Help recipients took their own security action when they tried new hobbies or felt a sense of ownership. We classify the Help recipients into four group based on the level of literacy and experience, user occupation and so on. It is important that different approaches are used for each group. Based on these findings, we propose a personalized intervention that shows an optimized security action list according to user situation such as OS of device, service, and so on.

This paper describes related works in Sect. 2 and the interview method in Sect. 3. We describe the findings of our interviews in Sect. 4 and discuss appropriate approaches for the Help recipients and the limitations of this research in Sect. 5. Finally, the conclusion and future works are discussed in Sect. 6.

2 Related Works

In this section, we describe approaches to bridge the digital divide and difference of security awareness between user characteristics.

2.1 Approaches to Bridge Digital Divide

Various approaches are proposed to bridge the digital divide. It is reported that two approaches, improving digital literacy and improving UI/UX are required for digital utilization [2]. We describe related works in terms of the two points above.

First, we focus on improving digital literacy. Telecommunication Carriers [5–8] conducted several seminars on how to use smartphones. Local governments [9] and schools [10] held smartphone classes for elderly people.

Next, we describe improving UI/UX. To realize this, it is important to provide personalized intervention according to user characteristics. For example, appropriate UI design differs depending on gender, age, and years of device use such as expert or beginner [11, 12]. In addition, there are several works [13, 14] on approaches for elderly people. Salman et al. [13] described the research in which five experts evaluated the usability of smartphones for elderly people. As a result, "appearance", "language", "dialogue" and "information" have an effect on improving the usability. Olwal et al. [14] proposed the idea of UI design of smartphones for elderly people and this design improved usability. From the above, we expect that UI/UX will be improved to create design that takes into consideration the gender, age, and years of device use of target users.

2.2 Difference of Security Awareness Between User Characteristics

Related works [15–18, 20, 21] describe that security awareness differs according to age, gender, culture, and the level of IT skill. Frik et al. [15] conducted interviews of elderly people and analyzed their security and privacy awareness. As a result, services that are difficult to use or the lack of knowledge and skill decrease their self-efficacy. Garg et al. [16] conducted an experiment with two groups (video-based risk communication or text-based risk communication) of elderly people. They showed that leveraging episodic memory using video-based risk communication can improve the ability to avoid phishing attacks and downloading malware. McGill et al. [17] describe the difference of security awareness and behavior between genders. Females have lower levels of security action than males because they have a lack of knowledge and skill. Sawaya et al. [18] conducted a survey on seven countries by using the revised Security action Intentions Scale (SeBIS [19]). SeBIS is a scale to measure the level of taking security action of users. The survey revealed the different characteristics that exist between the seven countries. Ndibwile et al. [20] investigated the difference of security action between Japan and Tanzania. They state that demographic factors such as culture, income, and geographic location have a major impact on OS update behavior. Wash et al. [21] found that there is a demographic difference of beliefs regarding security and security actions. They describe that a one-size-fits-all education approach is not appropriate for security. From the above, user awareness and behaviors are different according to characteristics, surrounding environment, past experiences, and so on.

3 Methodology

As described in Sect. 2.2, the level of literacy and skill differ among each user. To promote security action, personalized intervention is required for each user. In particular, the Help recipients cannot take security action by themselves. There is a risk that the Help recipient may incur security damage because the Helper does not always have a close connection with the Help recipient. In addition, it is important to bridge the digital divide. Therefore, we aim to support the Help recipients to take appropriate defensive actions against security threats by themselves.

In this paper, we investigated human factors which impact on taking security action of the Help recipients as a first step. We conducted online in-depth interviews from March 21st to March 29th 2022 by using the services of a Japanese research company (Cross Marketing, Inc. [22]). We recruited 10 participants who delegated security action to a Helper before, but who now take security action by themselves. However, as 1 participant did not delegate to others, we focused on the interviews of the remaining 9 participants. They had past experience which made it difficult for them to take their own security action. The related works [23, 24] state that more than 5 participants are required to ensure completeness at the usability test. Therefore, we set 9 participants as the number of this interview. The participant demographics are shown in Table 1. We judged their level of literacy subjectivity based on their past experience and the status of understanding technical terms.

We conducted 1-h, in-depth interviews online for each participant. The main content of the interviews is as follows.

1. What causes you anxiety when using a computer or/and smartphone?
2. What kinds of security actions do you or Helpers take on your computer or/and smartphone now?
3. Why did you delegate security action to Helpers before?
4. What kinds of security actions do you now take by yourself?
5. How did you come to take security actions by yourself?
6. Who are Helpers for you?

In this interview, we did not collect any information that could identify the participants. All responses from the participants were anonymized. In accordance with the regulations of the research company, the participants received a fee after we conducted the interviews. This research was exempt from ethical review in accordance with the guidelines of our office.

Table 1. Participant demographics.

No	Gender	Age	Occupation	Literacy
1	Female	70–79	Homemaker	Low
2	Male	50–59	Employee	High
3	Female	20–29	Homemaker	Middle
4	Male	40–49	Employee	Middle
5	Male	40–49	Self-employment	Middle
6	Female	20–29	Employee	Low (PC), High (Smartphone)
7	Female	50–59	Part-time job	Low
8	Female	20–29	Student	Low (PC), High (Smartphone)
9	Male	40–49	Employee	Middle

4 Result

In this section, we describe the interview results.

4.1 Causes of Anxiety in Users

We describe the causes of anxiety in users. The numbers of participants are shown in parentheses.

- Spam/Phishing attack (1, 5)
- Communication with a person whose face the user does not know (1)
- Unauthorized use of money (1, 2, 3, 5, 6, 8, 9)
- Leakage of personal data (1, 2, 3, 4, 5, 6, 7, 9)
- Influence on other people in the case of incurring security damage (1, 2, 9)
- Unauthorized use of my computer (4, 7)
- Fraud (7)
- Be Identified oneself when using SNS (8)

We found that many users feel anxious when unauthorized use of money and leakage of personal data occur because phishing attacks are increased near them recently. Users recognize the importance of taking security action to prevent damages from phishing attacks.

4.2 Status of Taking Security Action

We describe the security actions taken by the Help recipients or Helpers as follows. The numbers of participants are shown in parentheses. Security action taken by the Help recipients is indicated as "R", and security action taken by the Helpers is indicated as "H". "PC" indicates those users that take security action only on a PC.

- Defense against spam/phishing attack (1-R, 2-R, 3-R, 4-R, 5-R, 6-R,7-R, 8-R, 9-R)
- Set password complexity (1-R, 3-R, 5-R, 7-R, 8-R)
- Change password regularly (1-R, 2-R, 3-R, 5-R, 6-R, 7-R, 9-R)
- Use biometric authentication/one-time password/two-factor authentication (1-R, 3-R, 4-R, 5-R, 7-R, 8-R, 9-R)
- Install anti-virus software (1-H, 2-R, 3-R, 4-H, 5-H, 6-H: PC, 7-H, 8-H: PC, 9-H)
- Update OS and applications (1-H, 2-R, 3-R, 4-R, 5-R, 6-R,7-R, 8-R: PC, 9-R)
- Virus scan (1-R, 2-R, 3-R, 4-R, 7-R)
- Get security protection service (1-H, 2-R, 4-H, 5-R, 9-R)
- Do not carry device (1-R)
- Do not save personal data on my device (2-R)
- Do not use personal data (2-R, 6-R)
- Do not use public Wi-Fi (1-R, 2-R, 3-R, 7-R, 8-R)
- Download application after investigating it (2-R, 4-H)
- Set network configuration (3-H)
- Use VPN (3-R)
- Do not use auto login service (6-R)
- Use private browser (8-R)
- Share with close friends only (8-R)

All users defend themselves against phishing attacks because such attacks are increased near them and they feel anxious about phishing attacks as described in Section 4.1. Moreover, we assume that the rate of defense against phishing attacks is high because security damage from phishing attacks is constantly in the news and the degree of recognition of this security action is high. We assume that increasing the degree of recognition of other security actions is required.

The rate of installing anti-virus software is high; however, the users cannot do it by themselves. Some users do not recognize that updating OS and applications are types of security actions. However, they recognize that such actions are necessary to improve the usability of OS and applications.

4.3 Reasons Why the Help Recipients Delegated Security Action to Helpers

The reasons why the Help recipients delegated security action to Helpers are shown in Table 2. In Table 2, many users stated their lack of knowledge. In particular, they cannot distinguish proper security action because there are much of information and many items such as anti-virus software. Any users said that they have no adviser and seminar by which they can learn about security action. Therefore, it is important for the Help recipients to attend seminars and show them security actions that are suitable for their devices. Moreover, any users do not feel that their device is dangerous. We recognize that it is important for the Help recipients to increase their sense of ownership.

Table 2. Reasons why the Help recipients delegated security action to Helpers.

Category	No	Comment
The lack of knowledge	1	I did not understand the technical terms
	2	I cannot distinguish the difference between correct mail and suspicious mail
	3	I did not know what anti-virus software is appropriate for my computer when checking various sites. I did not understand about computer configuration
	4	There are many applications of Android smartphone. I cannot distinguish legitimate applications
	5	I did not distinguish appropriate security action. I did not understand the technical terms
	6	As it was the first time for me to use the device, I felt that it was safe to delegate security to the store clerk of an electronics store
	7	I do not know about anti-virus software in detail. I do not know what security action is appropriate and how to take it
	9	I have no knowledge and skills about security. There is much anti-virus software, but I cannot distinguish suitable software on my device. It is difficult for me to read the user manual because there are many technical terms

(*continued*)

Table 2. *(continued)*

Category	No	Comment
Busyness	2	I was busy, so I did not have enough time
Interest	3	I was not interest in security action. I did not feel dangerous not to take security action
Necessity	4	According to the news, cyberattacks happen all the time, but I feel that it is safe
	5	I feel that security action is not necessary because I can ask to an engineer
	7	I do not feel that it is dangerous until showing warning messages
	8	I share a computer with my family. My family takes security action, so I do not need to take it
The lack of educational opportunity	7	I want to learn about security action at a seminar, but I do not have the opportunity to do so

4.4 Security Actions that the Help Recipients Came to Take by Themselves

We describe the security actions that the Help recipients came to take by themselves as follows. The numbers of participants are shown in parentheses.

- Defense against spam/phishing attack (1, 2, 4, 5, 6, 7, 8, 9)
- Set password complexity (1, 7, 8)
- Change password regularly (1, 7)
- Update OS/applications (2, 3)
- Do not save personal data on device (2)
- Install anti-virus software (3)
- Virus scan (3, 4)
- Use VPN (3)
- Do not use personal data (6)
- Do not use auto login service (6)
- Do not use public Wi-Fi (7)

Defense against spam/phishing attack, setting password complexity, changing password regularly, and updating OS and applications are relatively easy for Help recipients to do. Therefore, the Help recipients came to take security action when they felt that it was easy for them to do so. As described in Sect. 4.3, installing anti-virus software is difficult for them because they cannot distinguish software that is suitable for their devices. We consider that the rate of these security actions may increase if we tell the Help recipients that they are easy to take.

4.5 Opportunities that Enabled the Help Recipients to Take Security Action by Themselves

The opportunities that enabled the Help recipients to take security action by themselves are shown in Table 3. In Table 3, many users stated past experience by which the Help recipient and someone close to them damaged from security attacks. We assume that the sense of crisis and ownership increased in the Help recipients. Recently, smartphone seminars are conducted at a smartphone shop [5–8], government office [9], school, and office. The Help recipient could understand the necessity of taking security action and learned how to take it. News announcements and notifications by trusted companies (e.g., large company, famous company) enable the Help recipient to take their own security actions. Therefore, regular information campaigns are required to increase their awareness of the Help recipient.

A Helper that can teach about security action is required until the Help recipient can take security action by themselves according to the comments of participant no. 2, 4, and 9 in the "Helper" category. The Help recipient take security action by themselves because the Help recipient recognizes that the Helper does not always have a close connection with them according to the comment of participant no.7 in the "Helper" category. Therefore, we consider that support by a Helper is required until the Help recipient reaches a certain competency level, but it also helps the Help recipient to recognize that they cannot always delegate their security actions to a Helper.

In addition, the Help recipients take actions by themselves when the Help recipients begin to use new services according to the comments of participants no. 1 and 3 in the "Usability" category. We assume that it is easy for the Help recipients to feel a sense of responsibility which is required to take security action in the case of beginning new interests because they have high security awareness. However, the support for them is insufficient currently. For example, much information is provided for them and it is difficult to take security action by themselves. It is urgently necessary that we consider the solutions to provide suitable information for each user.

Table 3. The opportunities that enabled the Help recipients take security action by themselves.

Category	No	Comment
Past experience	1	I was infected by a trojan horse before
	2	I heard from my friend about the experience where her computer was infected by a virus
	4	I heard from my colleague about the experience of damage incurred by phishing mail. I heard from my customer about the experience of a cyberattack. I have been receiving phishing mails for the last 2 ~ 3 years

(*continued*)

Table 3. (*continued*)

Category	No	Comment
	5	I incurred security damage from ransomware and spam mails. I fell victim to a fraud by dishonest businessman
	7	I was incurred by phishing attack before
	8	I was incurred by unauthorized using money. I accessed a suspicious site before
	9	I got phone fraud before
Notification	1	The financial institution advises me to change my password frequently
	2	The game office tells me not to upload personal data on games
	3	The site advises me to change password to protect myself from attackers
Anxiety	1	I worry about leakage of personal data
Education	1	I attended a smartphone seminar. I got information from a television show
	2	I attended seminars at the office. I got information from newspapers and TV news
	3	I was taught by my family, friend, news, and YouTube
	6	I was taught by school, my family, and news
	7	I was taught by my family and news
	8	I learn at school seminars
	9	I learn about security action at a seminar at my office
Usability	1	I want to broaden my world
	3	I want to use my computer thoughtfully to begin blog
Privacy	1	I have secrets that I do not want my family to know
Helper	2	I consult an adviser (e.g., store clerk, office worker) when I am in trouble
	4	I consult my friend who is knowledgeable about security
	5	I have no one else to consult about security near me
	7	I cannot always consult my family in the case they go out
	9	I consult my friend who is knowledgeable about IT literacy. I am taught by him to take security action by myself
Necessity	3	I need to protect my computer and blog
Cost	6	I need to pay the fee for setting up my computer to a store clerk in the case of delegating to them
Easiness	6	I feel that taking security action is easy for me if I am taught by others how to do so

4.6 Helpers

We describe the Helpers to whom the Help recipients delegate their security action as follows. The numbers of participants are shown in parentheses.

- Family (1, 3, 7, 8)
- Store clerk of electronics store/smartphone shop (1, 2, 4, 5, 6, 7)
- Friend (4, 9)
- Colleague at office (2, 7)

Many users answered that they delegate their security action to their family, friend and store clerk of an electronics store/smartphone shop. We found that the Help recipients delegate their security action to the store clerk of an electronics store/smartphone shop in the case of asking about technical content such as the functions of each anti-virus software.

5 Discussion and Limitation of Works

In this section, we discuss the characteristics of the Help recipients, proper approaches to promote security action for the Help recipients, and the limitations of this research.

5.1 Characteristics of Help Recipient

We discuss the characteristics of the Help recipients based on the result of interviews. As described in Sect. 4.3, the reason why the Help recipients delegated their own security actions to Helpers is that they have insufficient knowledge and skills to take security action by themselves. However, their security awareness and the motivation to take security action are high because they or someone close to them has incurred security damage in the past. Therefore, it is important to support them to increase their knowledge and skill by using unconventional methods.

In addition, as described in Sect. 4.5, the Help recipient is easy to feel a sense of responsibility to take security action in the case of beginning new interests such as a blog. If the company announces the importance of taking security action when using a service, the users may feel a sense of ownership of it.

Based on the above, we classify the Help recipients into four group as shown in Table 4. Each group differs in terms of age, gender, occupation, the level of literacy, experience, and opportunity that enabled them to take security action by themselves. Therefore, different approaches are required to promote security action for each group.

Table 4. Classification of Help recipients.

Group	No	Characteristics
A	1, 3, 7	• Users delegate to family • Occupation is mainly housemaker • The level of literacy is low • The motivation to take security action is high when users begin new interests
B	2, 4, 9	• Users delegate a part of security action • Occupation is employee • The level of literacy is middle or high • The motivation to take security action is high when users take seminars at the office
C	6, 8	• Users who are young • Occupation is student or freshman employee • Users delegated to family when they were children, but they now take security action by themselves when they become independent • Users took a seminar at school before • The frequency of PC is low, but the frequency of smartphone is high
D	5	• Users delegate to a vendor • Occupation is self-employment • Users have no experience of taking a seminar • Users want to take security action that does not to cost money

5.2 Approach for Help Recipients

In Table 2, many users find it difficult to identify suitable security actions on their devices because there are many security services and anti-virus software. Any users do not understand the technical terms on the website, and in the manual. Therefore, we consider that it is appropriate to provide for the Help recipients only the information they need to take security action. For example, we propose an optimized system to show the personalized list of the security actions for each user. First, users input their device information such as model, OS type, services used, and so on. Next, the system extracts the information they need to take security action. Finally, the system shows the information for users.

As described in Sect. 5.1, different approaches are required to promote security action in terms of the characteristics of the Help recipients. For example, for Group A in Table 4, it is appropriate that the Helper teaches security action until the Help recipient attains a certain level of comprehension about security. For Group C in Table 4, it is appropriate to raise the degree of recognition about other security action with the exception of spam mail, phishing attack and SNS privacy. The current seminar contents are mainly about defense against spam and phishing attacks and protection of SNS privacy, so users do not recognize the importance of taking other security action. In the future, we will consider suitable approaches for each group in detail. We will reveal whether the personalized list described above is appropriate for each group.

5.3 Limitation

In this research, we conducted online interviews in Japan. We assume that there are different characteristics of Help recipients in other countries. The result described in Sect. 4 is obtained from the qualitative research of 9 users. We will conduct quantitative research with a larger number of people to verify the validity of the classification of Help recipients, and approaches for them.

6 Conclusion and Future Works

We describe the risk of security damage that the Help recipients incur in the case that the Helper does not have a close connection with the Help recipient. Therefore, we aim to support appropriate approaches that enable the Help recipients to take security action by themselves. In this paper, we conducted online interviews to reveal the user characteristics of Help recipients as a first step. As a result, the reason why the Help recipients delegated security action to Helpers is their lack of literacy, skill, time and security awareness. We understand that Help recipients come to take security action by themselves when they try new hobbies or feel a sense of ownership. We classify Help recipients into four groups based on the level of literacy and experience, user occupation, and so on. It is important for each group to promote different approaches based on user characteristics, and we described an example of personalized intervention in which the optimized system shows a security action list according to user situation such as device and services.

In the future, we will conduct quantitative research to verify the effectiveness of the classification of Help recipients. We will develop the proposed method and verify the effectiveness of this method.

References

1. Sano, A., Sawaya, Y., Yamada, A., Kubota, A.: SeBeST: security behavior stage model and its application to OS update. In: Proceedings of the 35th International Conference on Advanced Information Networking and Applications, AINA, vol.2, pp. 552–566 (2021)
2. Ministry of Internal Affairs and Communications: WHITE PAPER information and Communications in Japan. https://www.soumu.go.jp/johotsusintokei/whitepaper/eng/WP2021/chapter-3.pdf. Accessed 17 May 2023
3. United Nations Development Programme: Bridge the digital divide. https://www.undp.org/digital/standards/2-bridge-digital-divide. Accessed 17 May 2023
4. IEEE: Solutions to the digital divide moving toward a more equitable future. https://ctu.ieee.org/solutions-to-the-digital-divide-moving-toward-a-more-equitable-future/. Accessed 17 May 2023
5. NTT Docomo: Safety and Security for Children; https://www.docomo.ne.jp/english/corporate/csr/safety/. Accessed 07 June 2023
6. KDDI: Creating a Digital Society for Affluence and Peace of Mind (Connecting Hearts and Minds). https://www.kddi.com/english/corporate/sustainability/sdgs/security/. Accessed 07 June 2023
7. Softbank: Responding to the Digital Divide; https://www.softbank.jp/en/corp/sustainability/esg/social/digital-divide/. Accessed 07 June 2023

8. Rakuten Senior: Smartphone Class. https://senior.rakuten.co.jp/sp_class/movie/. (In Japanese). Accessed 07 June 2023
9. Kyodo News: FEATURE: Japan local gov'ts helping aged people learn how to use smartphone. https://english.kyodonews.net/news/2021/11/a52b5f784700-feature-local-govts-helping-aged-people-learn-how-to-use-smartphone.html. Accessed 08 June 2023
10. The Mainichi: Smartphone classes taught by teens attract flood of elderly applicants in central Japan. https://mainichi.jp/english/articles/20220816/p2a/00m/0na/020000c. Accessed 08 June 2023
11. Weinschenk, S.: 100 MORE Things Every Designer Needs to Know About People. O'REILLY, Japan (2016)
12. Weinschenk, S.: 100 Things Every Designer Needs to Know about People 2nd edn. O'REILLY Japan (2021)
13. Salman, H.M., Ahmad, W.F.W., Sulaiman, S.: Usability evaluation of the smartphone user interface in supporting elderly users from expert's perspective. IEEE Access **6**, 22578–22591 (2018)
14. Olwal, A., Lachanas, D., Zacharouli, E.: OldGen: mobile phone personalization for older adults. In: Proceedings of the SIGCHI Conference on Human Factors in Computing Systems, pp. 3393–3396 (2011)
15. Frik, A., Nurgalieva, L., et al.: Privacy and security threat models and mitigation strategies of older adults. In: Proceedings of the Fifteenth Symposium on Usable Privacy and Security, SOUPS, pp. 21–40 (2019)
16. Garg, V., Jean Camp, L., Connelly, K., Lorenzen-Huber, L.: Risk communication design: Video vs. text. In: Proceedings of the 12th Privacy Enhancing Technologies Symposium, PETS, pp. 279–298 (2012)
17. McGill, T., Thompson, N.: Gender Differences in Information Security Perceptions and Behaviour. Australasian Conference on Information Systems (2018)
18. Sawaya, Y., Sharif, M., et al.: Self-confidence trumps knowledge: a cross-cultural study of security behavior. In: Proceedings of the CHI Conference on Human Factors in Computer Systems, CHI, pp. 2202-2214 (2017)
19. Egleman, S., Peer, E.: Scaling the security wall: developing a security behavior intentions scale (SeBIS). In: Proceedings of the 33rd Annual ACM Conference on Human Factors in Computer, CHI, pp. 2873–2882 (2015)
20. Ndibwile, J.D., Luhanga, E.T., et al: A demographic perspective of smartphone security and its redesigned notifications. J. Inf. Process. **27**, 773–786 (2019)
21. Wash, R., Rader, E.: Too much knowledge? Security beliefs and protective behaviors among United States Internet Users. In: Proceedings of the Eleventh USENIX Conference on Usable Privacy and Security, SOUPS, pp. 309–325 (2015)
22. Cross Marketing Inc. https://www.cross-m.co.jp/en/. Accessed 17 May 2023
23. Nielsen, J., Landauer, K.T.: A mathematical model of the finding of usability problems. In: Proceedings of INTERACT'93 and CHI'93 Conference on Human Factors in Computing Systems , CHI, pp. 206–213 (1993)
24. Nielsen Norman Group: Why You Only Need to Test with 5 Users. https://www.nngroup.com/articles/why-you-only-need-to-test-with-5-users/. Accessed 17 May 2023

Proposal for Approaches to Updating Software on Android Smartphone

Ayane Sano[1,2](✉), Yukiko Sawaya[1], Takamasa Isohara[1], and Masakatsu Nishigaki[2]

[1] KDDI Research, Inc., Fujimino 356-8502, Saitama, Japan
ay-sano@kddi.com

[2] Graduate School of Science and Technology, Shizuoka University, Hamamatsu, Shizuoka 432-8011, Japan

Abstract. It is important to provide personalized interventions that focus on the different security awareness of each user. We focus on updating operating system (OS) and seven major types of applications (Communication, Finance, Lifestyle, Games, Utility, Health, Entertainment) for smartphone users and aim to provide approaches that lead to updating of OS and seven types of applications (hereinafter, this is called "software"). We consider that user awareness of updating software may differ, and this relates to user understanding of the updating procedure. In this paper, we propose intervention methods according to users' knowledge about update procedures. We conduct an online survey to evaluate effective approaches such as dialog message and type of incentive that increases the intention of smartphone users to update. We found that effective phrases of dialog messages differ according to the users' understanding of the update procedures and that reward points are the best incentive for many users.

1 Introduction

Users are required to take their own security measures to prevent cyberattacks. However, it is reported that more than half of users do not continually take security action [1], because the same approaches are provided for all users even though security awareness differs from each user. Therefore, it is important to provide personalized interventions that focus on the different security awareness of target users. As one of the problems regarding the rate of taking security action, related works [2–4] have reported that many users do not update applications immediately. To improve the rate of immediate updating, we focus on updating operating system (OS) and applications for smartphone users and aim to provide appropriate approaches for them. We target seven types of applications (Communication, Finance, Lifestyle, Games, Utility, Health, Entertainment) which are commonly provided in Google Play [5].

As one of the existing intervention methods, it is reported that a personalized user interface (UI) is required to promote the updating of computer applications [6]. In the previous work, we stated that effective dialog messages for promoting Windows Update differ according to user awareness [1, 7]. Similar to personal computers, we consider that personalized dialog messages are required for the smartphones. However, computers and smartphones have aspects of both commonality and difference [6, 8]. Therefore, effective

L. Barolli (Ed.): BWCCA 2023, LNDECT 186, pp. 94–108, 2024.
https://doi.org/10.1007/978-3-031-46784-4_9

dialog messages for smartphone users do not necessarily correspond to effective dialog messages for computer users. In this paper, we evaluate personalized dialog messages for smartphone users. We consider that user awareness of updating software may differ according to the users' understanding of the update procedure. As a first step, we analyze effective phrases of dialog messages in terms of users' knowledge about the update procedure. In addition, since the related works [9, 10] have reported that incentives are appropriate for promoting certain behaviors, we verify that effectiveness of incentives. We assume that these approaches may differ between software and compare them.

We evaluate effective dialog messages and incentives that promote the updating of software for Android users by an online survey. First, we classify users into two groups based on the status of understanding the update procedure. The two groups consist of the group that understands the update procedure (UUP), and the group that does not understand the update procedure (NUUP). Second, we conduct the online survey to evaluate effective phrases of messages for the two groups and the best incentive from among several types of incentives such as reward points, coupons, mobile data and so on. We consider the following two research questions.

- RQ1: What are effective phrases of messages for each group (UUP and NUUP)?
- RQ2: What is the best incentive to promote the updating software?

We recruited 2400 participants to verify the above research questions. We allocated 300 participants to eight groups who answered about updating each software. We analyzed the difference in the rate of immediate updating between UUP and NUUP, and between software. We found the following results.

- Effective phrases of dialog messages differ between UUP and NUUP, and between software. Loss framed messages (LFM) are the most effective phrases for many groups.
- Reward points are the best incentive to update software.

These findings show that personalized interventions are effective in encouraging the updating of software for smartphone users.

2 Related Works

In this section, we describe user awareness of updating OS and applications, interventions for behavior change, and approaches that promote the taking of security actions.

2.1 User Awareness of Updating OS and Application

We describe related works [2–4, 11–13] about user awareness of updating OS and applications. Mathur et al. [2] described the characteristics of users who do not automatically update applications. For example, these users had negative experiences against updating applications in the past, they tend not to take risks, and they have high security awareness. Rajivan et al. [3] researched the characteristics of users who delay updating software. Experience, cost, and risk preference affect user behavior. Vaniea et al. [4] interviewed Windows users about the reason for avoiding the updating of OS. They found that negative experiences in the past (such as user interface changes) lead to update avoidance. Vaniea et al. [11] described user behavior toward updating software. They stated that there are six stages until updating of software: awareness, deciding to update, preparation, installation, troubleshooting, and post state. Wash et al. [12] researched Windows Update and analyzed it with log data. They reported that users may misunderstand the updating of OS (e.g., setting and automatically updating). Fagan et al. [13] researched existing application update messages of computers. They showed that users are commonly annoyed and sometimes confused by these messages.

2.2 Interventions for Behavior Change

Several works [9, 10, 15] have described some interventions that are excellent for changing user behavior. Vlaev et al. [9] reported that financial incentives influence the change to healthy behaviors. Acquisti et al. [10] claim that nudges (such as framing effects [14], education, and incentives) affect user behaviors regarding security and privacy. The framing effect [14] is where each user recognizes different outcomes from two messages: a loss framed message (LFM) and a gain framed message (GFM). LFM shows a disadvantageous message and GFM shows an advantageous message. However, these interventions are not necessarily appropriate for all users. Related work [15] reported that appropriate interventions differ depending on the status of user awareness and behavior. Therefore, personalized interventions are required in terms of user awareness and behavior.

2.3 Approaches that Promote the Taking of Security Actions

There are many related works [1, 6–8, 16–19] about approaches that promote the taking of security actions. Mathur et al. [6] evaluated the prototype of updating computer applications. As a result, they proposed personalizing interfaces, minimizing update interruptions, improving information for decision making, and centralizing update management. We focused on Windows Update and analyzed effective messages for different types of users [1, 7]. We created some phrases and UI that are based on the framing effects [14] and UI design, and conducted online surveys to verify suitable dialog messages for each user. Ndibwile et al. [8] proposed UI of update messages for smartphone users. They reported that color is important for persuading a user to update. Turland et al. [16] conducted experiments on Wi-Fi network selection for Android users by using their application prototype. They found that users' security action was influenced by color coding. Tahaei et al. [17] conducted experiments on privacy for participants with an app development experience. They analyzed user choices between personalized ads and non-personalized ads for six conditions. They found that users' choices differed for each condition.

Goel et al. [18] conducted experiments for employees to evaluate the effectiveness of financial incentives for promoting security action. The result of the experiments showed that incentives are effective for using safe passwords and reducing the use of non-work email. Goel et al. [19] conducted experiments on user behavior toward phishing emails in the case of using financial incentives. They set up two groups: positive frame group and negative frame group. The positive frame group gained up to a certain quantity of incentives for compliance. The negative frame group was awarded a certain quantity of incentives but lost a portion with each compliance failure. As a result, a negative frame is effective for raising user awareness toward phishing emails. From the above, the framing effect, UI and incentives are effective for promoting security action.

3 Proposed Method

As described in Sect. 2.2, it is important to promote security action for users by personalized interventions. In the previous work [1, 7], effective dialog messages for promoting Windows Update differ according to user awareness. Similar to personal computers, we consider that personalized dialog messages are required for the smartphones. However, these related works did not reveal the personalized approaches of OS and applications for smartphone users in terms of user awareness. Moreover, computers and smartphones have aspects of commonality and difference [6, 8]. Therefore, effective intervention methods for smartphone users are not necessarily the same as effective intervention methods for computer users. In this paper, we evaluate personalized intervention methods for smartphone users. We consider that user awareness of updating OS and applications may differ according to the users' understanding of the update procedure and evaluate appropriate approaches with this difference in mind. In Sect. 2.3, we described that the framing effect, UI, and incentives are effective for promoting security action. Therefore, we analyze effective phrases of dialog messages based on the knowledge of the update procedure as a first step. We set evaluation of UI design as the second step because we need to evaluate using the effective phrases. The second step is future work. In addition, we verify the effectiveness of incentives in the same way as dialog messages. We assume that these approaches may differ between software and compare them.

We propose 26 phrases as shown in Table 1 in reference to related work [1]. Many phrases in Table 1 are applicable to the cases of both OS and applications. In Table 1, the phrases which are applicable only to the case of OS are indicated as "(OS)", and the phrases which are applicable only to the case of applications are indicated as "(App)". However, LFM-1 and LFM-2 are applicable only to the case of OS because there is a possibility of ransomware infection of devices only in the case of not updating OS. These phrase types consist of usual, framing effect (LFM, GFM), increase security effect (ISE), increase security interest (ISI), receive other request (ROR), recognize about contribution (RAC), decrease security cost (DSC), Wi-Fi and auto update setting (AUS). The usual type is the phrase that we usually see when we update OS or applications. The LFM type shows the disadvantages of not updating. The GFM type shows the advantages of updating. Most of the phrases are created in reference to related work [1]; however, LFM-5, LFM-6, GFM-3, GFM-4, Wi-Fi-1, Wi-Fi-2 and AUS-1 are newly created by us. Figure 1 shows an example of usual messages and these phrases are the same user interface as that in Fig. 1. These dialogs are shown in the Japanese language. We evaluate the effectiveness of phrases in an online survey in Sect. 4.

Table 1. 26 phrases of message.

Type	Phrases
Usual	Updates are available. Please update right now
LFM-1	(OS only) Updates are available. If you do not update, you may be infected by ransomware and not be able to access data in your smartphone. Please update right now
LFM-2	(OS only) Updates are available. If you do not update, you may be infected by ransomware and need money to use your smartphone. Please update right now
LFM-3	Updates are available. If you do not update, others may access your smartphone dishonestly and leak your personal data. Please update right now
LFM-4	Updates are available. If you do not update, others may access your smartphone dishonestly and damage people around you. Please update right now
LFM-5	Updates are available. If you do not update, your smartphone bugs may not be fixed. Please update right now
LFM-6	Updates are available. If you do not update, you cannot use applications. Please update right now
LFM-7	Updates are available. If you do not update, you cannot use new functions of usability. Please update right now
GFM-1	Updates are available. If you update, you can solve security problems and improve security. Please update right now
GFM-2	Updates are available. If you update, you can use good user interfaces and improve usability. Please update right now
GFM-3	(OS) Updates are available. If you update, you can fix smartphone bugs. Please update right now. (App) Updates are available. If you update, you can fix application bugs. Please update right now
GFM-4	Updates are available. If you update, you can use your smartphone in the same way as before. Please update right now
GFM-5	(OS) Updates are available. According to the NISC[1], if you always update OS, your smartphone will be safe. Please update right now. (App) Updates are available. According to the NISC, if you always update application, your smartphone will be safe. Please update right now
ISE-1	Updates are available. If you update, you can defend yourself from various attacks. Please update right now
ISE-2	Updates are available. According to recent research, more than 60% of people have incurred security damages in the past one year. Please update right now
ISI-1	Updates are available. According to recent research, about 90% of people update now. Please update right now
ISI-2	Updates are available. It is important for you to update now because the attacks are sophisticated. Please update right now

(*continued*)

[1] NISC is an abbreviation for the National center of Incident readiness and Strategy for Cybersecurity.

Table 1. *(continued)*

Type	Phrases
ROR-1	Updates are available. According to the office or school, you need to update now. Please update right now
ROR-2	Updates are available. Your friend who is knowledgeable about security advises you to update now. Please update right now
RAC-1	Updates are available. If you update, you can defend not only yourself but also others from security damage. Please update right now
RAC-2	Updates are available. If you update, the rate of incurring security damage decreases by 50%. Please update right now
DSC-1	(OS) Updates are available. If you update once every few months, your smartphone will be safe. Please update right now. (App) Updates are available. If you update once every few weeks, your application will be safe. Please update right now
DSC-2	(OS) Updates are available. If you only reboot your smartphone, that is all you need to do. Please update right now. (App) Updates are available. If you only reboot your application, that is all you need to do. Please update right now
Wi-Fi-1	Updates are available. If you use Wi-Fi, you will not consume mobile data. Please update right now
Wi-Fi-2	Updates are available. You are using Wi-Fi now. If you update now, you will not consume mobile data. Please update right now
AUS-1	Updates are available. If you set auto update, you do not need to see the notification every time and the system updates automatically. Please update and set auto update right now

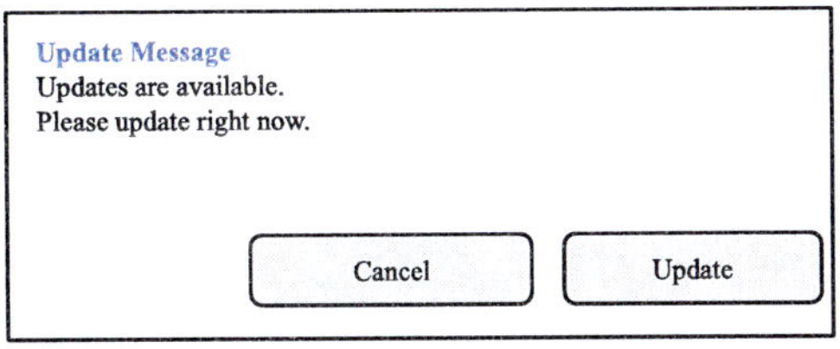

Fig. 1. Example of a usual message.

4 Evaluation

4.1 Methodology

We conducted an online survey by using the services of Cross Marketing, Inc. [20]. Cross Marketing, Inc. is a Japanese research company. We conducted the survey from February 10th to February 14th 2022. We recruited 2400 participants who use Android smartphones of one communication carrier. We focus on Android users because many people use Android smartphones around the world [21]. In addition, we narrowed it down to one communication carrier to analyze the participants who use the same service. Android smartphones cover models made by Sony, Samsung, Sharp, and Kyocera. The age of participants ranged from 20 to 69 years of age. They do not set auto update and update manually or do not update. Table 2 shows the participant demographics. We allocated 300 participants to eight groups who answered about updating OS or each application (Communication, Finance, Lifestyle, Games, Utility, Health or Entertainment). We consider that effective phrases differ according to the frequency of application use. Each group is organized into three types of 100 participants who were classified by frequency of application use; sometimes, often, and every day. We compare the difference between UUP and NUUP, between software, and among frequency of application use. In this paper, we do not describe about the difference among frequency of application use due to the limit of space.

Table 2. Demographics of the 2400 participants.

Age	20–29	30–39	40–49	50–59	60–69
Male	90	258	296	315	346
Female	101	212	228	278	276

The questions are shown in the Appendix. In this survey, we did not collect any information that could identify the participants. All responses from the participants were anonymized. In accordance with the regulations of the research company, the participants received a fee for answering the survey. This research was exempt from ethical review in accordance with the guidelines of our office.

4.2 Analysis

We analyzed the following items and describe details in the next sub-subsection. We analyzed the data of this survey by using HAD [22] for statistical analyses. We compared the difference between software.

1. Effective phrases of dialog message for the UUP group and NUUP group (RQ1)
2. Effective incentives (RQ2)
3. Relationship between incentives that users are interested in and incentives that users want to get

4.2.1 Phrases

We classified the participants into two groups (UUP and NUUP) by using user answers to Question 1 ("Status of understanding the update procedure"). For example, UUP includes participants who answered "I understand" to Question 1, and NUUP includes participants who answered "I do not understand" to Question 1. In the same way as related work [1], we calculated the ratio of those who answered "I update right now" to Question 4 ("User behavior toward each phrase in Table 1") because it is important that users update now after receiving the message. In the case of OS, we used a t-test to compare 25 phrases (LEM-1 ~ AUS-1) with the usual phrase. In the case of applications, we used a t-test to compare 23 phrases (LEM-3 ~ AUS-1) with the usual phrase. If there was a significant difference between two phrases, an effective phrase was one in which the ratio of those who answered "I update right now" was higher than another phrase.

4.2.2 Incentives

We calculated the ratio of selecting each incentive for Question 2 ("Incentives that users are interested in") and Question 5 ("Incentives that users want to get"). We define the top three incentives that are of a higher rate for Question 5 as effective incentives. We consider that effective incentives match the incentives that users are interested in. Therefore, we used correlation analysis to analyze the relationship between user answers to Question 2 and Question 5. In this paper, if the correlation coefficient is more than 0.4, there is a correlation between two answers.

4.3 Result

Table 3 shows the ratio of those who answered "I update right now" to Question 4 for the UUP group in the upper part, and for the NUUP group in the lower part. In this paper, the level of significance is 5%. If the p-value is less than 0.05, the value is statistically significant. If the value is statistically significant and positive, the phrase for comparison is an effective phrase; on the other hand, if the value is statistically significant and negative, the phrase for comparison is a backfire phrase. Effective phrases are shown in red and backfire phrases are shown in blue in Table 3.

In Table 3, we found that different phrases are effective or backfire between the UUP and the NUUP groups. For example, in the case of the Communication application, effective phrases are LFM-3, LFM-5, LFM-6, GFM-1, GFM-3, GFM-4, GFM-5, ISE-1, ISI-2, and DSC-2 in the UUP group, but they are LFM-3 and LFM-6 in the NUUP group. Therefore, it is important to show different appreciated phrases for the two groups. We confirmed that the ratio of those who answered "I update right now" of the UUP group is higher than that of the NUUP group in the case of most of the phrases. The more the users understand the update procedure, the more the users update OS and applications.

In addition, effective phrases differ between software. For example, we focus on the UUP group. Effective phrases are LFM-6 in the case of OS; LFM-3, LFM-5, LFM-6, GFM-1, GFM-3, GFM-4, GFM-5, ISE-1, ISI-2 and DSC-2 in the case of the Communication application; LFM-3 and GFM-1 in the case of the Games application. As we expected, personalized phrases are required for updating software.

We calculated the ratio of selecting each incentive for Question 2 ("Incentives that users are interested in") and Question 5 ("Incentives that users want to get") in Table 4.

In Table 4, many users select Ponta points that one communication carrier provides (hereinafter, this is called "points"), mobile data, and coupons for Question 2 and Question 5. In particular, more than 50% of users selected points for both questions. Commonly, points are the best incentive for Android smartphone users. There is not much difference between software. Therefore, effective incentives are points, mobile data, and coupons in all cases. However, there are major differences between two questions in the case of games, music, and movies. No effect of providing these incentives

Table 3. Ratio of those who answered "I update right now" for the UUP and NUUP group.

UUP	OS	Comm-unicati-on	Finance	Lifesty-le	Games	Utility	Health	Enterta-inment
Partici-pants	177	183	201	214	199	228	237	207
Usual	0.175	0.142	0.204	0.248	0.196	0.254	0.186	0.193
LFM-1	0.215	-	-	-	-	-	-	-
LFM-2	0.220	-	-	-	-	-	-	-
LFM-3	0.215	0.273	0.244	0.252	0.271	0.316	0.278	0.246
LFM-4	0.198	0.186	0.214	0.229	0.226	0.272	0.262	0.232
LFM-5	0.203	0.230	0.209	0.257	0.211	0.294	0.249	0.251
LFM-6	0.237	0.235	0.259	0.252	0.246	0.276	0.278	0.300
LFM-7	0.198	0.175	0.189	0.215	0.206	0.237	0.236	0.169
GFM-1	0.220	0.213	0.229	0.271	0.251	0.298	0.253	0.222
GFM-2	0.186	0.137	0.194	0.206	0.191	0.237	0.177	0.174
GFM-3	0.203	0.197	0.209	0.238	0.216	0.259	0.253	0.237
GFM-4	0.198	0.197	0.224	0.215	0.216	0.254	0.219	0.193
GFM-5	0.198	0.224	0.214	0.206	0.181	0.232	0.236	0.188
ISE-1	0.209	0.202	0.214	0.243	0.221	0.246	0.232	0.208
ISE-2	0.175	0.175	0.169	0.234	0.201	0.219	0.203	0.184
ISI-1	0.153	0.148	0.169	0.196	0.141	0.193	0.177	0.169
ISI-2	0.215	0.219	0.194	0.234	0.221	0.272	0.241	0.227
ROR-1	0.147	0.137	0.169	0.196	0.166	0.175	0.160	0.164
ROR-2	0.124	0.109	0.144	0.182	0.136	0.197	0.131	0.159
RAC-1	0.198	0.158	0.194	0.220	0.191	0.228	0.219	0.184
RAC-2	0.181	0.169	0.174	0.201	0.196	0.215	0.198	0.174
DSC-1	0.198	0.158	0.179	0.201	0.181	0.215	0.190	0.174
DSC-2	0.226	0.230	0.219	0.192	0.201	0.241	0.211	0.237
Wi-Fi-1	0.215	0.158	0.214	0.234	0.201	0.272	0.207	0.179
Wi-Fi-2	0.226	0.164	0.234	0.285	0.201	0.294	0.219	0.227
AUS-1	0.192	0.153	0.179	0.252	0.171	0.211	0.177	0.164

NUUP	OS	Comm-unicati-on	Finance	Lifesty-le	Games	Utility	Health	Enterta-inment
Partici-pants	123	117	99	86	101	72	63	93
Usual	0.106	0.128	0.182	0.151	0.168	0.181	0.111	0.129
LFM-1	0.106	-	-	-	-	-	-	-
LFM-2	0.089	-	-	-	-	-	-	-
LFM-3	0.098	0.205	0.232	0.163	0.198	0.292	0.159	0.215
LFM-4	0.122	0.188	0.212	0.128	0.198	0.250	0.127	0.151
LFM-5	0.089	0.120	0.202	0.140	0.158	0.208	0.175	0.129
LFM-6	0.114	0.188	0.242	0.140	0.198	0.236	0.190	0.140
LFM-7	0.073	0.120	0.172	0.128	0.158	0.153	0.143	0.108
GFM-1	0.106	0.137	0.242	0.140	0.158	0.236	0.190	0.151
GFM-2	0.065	0.111	0.121	0.128	0.139	0.153	0.127	0.086
GFM-3	0.081	0.128	0.182	0.174	0.168	0.250	0.143	0.118
GFM-4	0.081	0.145	0.222	0.116	0.158	0.208	0.175	0.118
GFM-5	0.106	0.154	0.141	0.128	0.139	0.194	0.127	0.151
ISE-1	0.098	0.162	0.202	0.128	0.168	0.250	0.175	0.151

(continued)

Table 3. *(continued)*

ISE-2	0.081	0.128	0.172	0.151	0.158	0.153	0.095	0.118
ISI-1	0.073	0.111	0.152	0.105	0.149	0.167	0.063	0.108
ISI-2	0.106	0.145	0.192	0.140	0.168	0.278	0.143	0.161
ROR-1	0.073	0.128	0.172	0.116	0.149	0.125	0.095	0.108
ROR-2	0.081	0.094	0.152	0.128	0.149	0.111	0.063	0.108
RAC-1	0.081	0.128	0.202	0.105	0.158	0.208	0.127	0.118
RAC-2	0.081	0.154	0.182	0.151	0.158	0.194	0.127	0.140
DSC-1	0.073	0.137	0.162	0.105	0.149	0.222	0.143	0.161
DSC-2	0.081	0.162	0.192	0.093	0.198	0.250	0.143	0.118
Wi-Fi-1	0.114	0.137	0.192	0.128	0.158	0.167	0.159	0.065
Wi-Fi-2	0.106	0.137	0.172	0.163	0.149	0.222	0.159	0.118
AUS-1	0.073	0.120	0.172	0.105	0.168	0.194	0.111	0.118

is seen for users who are interested in games, music, and movies. A certain number of users answered N/A. Therefore, other incentives or other approaches are required for those users.

Table 4. Ratio of users selecting each incentive for Question 2 and Question 5.

Question2	OS	Communication	Finance	Lifestyle	Games	Utility	Health	Entertainment
Points	51.3%	55.7%	68.7%	75.3%	66.7%	78.3%	69.7%	63.3%
Mobile data	27.0%	37.3%	15.3%	55.7%	19.7%	53.0%	55.0%	15.7%
Games	17.7%	17.0%	24.0%	23.0%	54.3%	29.3%	24.7%	24.0%
Music	23.7%	28.0%	30.0%	39.3%	37.3%	36.3%	41.3%	41.3%
Movies	23.7%	28.3%	37.3%	45.7%	45.3%	43.3%	42.7%	58.7%
Coupons	34.0%	40.3%	45.0%	65.7%	45.0%	55.3%	52.7%	46.0%
N/A	25.0%	19.7%	10.3%	7.0%	6.7%	6.7%	7.7%	7.7%
Question5	OS	Communication	Finance	Lifestyle	Games	Utility	Health	Entertainment
Points	50.3%	58.7%	70.3%	71.3%	68.0%	80.3%	72.7%	61.3%
Mobile data	23.7%	33.3%	32.3%	36.3%	34.7%	37.7%	35.7%	33.7%
Games	4.7%	1.7%	3.0%	6.3%	9.3%	4.0%	5.3%	3.0%
Music	7.7%	7.7%	10.3%	12.3%	13.0%	13.3%	14.0%	14.7%
Movies	3.7%	3.7%	4.3%	8.7%	6.0%	7.0%	4.7%	8.3%
Coupons	26.3%	31.7%	37.3%	53.7%	28.0%	39.3%	44.0%	34.0%
N/A	34.7%	25.7%	13.7%	12.0%	16.0%	10.3%	12.0%	21.3%

Table 5 shows the result of correlation analysis. If the correlation coefficient is more than 0.4, the correlation coefficient is shown in red in Table 5. The correlation coefficient of points and coupons as two of the top three incentives are higher than the other incentives. However, the correlation coefficient of mobile data as one of the top three incentives is not high. We found that effective incentives do not always match the incentives that users are interested in. There are no differences between software. Commonly, points have the highest correlation coefficient in all cases. Therefore, to giving the users points is an appreciated intervention.

Table 5. Correlation coefficient between Question 2 and Question 5. (** p < .01, * p < .05, + p < .10)

	OS	Comm-unicati-on	Finance	Lifesty-le	Games	Utility	Health	Enterta-inment
Points	0.58**	0.72**	0.65**	0.71**	0.59**	0.45**	0.52**	0.52**
Mobile data	0.42**	0.39**	0.33**	0.30**	0.31**	0.39**	0.37**	0.31**
Games	0.35**	0.15**	0.22**	0.25**	0.16**	0.17**	0.21**	0.08
Music	0.34**	0.41**	0.35**	0.42**	0.28**	0.27**	0.36**	0.26**
Movies	0.27**	0.23**	0.17**	0.29**	0.16**	0.21**	0.19**	0.16**
Coupons	0.42**	0.45**	0.53**	0.47**	0.45**	0.48**	0.44**	0.50**

5 Discussion

In this section, we discuss appropriate phrases and incentives for each user and the limitations of this research.

5.1 Appropriate Phrases

As described in Sect. 4.3, effective phrases differ between UUP and NUUP. In other words, appropriate phrases are different in terms of the status of understanding the update procedure. LFM and GFM are more effective than other categories in the case of the UUP group, and LFM is more effective than other categories in the case of the NUUP group. However, in Table 3, there are different effects among the same type (such as LFM-3 and LFM-4). Therefore, it is important for users to show effective phrases. For example, in the case of the UUP group, LFM-6 is the most effective phrase for updating of OS, and the Finance, and Entertainment applications; LFM-3 is the most effective phrase for updating of Communication, Games, and Utility applications; LFM-3 and LFM-6 are the most effective phrases for updating of Health application. In the case of the NUUP group, LFM-3 is the most effective phrase for updating of Communication, Utility, and Entertainment applications; GFM-4 is the most effective phrase for updating of the Health application.

Moreover, there are more effective and backfire phrases of the UUP group than there are those of the NUUP group. Showing appropriate phrases to the UUP group is expected to have a strong effect. However, there are some cases where there are no effective phrases for the NUUP group in the case of OS, and the Finance, Lifestyle, and Games applications. We consider that other elements such as UI design with added dialog message or other approaches are required for them.

Different phrases are effective according to software. In the case of UUP, Wi-Fi-2 is an effective phrase for updating OS, but this phrase is not effective in encouraging updating of applications. This is because that the mobile data consumed by updating OS is greater than that of applications because the users in the UUP group understand that updating OS and applications do not consume mobile data in the case of using Wi-Fi. However, it is possible that the users in the NUUP group do not understand Wi-Fi and/or mobile data. Therefore, it is necessary for the users in the NUUP group to understand aspects of the update procedure such as Wi-Fi and mobile data. Moreover, effective phrases differ among software. For example, GFM are effective for Communication, Games, Utility, Health, and Entertainment applications, but backfire for the Lifestyle application in the case of the UUP group. This is because the purpose of use differs according to the type of application. Therefore, it is important for each application to show different appropriate phrases.

5.2 Appropriate Incentives

As shown in Table 4, most users want to get points as incentives. In particular, 70% of the users want to get points in the case of Finance, Lifestyle, Utility, and Health applications. Considering of Table 5, if we give points to Finance and Lifestyle application users, strong positive effect is expected on updating these applications. It is necessary to set an appreciated volume of incentives for updating OS and applications, and this is future work. However, the correlation coefficient of mobile data is not higher than points and coupons. This is because some users do not understand mobile data as described in Sect. 5.1. Many users want to get cash in the form of points, mobile data, and coupons. On the other hand, few users want to get contents such as games, music, and movies. Therefore, effective incentives are cash in the form of points, mobile data, and coupons, and provision of contents has a small effect for users.

As described in Sect. 4.3, there is not much difference between software. Before this survey, we expected the correlation coefficient of games to be high in the case of the Games application. However, it is not high because free trial of a game that comes with a charge does not always match the game that the user is interested in. In the case of entertainment applications, there is a similar phenomenon for the movies. Therefore, it is effective to give points, mobile data, and coupons to users.

5.3 Limitations

In this research, we conducted an online survey for Android smartphone users. We will obtain different results in the case of conducting surveys for different devices (e.g., iPhone, personal computer). Moreover, we focused on seven types of applications (Communication, Finance, Lifestyle, Games, Utility, Health, Entertainment), but there are other applications (e.g., News, Education) around the world. However, the use of these seven types of applications has a high frequency in Japan [23], and the comprehensiveness of applications is satisfied. We conducted this survey in Japan, and there is a possibility that other phrases of messages and incentives are appropriate in other countries. Evaluating the effectiveness of approaches in other countries is future work.

6 Conclusion and Future Works

We analyzed effective dialog messages and incentives for encouraging updating of software for smartphone users. After conducting an online survey, we found that effective phrases differ according to users' understanding of the update procedure, and each software. Therefore, we understood that personalized intervention is required for each user according to the users' level of literacy. In addition, points, mobile data and coupons are effective incentives for many users, and there is not much difference between software. In the future, we will discuss effective UI design of the dialog message based on users' understanding of the update procedure and each software. We will verify the appropriate volume of effective incentives. After revealing them, we will develop a system and conduct demonstration experiments.

Appendix: Questionnaire

- Q1: Status of Understanding the Update Procedure (Do You Know What Updates OS and Applications of Smartphone and How?)
  1: I understand, 2: I do not understand.
- Q2: Incentives that Users Are Interested in (Multiple Choice Allowed)
  1: Ponta points that one communication carrier provides, 2: Mobile data, 3: Games, 4: Music, 5: Movies, 6: Coupons, 7: N/A
- Q3: Frequency of checking incentives
  Users answer only item selected in the question above ("Incentives that Users Are Interested in")
  1: I do not check, 2: I rarely check (two or three times per year), 3: once a month, 4: once every couple of weeks, 5: once a week, 6: once every two or three days
  7: every day
- Q4: User Behavior Toward Each Phrase in Table 1
  1: I ignore the notification, 2: I check the notification, but I do not update, 3: I update later, 4: I update right now
- Q5: Incentives that Users Want to Get (Multiple Choice Allowed)
  1: Granting of Ponta points that one communication carrier provides, 2: Granting of mobile data, 3: Granting of a free trial of a game that comes with a charge, 4: Granting of favorite music, 5: Granting of a movie that comes with a charge, 6: Granting of coupons, 7: N/A

References

1. Sano, A., Sawaya, Y., et al.: SeBeST: security behavior stage model and its application to OS update. In: Proceedings 35th International Conference on Advanced Information Networking and Applications, AINA (2021), vol. 2, pp. 552–566 (2021)
2. Mathur, A., Chetty, M.: Impact of user characteristics on attitudes towards automatic mobile application updates. In: Proceedings Thirteenth USENIX Conference on Usable Privacy and Security, SOUPS (2017), pp. 175–193 (2017)

3. Rajivan, P., Efrat, A., Gonzalez, C.: Update now or later? Effects of experience, cost, and risk preference on update decisions. J. Cybersecurity, 1–12 (2020)
4. Vaniea, K., Rader, E., Wash, R.: Betrayed by updates: how negative experiences affect future security. In: Proceedings SIGCHI Conference on Human Factors in Computing Systems (2014), pp. 2671–2674 (2014)
5. Google Play: https://play.google.com/store/games?hl=en-JP. Accessed 27 June 2023
6. Mathur, A., Engel, J., et al.: "They keep coming back like zombies": improving software updating interfaces. In: Proceedings Twelfth USENIX Conference on Usable Privacy and Security, SOUPS (2016), pp. 43–58 (2016)
7. Sano, A., Sawaya, Y., et al.: Designing personalized OS update message based on security behavior stage model. In: Proceedings 18th Annual International Conference on Privacy, Security and Trust, PST (2021)
8. Ndibwile, J.D., Luhanga, E.T., et al.: A demographic perspective of smartphone security and its redesigned notifications. J. Inf. Process. **27**, 773–786 (2019)
9. Vlaev, I., King, D., et al.: Changing health behaviors using financial incentives: a review from behavioral economics. BMC Public Health **19**(1059) (2019)
10. Acquisti, A., Adjerid, I., et al.: Nudges for privacy and security: understanding and assisting user's choices online. ACM Comput. Surv. **50**(44), 1–41 (2018)
11. Vaniea, K., Rashidi, Y.: Tales of software updates: the process of updating software. In: Proceedings 2016 CHI Conference on Human Factors in Computing Systems, CHI (2016), pp. 3215–3226 (2016)
12. Wash, R., Rader, E., et al.: Out of the loop: how automated software updates cause unintended security consequences. In: Proceedings of the Tenth USENIX Conference on Usable Privacy and Security, SOUPS (2014), pp. 89–104 (2014)
13. Fagan, M., Khan, M.M.H., Buck, R.: A study of users' experiences and beliefs about software update messages. Comput. Hum. Behav. **51**, 504–519 (2015)
14. Tversky, A., Kahneman, D.: The framing of decisions and the psychology of choice. Science **211**, 453–458 (1981)
15. Prochaska, J.O., Velicer, W.F.: The transtheoretical model of health behavior. Am. J. Health Promot. **12**, 38–48 (1997)
16. Turland, J., Coventry, L., et al.: Nudging towards security: developing an application for wireless network selection for android phones. In: Proceedings 2015 British HCI Conference (2015), pp. 193–201 (2015)
17. Tahaei, M., Frik, A., Vaniea, K.: Deciding on personalized ads: nudging developers about user privacy. In: Proceedings of the Seventeenth Symposium on Usable Privacy and Security, SOUPS (2021)
18. Goel, S., Williams, K., et al.: Understanding the role of incentives in security behavior. In: Proceedings 53rd Hawaii International Conference on System Sciences (2020), pp. 4241–4246 (2020)
19. Goel, S., Williams, K., et al.: Can financial incentives help with the struggle for security policy compliance? Inf. Manag. **58**(2) (2021)
20. Cross Marketing Inc. https://www.cross-m.co.jp/en/. Accessed 27 June 2023
21. Statcounter Global Stats: https://gs.statcounter.com/os-market-share/mobile/worldwide. Accessed 28 June 2023

22. Shimizu, H.: An introduction to the statistical free software HAD: suggestions to improve teaching, learning and practice data analysis. J. Media Inf. Commun. 59–73 (2016)
23. Ministry of Internal Affairs and Communications: WHITE PAPER information and Communications in Japan. https://www.soumu.go.jp/johotsusintokei/whitepaper/eng/WP2021/chapter-1.pdf. Accessed 27 June 2023

Scheme for Selection of Deceptions as a Countermeasure for Insider Threats

Sana Okumura[1], Tomoya Amagasa[1], Tsubasa Shibata[1], Takumi Yamamoto[2], Tadakazu Yamanaka[2], Tetsushi Ohki[1], and Masakatsu Nishigaki[1](✉)

[1] Graduate School of Integrated Science and Technology, Shizuoka University, Hamamatsu, Shizuoka, Japan
nisigaki@inf.shizuoka.ac.jp

[2] Information Technology R&D Center, Mitsubishi Electric Corp, Chiyoda, Tokyo, Japan

Abstract. The number of insider threats and the expense of handling them increase every year, rendering it imperative to adopt measures against insider threats. In particular, insiders must adopt a psychological approach. Psychological approaches can be classified into three categories: deceiving insiders, demoralizing insiders, and luring insiders. The deception simultaneously accomplishes these three approaches. The disruption of enterprise systems by deception mechanisms causes a significant decrease in usability for users. The application of deception mechanisms requires careful consideration. Therefore, when designing deception as a countermeasure against insider threats, usability and security must be balanced from the viewpoint of cost-effectiveness. For evaluating cost-effectiveness of security measures, a method that models the relationship between "assets", "threats", and "countermeasures" as well as formulates the countermeasure selection problem as a discrete optimization problem has been proposed. However, these methods assume an external intruder, and to the best of our knowledge, no existing research explicitly covers the selection of countermeasures against insiders (malicious insiders). This paper proposes a scheme to quantitatively evaluate the effectiveness of deception against insider threats and determine the optimal deception mechanism. In the existing method, the relationship between "assets", "threats", and "countermeasures" is formulated as a discrete optimization problem, but the proposed method explicitly includes insiders as "threats" and deception as "countermeasures". In addition, when evaluating the effectiveness of insider-threat countermeasures in the model, the usability of users must be considered. Therefore, by adding "operation" to "assets", "threats", and "countermeasures", the proposed method incorporates the impact of selected countermeasures on the "usability" of business into the formulation of existing methods. Specifically, the existing method is sublimated into a security measure selection method that includes insider threat countermeasures (deceptions) by formulating the objective function of the existing method to be maximized with "usability" as a constraint.

1 Introduction

The number of insider threats and the expense of handling them increase every year [1], rendering it imperative to adopt measures against insider threats. Those involved in insider threats can be classified as careless/negligent employees, credential

L. Barolli (Ed.): BWCCA 2023, LNDECT 186, pp. 109–121, 2024.
https://doi.org/10.1007/978-3-031-46784-4_10

thieves/imposters, and malicious/criminal insiders [1]. For careless employees, human error countermeasures are effective [2, 3]. For a credential thief, applying anomaly detection (evaluating the risk value from the user's behavior and searching for high-risk users) is effective. In the current zero-trust environment, because each user activity is traced in detail by the enterprise system, accumulated log information can be used to calculate the risk value [3]. However, these countermeasures do not work effectively against malicious insider threats. Therefore, this study focuses on countermeasures against malicious insider threats (hereafter referred to as "insiders").

Human error countermeasures are ineffective against insiders who intentionally commit insider threats. Because insiders have legitimate authority, user authentication does not function as a countermeasure. In anomaly detection, such as risk-based authentication, it is not easy to calculate the risk value accurately because employees' work content varies widely, and even when the same user performs the same task, the work can change from situation to situation. Therefore, a psychological approach must be adopted against insiders [4]. In our interpretation of the literature [14], psychological approaches can be classified into three categories: deceiving insiders, demoralizing insiders, and luring insiders. The deception is a countermeasure that simultaneously accomplishes these three approaches. However, deception also works against users who do not have malicious intent (hereafter referred to as "harmless users"). The disruption of enterprise systems by deception mechanisms causes a significant decrease in usability for harmless users. Therefore, the deployment of deception mechanisms requires careful consideration.

More preciously, when designing deception as a countermeasure against insider threats, usability and security must be balanced from the viewpoint of cost-effectiveness. For evaluating the cost-effectiveness of security measures, a method that models the relationship between "assets", "threats", and "countermeasures" to formulate the countermeasure selection problem as a discrete optimization problem has been proposed [5]. This approach quantitatively evaluates the effectiveness of a security countermeasure by subtracting the cost of implementing the countermeasure from the value of the remaining assets. The security officer then selects an appropriate countermeasure that maximizes its effectiveness. Another method that dynamically deploys a deception mechanism based on the network state and attacker model and selects the optimal deception for each case has also been proposed [6]. However, these methods assume an external intruder, and to the best of our knowledge, no existing research explicitly covers the selection of countermeasures against insiders (malicious insiders).

This paper proposes a scheme to quantitatively evaluate the effectiveness of the deception mechanism against insider threats and determine the optimal deception deployment. Deceptions that are cost-effective against insider threats can vary depending on the information assets of an enterprise and the anticipated attacks. Therefore, similar to the existing method [5], the relationship between "assets", "threats", and "countermeasures" is formulated as a discrete optimization problem in the proposed method, but the objective function explicitly includes insiders as "threats" and deception as "countermeasures". In addition, when evaluating the effectiveness of insider-threat countermeasures, the usability of harmless users must be considered in the model. Therefore, by adding "business operation" to "assets", "threats", and "countermeasures", the proposed method incorporates the impact of selected countermeasures (deceptions) on the "usability" of business operations into the formulation of existing methods. Specifically, a security measure selection method that includes insider threat countermeasures (deceptions) is formulated into the objective function to be maximized with "usability" as a constraint.

Section 2 introduces related works about insider threats, deceptions, and selection of security measures. Sects. 3 and 4 describe our concept and deception selection scheme of the proposal. Section 5 summarizes the paper.

2 Related Work

2.1 Insider Threats

2.1.1 Classification of Insiders

Insiders are classified into careless/negligent employees, credential thieves/imposters, and malicious/criminal insiders [1]. Careless insiders are harmless users without malicious intent and are not motivated to commit fraud. Improvements in user interfaces and operational processes [2], and the introduction of human error countermeasures represented by education and training [3], are expected to deter inadvertent insider threats by reducing the possibility of erroneous operations. Because credential thieves and malicious insiders possess legitimate authentication information, checking user authentication information (something you know, something you have, something you are) is not effective.

Therefore, it is necessary to expose insiders through "anomaly detection", in which risk values are evaluated based on user behavior and high-risk users are detected by introducing User and Entity Behavior Analytics (UEBA) and zero-trust defense, as described in detail in Sect. 2.1.2 [7]. In the case of credential thieves, the difference between the normal behavior of a harmless internal employee and that of a malicious external imposter is clear. However, in the case of an insider, the attacker is also an authenticated internal user who is performing his/her regular duties most of the time at work; therefore, an anomaly is rarely expected to occur in his/her behavior. Besides, normal operations in regular duties are basically diverse and not constant; therefore, deviations from "normal operations" are not easy to be precisely detected.

This study focuses on countermeasures against insiders who intentionally commit threats. In this paper, the term "insider" is used in the sense of an insider who maliciously commits fraud, unless otherwise specified.

2.1.2 User and Entity Behavior Analytics (UEBA)

UEBA evaluates risk values based on user behavior and device conditions, and detects the occurrence of attacks based on those risk values [8]. The risk value in the UEBA is a numerical value that represents the deviation from the authenticated user/legitimate operation pattern. Approaches for calculating risk values can be broadly classified into two types: signature-based [9] and anomaly- based [10], both of which are generally used together. The signature-based method defines regular patterns as white lists/black lists and detects violations from them. This is similar to the technologies used in signature-based intrusion detection systems (IDS) or intrusion prevention systems (IPS). The anomaly-based method collects "normal operation patterns of authenticated users" and detects deviations from them. This is similar to the technologies used in risk-based authentication and anomaly based IDSs/IPSs.

In the case of insiders, the attacker is also an authenticated user; therefore, no difference exists in "identity" in their behavior. Thus, to achieve insider detection by the UEBA, the "legitimate work of authenticated users" must be defined and the deviation from legitimate operations must be quantified. However, it is unrealistic to define a pattern that covers all "legitimate operations" owing to its wide range of operations. Even when the same user performs the same operation, the task details vary from situation to situation. Therefore, it is not easy to determine the "anomaly" to a normal operation pattern.

In recent years, many attempts have been made to use Artificial Intelligence (AI) to calculate the risk value of insider behaviors [11]. However, insider threat is not a frequent occurrence. In addition, enterprises tend to conceal incident information about insider threats (more than incidents caused by external cyberattacks) for fear of reputational risk [12]. Therefore, preparing the training data required for AI training is not straightforward.

2.1.3 Zero Trust Architecture

As remote work has expanded in many enterprises, corporate security measures have shifted from perimeters to zero-trust defenses. The zero-trust defense assumes that attackers can steal user IDs and passwords. In other words, zero-trust protection is a countermeasure that assumes a credential thief from outside.

In zero-trust protection, user authentication is applied each time a user or device accesses information assets in an enterprise system. Various tools continuously monitor the behavior of users and devices and alerts are raised when a behavior exceeds a predefined risk value [3]. This implies that zero-trust defense is an enabler for creating the UEBA described in Sect. 2.1.2. As explained previously, in the case of a credential thief from the outside, the difference in behavior is clear because the authenticated user and attacker are different. However, in the case of a malicious insider threat, the attacker is also an authenticated user; therefore, the UEBA must be difficult to detect deviations from legitimate operations.

2.2 Deceptions

A deception mechanism is a security measure for deceiving attackers by providing false information (data, files, servers, networks, etc.) [13]. Cyberattacks are becoming increasingly sophisticated and diverse, and complete prevention of intrusions into enterprise systems is difficult. Therefore, it is necessary to use a combination of countermeasures based on the assumption that intrusion will occur and that deception mechanisms are used as one of the countermeasures. Deceptions have three basic missions: "Seduction", "Saturation", and "Detection" [14].

- Seduction: Causes an enemy to switch an attack from the intended target to the decoy.
- Saturation: Saturates enemy defenses.
- Detection: Causes an enemy to expose his offensive assets by preparing to attack a decoy.

The key to the deployment of a deception is the degree of imitation. If it is easy to distinguish real from fake, an attacker can bypass the deception mechanism, thereby rendering it less secure. If it is extremely difficult to distinguish genuine objects from counterfeit ones, harmless users will be confused, and usability will decrease. To address this problem, Aoike et al. tried to preserve usability by rendering deception invisible to harmless users [15]. However, designing deception is not a simple task because usability for harmless users must be traded-off with security against attackers.

Shuo et al. proposed a method that dynamically determines the optimal deception mechanism by monitoring the network state and combining this information with an attacker model [6]. However, usability has not been studied in the method, and it is insufficient as an insider countermeasure because it targets only external intruders.

2.3 Selection of Security Measures

Because each enterprise has different information assets and anticipated attacks, the security measures differ from enterprise to enterprise. Various methods have been proposed for selecting cost-effective countermeasures by comprehensively considering information assets, attacks on the assets, and security countermeasures against the attacks.

2.3.1 ISMS

In establishing an enterprise's information security management system (ISMS) [16], the steps to select security measures are as follows:

1. List all information assets and calculate the value of each asset.
2. List possible threats to each asset and estimate the frequency of occurrence of each threat.
3. List possible countermeasures to reduce the threats involved in the risk.
4. Evaluate cost-effectiveness based on the importance of the asset, severity of the threat, and cost of the countermeasure to determine the countermeasures to be adopted.

2.3.2 Security Countermeasure Selection Based on the Remaining Assets

Nakamura et al. formulated steps 1–4 of the security countermeasure selection described in Sect. 2.3.1, as a discrete optimization problem [5]. This method (hereinafter referred to as the "existing method") models the relationship among "assets", "threats", and "countermeasures" to formulate an objective function for security measure selection. The effectiveness of the countermeasures is evaluated quantitatively by subtracting the cost of implementing the selected countermeasures from the expected value of the remaining assets in a situation where security countermeasures are selected. By selecting countermeasures to maximize this objective function, an optimal security measure selection is achieved.

The procedure for the existing method is as follows.

1. Model the assets, threats, and countermeasures:
   - A_k: Each asset in the enterprise ($1 < k < K$, where K is the total number of assets)

- V_k: Asset value of the asset A_k
- T_j: Each possible threat ($1 < j < J$, where J is the total number of threats)
- E_{jk}: Flag indicating whether or not the asset A_k likely to be compromised by the threat T_j
- P_j: Probability of occurrence of the threat T_j
- CM_i: Each countermeasure that can be applied ($1 < i < I$, where I is the total number of countermeasures)
- R_{ji}: Effectiveness of the countermeasure CM_i against the threat T_j
- C_i: Cost of deploying the countermeasure CM_i
- S_i: Flag indicating whether the countermeasure CM_i is deployed or not

2. Calculate the remaining assets with no measures:

$$\sum_k \left\{ V_k \prod_j [1 - E_{jk} P_j] \right\}. \quad (1)$$

3. Evaluate the remaining assets when the measures are deployed:

$$\sum_k \left\{ V_k \prod_j [1 - E_{jk} P_j (1 - R_{ji} S_i)] \right\}. \quad (2)$$

4. Formulate the objective function by subtracting the costs incurred in deploying the measure:

$$\sum_k \left\{ V_k \prod_j \left[1 - E_{jk} P_j \prod_i (1 - R_{ji} S_i) \right] \right\} - \prod_i C_i S_i. \quad (3)$$

5. Select countermeasures so that formula (3) is maximized.

Existing methods are formulated assuming only external attackers. To select countermeasures against insiders, an objective function that can evaluate the psychological effects on insiders and the usability effects on harmless users must be introduced.

2.3.3 MITRE ATT&CK

MITRE ATT&CK (Adversarial Tactics, Techniques, and Common Knowledge) is a framework that collects many actual attacks that have occurred in the past, systematizes them as "knowledge about the tactics and techniques used by attackers", and shares them [17]. In MITRE ATT&CK, attacks are systematized by Tactics and Techniques, and for each Tactics, multiple Techniques are integrated to achieve the Tactics.

The MITRE ATT&CK is a model of an attacker's sequence of actions to achieve an attack objective. The attacker first executes "attack techniques to achieve the first attack tactic"; when the first attack tactic is achieved, the attacker moves on to the second attack tactic; and so forth. For each attack tactic, MITRE ATT&CK provides an overview of the attack tactic, the steps performed by previous attack groups when they used the attack tactic, examples of mitigating the attack tactic, and the ways to detect the attack. MITRE ATT&CK provides an overview of each attack method.

MITRE ATT&CK is used as a knowledge base to support various security technologies, products, and services such as attack detection systems [18], performance evaluations of security software, and assessments of security measures by enterprises [19]. However, MITRE ATT&CK is, at the current time, a knowledgebase that mainly aggregates external attacks and does not substantially analyze insider threats.

3 Selection of Security Measures with Deception Mechanisms

3.1 Effects of Deceptions

As described in Sect. 2.1, because an insider is an authenticated user, human-error countermeasures and user authentication do not function as countermeasures against insider threats. Because it is difficult to detect deviations from legitimate operations, even if anomaly detection is applied, false alerts may frequently occur. Therefore, a psychological approach must be a good strategy to address insider threats [4]. By deducing from the literature [14], we grasp that psychological approaches can be classified into three categories: deceiving insiders, demoralizing insiders, and luring insiders. The deception is a countermeasure that simultaneously accomplishes these three approaches. The three-deception missions (seduction/saturation/detection) described in Sect. 2.2 [14] can be redefined as follows:

- Misidentification (deceiving the insider): Deploying decoys of information assets to misidentify the insider's attack target from a true target to a false target.
- Abandonment (demoralizing the insider): Deploying decoys of information assets to drastically increase the number of attack steps required by the insider before reaching the true target.
- Inducement (luring the insider): Deploying decoys of information assets to lure the insider to a false target that can reveal the insider.

3.2 Extending Existing Method

To introduce a deception mechanism as an insider countermeasure, considering the fact that harmless users are disturbed by the deception mechanism is important. In deciding whether to deploy or not, deception mechanisms must be evaluated from the viewpoint of "usability for harmless users" in addition to "effectiveness against insiders". To achieve this, this study extended the existing method [5] and formulated the security measure selection problem using the relationship among "assets", "threats (including insiders)", "countermeasures (including deception mechanisms)", and " business operations".

The effectiveness of deception mechanisms against insiders is modeled as a relationship between threats (insiders) and countermeasures (deceptions). Namely, similar to the existing methods, the effectiveness is expressed by "the degree to which the implemented countermeasure (deception) reduces the occurrence of the threat (insiders)". However, as explained in Sect. 3.1, the missions of deceptions are classified as Misidentification, Abandonment, and Inducement; Misidentification and Inducement deceptions have the psychological effect of exposing insiders, whereas Abandonment deception has the psychological effect of causing insiders to withdraw. Therefore, in this study, the deception effectiveness is decomposed into two types: Countermeasure effectiveness based on the psychological effect caused by Misidentification and Inducement deceptions is called

"Explicit effectiveness", and countermeasure effectiveness based on the psychological effect caused by Abandonment deception is called "Implicit effectiveness". Each single deception countermeasure may have both explicit and implicit effects against insiders. As shown in Fig. 1, these Explicit/Implicit effectiveness are superimposed on the objective function in the existing method.

The usability degradation against harmless users caused by deception mechanisms is modeled as a relationship between countermeasures (deceptions) and business operations. In this study, the decrease in usability is newly expressed as "the degree to which the implemented countermeasure (deception) reduces the efficiency of business operation for harmless employees". However, it is difficult to directly incorporate the usability degradation into the objective function of the existing method. In the existing method, the objective function of security measure selection is formulated based on the value and the cost. By contrast, usability is an indicator of operation efficiency, and thus not easy to translate directly into value/cost. Therefore, as shown in Fig. 1, the decrease in usability is considered as a constraint condition on the objective function of the existing method.

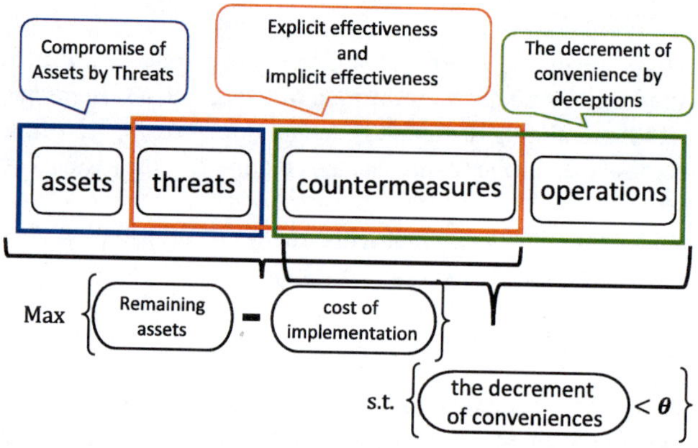

Fig. 1. Selection of security measures with deception countermeasures

3.3 Extending the MITRE ATT&CK

For selecting deceptions in the manner described in Sect. 3.2, a database that lists all available deception countermeasures and summarizes the Explicit effectiveness, Implicit effectiveness, decrease in usability, and cost of implementation of each deception countermeasure is needed. The explicit and Implicit effectiveness of countermeasures, even for the same countermeasure, can differ depending on the threat. In addition, even if the same countermeasures are proposed, the decrease in usability may differ depending on each business operation.

The MITRE ATT&CK framework is useful for identifying insider threats and evaluating the effectiveness of deception countermeasures. However, the current MITRE ATT&CK is a knowledge base that summarizes external attacks. In this study, we assumed that in the future, various case data on insider threats will be collected and aggregated in the same way as the MITRE ATT&CK to develop a knowledge base for insider threats. The framework classifies insider attack scenarios in terms of TTPs (tactics, techniques, procedures). Therefore, by listing the Explicit/Implicit effectiveness of each TTP and each deception countermeasure, the effectiveness of the deception mechanisms can be systematically summarized (Fig. 2 and Table 1).

Job descriptions provided by each enterprise can be used to identify the business operations necessary to evaluate the usability degradation. By listing each operation and each deception countermeasure, the decrease in usability of the deception mechanisms can be systematically summarized (Fig. 3 and Table 2).

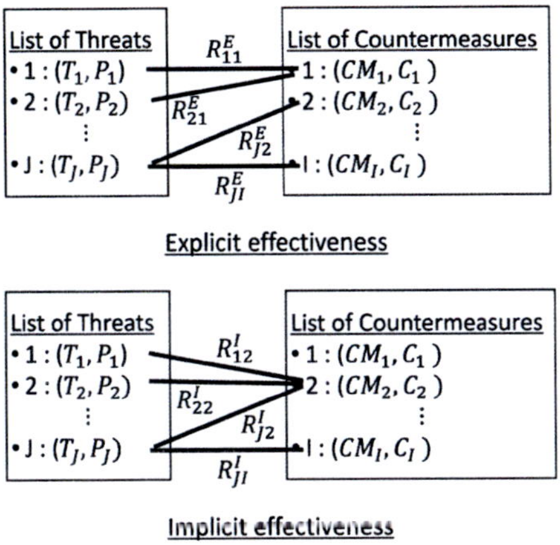

Fig. 2. Relationship between threats and countermeasures

Table 1. Relationship between threats and countermeasures

	CM_1 C_1		CM_2 C_2		...	CM_I C_I	
	explicit	implicit	explicit	implicit	...	explicit	implicit
T_1 P_1	R^E_{11}	R^I_{11}	R^E_{12}	R^I_{12}	...	R^E_{1I}	R^I_{1I}
T_2 P_2	R^E_{21}	R^I_{21}	R^E_{22}	R^I_{22}	...	R^E_{2I}	R^I_{2I}
⋮	⋮	⋮	⋮	⋮	⋱	⋮	⋮
T_{1J} P_J	R^E_{J1}	R^I_{J1}	R^E_{J2}	R^I_{J2}	...	R^E_{JI}	R^I_{JI}

The notations for modeling the Explicit/Implicit effectiveness and the decrease in usability are as follows:

- T_j: Each possible insider threat (TTP) ($1 < j < J$, where J is the total number of threats)

Table 2. Relationship between business operations and countermeasures

	$O_1\ V_1{}'$	$O_2\ V_2{}'$	$\cdots$	$O\ V_L{}'$
$CM_1\ C_1$	D_{11}	D_{12}	$\cdots$	D_{1L}
$CM_2 C_2$	D_{21}	D_{22}	$\cdots$	D_{2L}
$\vdots$	$\vdots$	$\vdots$	$\ddots$	$\vdots$
$CM_I\ C_I$	D_{I1}	D_{I2}	$\cdots$	D_{IL}

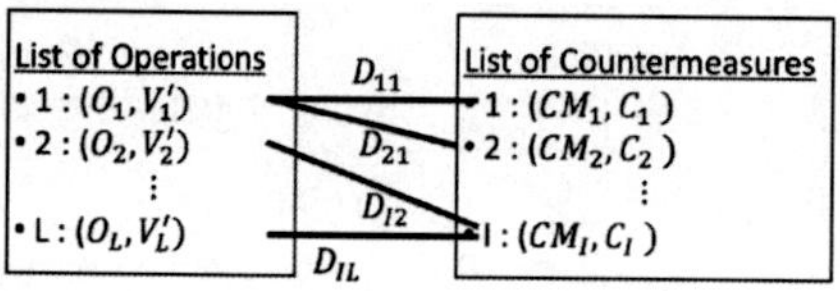

Fig. 3. Relationship between business operations and countermeasures

- CM_i: Each deception countermeasure ($1 < i < I$, where I is the total number of countermeasures)
- R^E_{ji}: Explicit effectiveness of countermeasures CM_i against threats T_j
- R^I_{ji}: Implicit effectiveness of countermeasures CM_i against threats T_j
- C_i: Cost of implementing countermeasures CM_i
- O_l: Each business operation owned by an enterprise ($1 < l < L$, where L is the total number of operations).
- F_l: Frequency of occurrence of the operation O_l
- D_{il}: Decrease in usability of the operation O_l caused by the deployment of countermeasures CM_i

Explicit and Implicit effectiveness can be tabulated by conducting the following questionnaire with a sufficient number of general users: "You are an insider and are attempting to commit an insider threat under a certain objective (TTP) T_j. If a deception countermeasure CM_i were deployed in your enterprise system, how much of a deterrent effect would that countermeasure have on your perception of the fraud being perpetrated?".

The decrease in usability can be measured by administering the following questionnaire to a sufficient number of general users: "You are a harmless user and are about to perform a certain business operation, O_l. If a deception countermeasure CM_i is deployed in your enterprise system, how much does the countermeasure decrease your usability in performing your operation?".

4 Formulation of Proposed Method

In this section, we describe the proposed deception selection scheme. It should be noted that this study focuses on the selection of deception countermeasures. As insider countermeasures in actual enterprise systems, it is recommended that not only deception mechanism but also data loss prevention, privileged access control (PAM), and UEBA [1] are adopted in combination.

4.1 Extending Objective Function by Implementing Explicit/Implicit Effectiveness

Formula (2) shown in step 2 of Sect. 2.3.2 is the amount of the "remaining asset" in the objective function of the existing method [5]. By contrast, in this study, the effectiveness R_{ji} of deploying countermeasure CM_i against threat T_j is extended to the Explicit effectiveness R^E_{ji} and Implicit effectiveness R^I_{ji}. The following is an example of the Explicit and Implicit effectiveness using the installation of dummy files. We assume that countless decoys (dummy files) are installed in an enterprise system. Insiders cannot determine the legitimate files among them. The Misidentification of dummy files as legitimate files or Inducement by dummy files results in access to dummy files by the insiders. Accessing any dummy file triggers insider detection. This is an Explicit effectiveness. Even when the insiders know that the enterprise system does not use dummy files as alarm triggers, the insiders may feel that they must check countless files one by one before finding the legitimate file, which may lead to Abandonment of the crime. This is Implicit effectiveness.

For simplicity, we assume that the Explicit effectiveness, R^E_{ji} and Implicit effectiveness, R^I_{ji} are uncorrelated, and the effects of each countermeasure are expected to be synergistic. Therefore, the remaining assets in the proposed method can be expressed as Formula (4).

$$\sum_k \left\{ V_k \prod_j \left[1 - E_{jk} P_j \left(1 - R^E_{ji} S_i \right) \left(1 - R^I_{ji} S_i \right) \right] \right\} \tag{4}$$

Therefore, the objective function results in Formula (5).

$$\sum_k \left\{ V_k \prod_j \left[1 - E_{jk} P_j \left(1 - R^E_{ji} S_i \right) \left(1 - R^I_{ji} S_i \right) \right] \right\} - \prod_i C_i S_i \tag{5}$$

4.2 Setting Constraints Based on Usability Degradation

The proposed method imposes the constraint that the decrease in usability is less than the acceptable value, while the objective function (Formula (5)) is to maximize. Here, the impact of usability on a business operation O_l is zero when no deception deployed. Then, the usability degradation yielded by a deception countermeasure CM_i is expressed as the degree of decline D_{il} in each single business operation of O_l, multiplied by the frequency F_l of the occurrence of the business operation O_l.

When multiple deception countermeasures are deployed, each deception decreases the usability of the business operation O_l in each degree. Therefore, the decrease in usability for business operation O_l is expressed as the sum of the effects of individual deceptions. By introducing flag S_i that indicates the deployment/non-deployment of a deception countermeasure CM_i, where $S_i = 1$ indicates deployment and $S_i = 0$ indicates non-deployment, the usability reduction caused in a business operation O_l through the implementation of deception countermeasures is shown in Formula (6).

$$F_l \sum_i D_{il} S_i \tag{6}$$

A decrease in usability was defined as the sum of the effects of each business operation on usability. Therefore, the constraint condition for selecting the deception countermeasures is that Formula (7) should not exceed the value set by each enterprise.

$$\sum_l F_l \sum_i D_{il} S_i \tag{7}$$

4.3 Determination of the Best Deceptions

Using Formula (5) derived in Sect. 4.1, as the objective function, and by searching for S_i with the constraint that Formula (7) derived in Sect. 4.2, is less than the permissible value θ, the optimal (maximizing the objective function within the range of conditions that satisfy the constraint) selection of the deception countermeasures is achieved. The formulation of the proposed method is Formula (8) and (9).

Maximize

$$\sum_k \left\{ V_k \prod_j \left[1 - E_{jk} P_j \left(1 - R_{ji}^E S_i \right) \left(1 - R_{ji}^I S_i \right) \right] \right\} - \prod_i C_i S_i \tag{8}$$

s.t.

$$\sum_l F_l \sum_i D_{il} S_i < \theta \tag{9}$$

5 Conclusion

In this study, we proposed a scheme to quantitatively evaluate the effectiveness of deception against insider threats and determined the optimal combination of deception mechanisms. To address insider threats, a psychological approach must be a good strategy. So, the effectiveness of deploying countermeasure against threat is extended to psychological effectiveness (the Explicit effectiveness and Implicit effectiveness). In addition, this study extended the existing security measure selection method and formulated the security measure selection problem using the relationship among "assets", "threats (including insiders)", "countermeasures (including deception mechanisms)", and " business operations". In the future, the questionnaire survey must be conducted as described in Sect. 3.3 and results must be used to verify the validity of the proposed scheme.

References

1. Proofpoint, Inc: 2022 COST OF INSIDER THREATS GLOBAL REPORT. Proofpoint., 2022 COST OF INSIDER THREATS GLOBAL REPORT. Proofpoint. https://www.proofpoint.com/us/resources/threat-reports/cost-of-insider-threats. Accessed 3 July 2023
2. Maxion, R.A.: Improving user-interface dependability through mitigation of human error. Int. J. Hum.-Comput. Stud. **63**(10), 25–50 (2005)
3. Carl, C.: Human factors in information security: the insider threat – Who can you trust these days? Inf. Secur. Tech. Rep. **14**, 189–196 (2009)
4. Schoenherr, J.R., Lilja-Lolax, K., Gioe, D.: Multiple Approach Paths to Insider Threat (MAP-IT): Intentional, Ambivalent and Unintentional Insider Threats, Counter-Insider Threat Research and Practice. https://www.proofpoint.com/us/resources/threat-reports/cost-of-insider-threats. Accessed 3 July 2023
5. Nakamura, I., Hyodo, T., Soga, M., et al.: A practical approach for security measure selection problem and its availability, IPSJ J. **1445**(8), 2022–2033(2004). (in Japanese)
6. Shuo, W, Qingpi, P, Jianhua, W, et al.: An intelligent deployment policy for deception resources based on reinforcement learning, IEEE, vol. 8, pp. 35792–35804 (2020)
7. Yousef, R.: Measuring the effectiveness of user and entity behavior analytics for the prevention of insider threats, Xi'an Jianzhu Keji Daxue Xuebao/J. Xi'an Univ. Archit. Technol. **13**(10), 175–181(2021)
8. Madhu, S., Min-Yi, S., Jisheng, W.: User and entity behavior analytics for enterprise security. In: IEEE International Conference on Big Data, pp.1867–1874 (2016)
9. N-able: Intrusion Detection System (IDS): Signature vs. Anomaly-Based, N-able. https://www.n-able.com/blog/intrusion-detection-system. Accessed 3 July 2023
10. Khan, M.Z.A., Khan, M.M., Arshad, J.: Anomaly detection and enterprise security using user and entity behavior analytics (UEBA). In: 2022 3rd International Conference on Innovations in Computer Science & Software Engineering, pp. 1–6 (2022)
11. Vasileios, K., Stavros, S., Bogdan, G., et al.: Detection of insider threats using artificial intelligence and visualisation. In: 2nd International Workshop on Cyber-Security Threats, Trust and Privacy Management in Software-defined and Virtualized Infrastructures (SecSoft), pp. 437–443 (2020)
12. Hunker, J., Probst, W.C.: Insiders and insider threats an overview of definitions and mitigation techniques. In: Journal of Wireless Mobile Networks, Ubiquitous Computing, and Dependable Applications, vol. 2, no. 1, pp. 4–27 (2011)
13. David, L., Surya, N., Kristen, M., et al.: Deception for cyber defence. In: Challenges and Opportunities, 2021 Third IEEE International Conference on Trust, Privacy and Security in Intelligent Systems and Applications, pp. 173–182 (2021)
14. David, A.: Ew101: a first coursein electronic warfare, pp. 223–227. Artech House Publishe (2000)
15. Aoike, Y., Kamizono, M., Eto, M., et al.: Decoy-file-based deception without usability degradation. In: 2021 IEEE Asia-Pacific Conference on Computer Science and Data Engineering10. (CSDE), pp. 1–7 (2021)
16. isms.online: Information Security Management System SaaS For ISO 27001., isms. https://www.isms.online/information-security-management-system-isms/. Accessed 3 July 2023
17. MITRE ATT&CK®: ATT&CK Matrix for Enterprise, MITRE ATT&CK. https://attack.mitre.org/. Accessed 3 July 2023
18. Microsoft: What is Microsoft 365 Defender? Microsoft. https://learn.microsoft.com/en-us/microsoft-365/security/defender/microsoft-365-defender?view=o365-worldwide&viewFallbackFrom=o365-worldwide. Accessed 3 July 2023
19. MITRE ATT&CK®: Open and fair evaluations based on ATT&CK®, MITRE ENGENUITY. https://attackevals.mitre-engenuity.org/. Accessed 3 July 2023

Device Classification via Passive Fingerprints with Clustering Algorithm

Masaki Ichino[1(✉)], Naoyuki Masuda[1], Takahiro Hayashi[1], Naoki Kodama[1], and Takamichi Saito[1,2]

[1] Meiji University, 1-1-1 Higashi-Mita, Tama-ku, Kawasaki-shi, Kanagawa, Japan
{ce225003,ce225029,ce225021,kodama}@meiji.ac.jp, saito@cs.meiji.ac.jp

[2] Rangeforce, Inc., Tokyo, Japan

Abstract. In browser fingerprinting, device identification becomes increasingly difficult as the number of accesses increases because the number of combinations to be predicted increases exponentially. This paper proposes a method that employs a clustering algorithm to minimize the number of prediction pairs. The proposed method also evaluates whether two access logs were sourced from the same device and groups the results using the union-find algorithm for each device. The proposed method was applied to the access logs of an actual website, and experiments were conducted to investigate the impact of the pair reduction process. To investigate the impact of the reduction, the proposed method and the proposed method excluding the reduction were applied to a dataset of about 100,000 access logs. As a result, the processing time of the proposed method was reduced to about 1% compared to the proposed method without the reduction, and the adjusted Rand index improved by about 0.30. The experimental results demonstrate the feasibility of rapidly grouping devices, even when the target data contains unlearned variables or patterns.

1 Introduction

In membership-based websites, malicious users can engage in various fraudulent activities, e.g., creating multiple accounts to post fake reviews or hoarding purchases. It is expected that such users can be detected by accesses by device. Thus, in this paper, we propose a method that uses passive fingerprinting, which a subset of browser is fingerprinting, to group access logs from websites according to the devices used.

In our previous work [7,12], we proposed a method to predict whether two access logs were obtained from the same device. We found that this method achieved an F-score accuracy of 0.99 or higher for mobile devices. In this study, we extended this approach by applying the union-find algorithm [5] to the prediction results for all possible pairs of access logs in order to group the access logs based on the devices used. The algorithm follows the predictions that suggest the access logs were obtained from the same device and then groups the logs

L. Barolli (Ed.): BWCCA 2023, LNDECT 186, pp. 122–134, 2024.
https://doi.org/10.1007/978-3-031-46784-4_11

accordingly. However, as the number of access logs increases, the combinations of access logs become quadratic ($O(n^2)$) in relation to the number of logs, thereby making it difficult to handle a large number of predictions. Thus, the proposed method implements a technique to reduce the combinations of access logs. With this approach, we aim to group logs using a smaller number of predictions. As a result, when the number of access logs was 100,000, the corresponding processing time was reduced by over 100 times, and the adjusted Rand index improved by approximately 0.30. Collectively, these results demonstrate that it is possible to quickly group devices even when the target data contain unlearned devices.

The rest of the paper is organized as follows. Section 2 describes the browser Fingerprinting and related works. Proposed method is described in Section 3. Section 4 lists datasets used for experiments and Sect. 5 describes experiments and their result. We discuss in Sect. 6 our method and limitations. Finally, we offer our conclusion in Sect. 7.

2 Related Work

2.1 Browser Fingerprinting

The browser fingerprinting technique is used to identify the device a user employs to access a website by combining information available from the web browser. The information that can be obtained from the browser is referred to as features, and the combination of these features is called a fingerprint.

Takahashi et al. [12] conducted experiments on passive fingerprinting and demonstrate that, among the available features, the global IP address and user-agent string are particularly useful for identification purposes. In addition, Vastel et al. [14] investigated the changes in fingerprints over time and confirmed that the target fingerprints gradually changed over the given period.

In 2010, Eckersley et al. [3] constructed websites for collecting browser fingerprints and demonstrated that 94.2% of the browser fingerprints are unique.On the other hand, Gómez-Boix et al. [6] conducted browser fingerprinting on a large-scale French website to examine the uniqueness of fingerprints in 2018. They found that browser fingerprints for desktops and similar devices were unique in 35.7% of cases, while those for mobile devices were unique in 18.5% of cases. This study revealed that the uniqueness of fingerprints is lower than in previous research.

Acar et al. and Iqbal et al. [1,8] proposed a method for detecting websites implementing browser fingerprinting. They achieve this detection by analyzing code such as JavaScript and Flash. Therefore, detecting passive fingerprinting is believed to be challenging.

2.2 Passive Fingerprinting

Passive fingerprinting is a subset of browser fingerprinting that only utilizes features obtained passively from the target browsers. The proposed method employed passive fingerprinting.

The references [7,12,13] conducted experiments on passive fingerprinting to attempt device identification. Masuda et al. and Fujii et al. [4,11] proposed the use of passive fingerprinting for creating a blacklist based on fingerprints of unauthorized devices that were accessed in the past.

3 Proposed Method

The proposed method comprises two main process, which are summarized as follows.

1. Pair reduction: Here, the number of pairs of access logs to be predicted are reduced using a clustering algorithm.
2. Grouping: Here, the proposed method predicts whether two access logs were obtained from the same device, and then the access logs for each device are grouped using the union-find algorithm.

Hereafter, in this paper, following terms will be defined and used. Figure 1 illustrates the prediction process of the proposed method and explains the terms.

Pair A set of two access logs
Pair Creation A process of creating these pairs
Pair Prediction A prediction of whether two sets of access logs were obtained from the same device
Teacher Class Sets of access logs grouped by device using teacher labels
Predicted Class Sets of access logs grouped by device using the proposed method
Cluster Sets of access logs generated via pair reduction process

3.1 Pair Reduction

In the pair reduction process, the proposed method reduces the number of pairs for the subsequent grouping process using a clustering algorithm.

Clustering algorithms are used to group similar data or find relationships between data. Fingerprints from the same device frequently exhibit similar characteristics. Thus, it is expected that clustering algorithms can be employed to classify access logs acquired from the same device into the same cluster.

The clustering algorithm we adopted has the characteristic of being unable to predict after learning the model. Alternatively, we employed a multiclass classification model which trained with the clustering result. In the proposed method, the light gradient boosting machine (LightGBM) framework [2] is used for model algorithm.

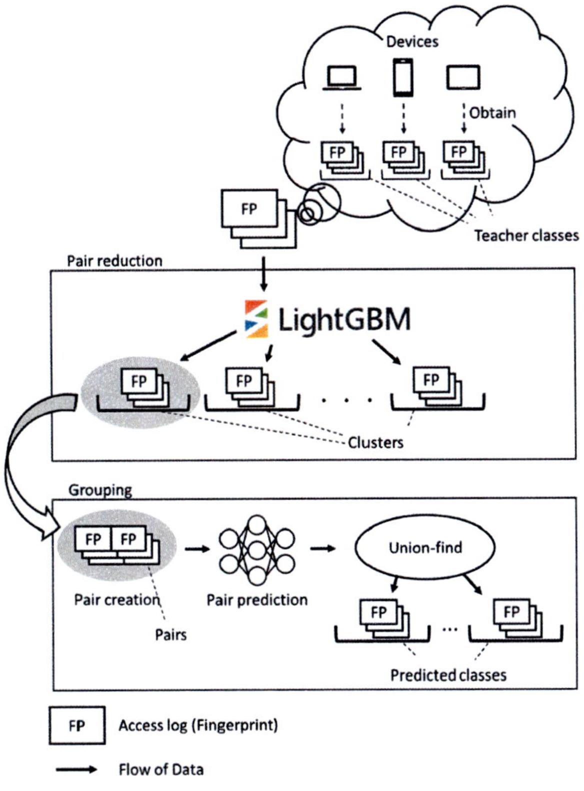

Fig. 1. Conceptual diagram of the proposed method

3.1.1 Training In the pair reduction process, a machine learning model is constructed using a clustering result. Based on the results of Experiment 3, Ward's method [9] was adopted with the cluster generation method, which is described in Sect. 3.1.2. However, this clustering algorithm cannot be used to predict new data after being trained; thus, a multiclass classification model using the LightGBM framework was constructed, where the clustering results served as the new ground truth information.

3.1.2 Cluster Generation Method We used Ward's method, which is an agglomerative clustering algorithm. However, by itself, agglomerative clustering algorithms can only measure the distances between data points, and another algorithm is required to group the data into clusters based on the measured distances. To address this issue, we propose the following procedure as the clustering algorithm. As an example, Fig. 2 shows the generation of six clusters from the dendrogram.

1. Apply Ward's method to the dataset and generate a dendrogram.
2. Generate clusters using the obtained dendrogram.
   a. Consider all leaf nodes in the root node as a single cluster.

b. If the specified number of clusters is reached, proceed to step 2c; otherwise, end the process.
c. Split the cluster based on the child nodes of the node associated with the largest cluster among the generated clusters.
d. Return to step 2b.

3.1.3 Prediction We perform prediction and generate clusters using the multiclass classification model trained on the generated data (Sect. 4.3). During the prediction phase, unlike the training process, we do not utilize the clustering algorithm. However, the model functions similarly to a clustering algorithm; thus, we refer to its results as a "cluster."

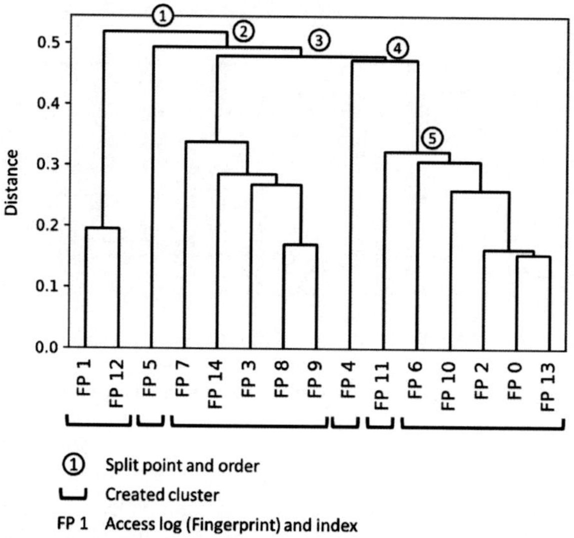

Fig. 2. Dendrogram and cluster generation process

3.2 Grouping

In the grouping, we perform pair creation for each cluster obtained from the results of the pair reduction process. This allows us to perform more detailed predictions and derive the final results. For pair prediction, a single deep learning model is used, and the union-find algorithm is employed to group the results for each device.

The machine learning model used for pair prediction is constructed following the approach described by Houjou et al.[7].

3.2.1 Prediction During the prediction process, pair creation is performed for each cluster, i.e., the results of the pair reduction process. The prediction procedure is described as follows.

1. Pair creation (Sect. 3.2.2) is performed for each access log in each cluster based on the results of the pair reduction process.
2. Pair prediction is applied using the constructed deep learning model.
3. The prediction results are grouped using the union-find algorithm:
   a. Each access log is initially considered as a separate device.
   b. For pairs with a prediction result of "acquired from the same device," the union operation is applied to create a corresponding prediction class.

The process of grouping the prediction results using the union-find algorithm is illustrated in Fig. 3; for example, when there are two predicted classes in the third row, i.e., FP1-FP2 and FP3-FP4. Here, if FP4 and FP1 are from the same device, the union operation is applied to these two predicted classes, thereby resulting in the creation of a joined prediction class, i.e., FP1-FP2-FP3-FP4, as shown in the fourth row.

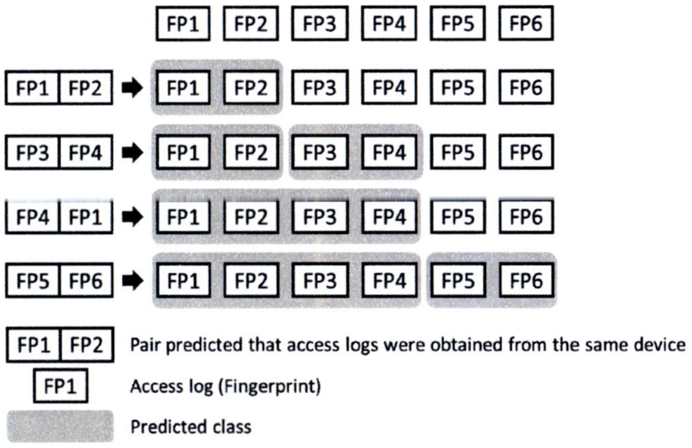

Fig. 3. Grouping access logs using union-find algorithm with prediction results

3.2.2 Pair Creation In the grouping process, new features are generated using the features shown in Table 2, and pairs of access logs are created following the method proposed by Houjou et al. [7]. The process is illustrated in Fig. 4.

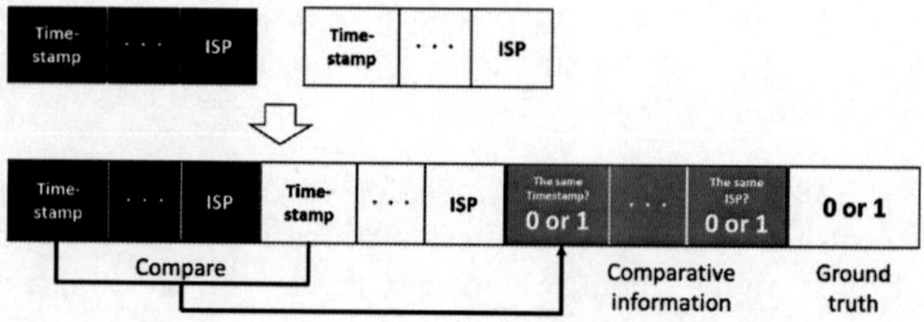

Fig. 4. Creation of Pairs of Access Logs and New Features

4 Dataset and Feature Engineering

4.1 Dataset Splitting

The data collection period was eight days. Here, devices that had access more than once every 10 min on average during the data collection period were considered to have failed to log correctly and were removed. In addition, logs with missing values for the timestamp, user-agent string, or global IP address were removed. As shown in Table 1, datasets were created by extracting samples randomly. The dataset for "Training for grouping and Experiment 3" (1st row) is further sampled randomly from the dataset for "Training for pair reduction" (2nd row).

Table 1. Datasets

Purpose	# of access logs	# of devices	Sampled from
Training for grouping and Experiment 3	100,000	98,342	from below dataset
Training for pair reduction	22,520,449	10,456,082	Days 1–7
Testing (Experiment 1 and 2)	10,000	8,157	Day 8
	50,002	40,948	
	100,002	80,805	

4.2 Features

The fingerprints used in the experiments are the features shown in Table 2. Here, the device identifiers were used as the ground truth labels, which were generated by assigning random values to HTTP cookies to distinguish the different devices.

Table 2. Examples of fingerprint and device identifier

Feature	Example
Timestamp	2022–01–01 12:00:00
User-agent string	Mozilla/5.0 (X11; Linux x86_64) AppleWebKit/537.36 (KHTML, like Gecko) Chrome/104.0.5112.81 Safari/537.36 Edg/104.0.1293.47
Global IP Address	133.26.81.168
Device identifier	glF0leOtAZEat...

4.3 Creation of Data for Pair Reduction Process

The data used in the pair reduction process comprised the user-agent strings, global IP addresses, and the values shown in Table 3. Note that the features listed in Table 3 were created in reference to the method proposed by Houjou et al. [7] using the feature values given in Table 2.

Table 3. Newly created features

Original feature	Newly created feature
User-agent string	Name of OS and web browser, Version, major, minor and maintenance versions Device model, device brand name
Global IP address	1st octet, 2nd octet, 3rd octet, 4th octet, ISP name, latitude, longitude, country name, city name, district name

5 Experiments

5.1 Evaluation Metrics

The evaluation metrics used in Experiments 1 and 2 included the adjusted Rand index, the perfect correct rate, the no-misclassified correct rate, the no-missed correct rate, and the partial correct rate. Here, the first metric is a clustering evaluation metric, and the last four metrics are defined for this study.

Perfect Correct Rate The perfect correct rate indicates the proportion of correctly classified devices. For example, as shown in Fig. 5a and 5b, when the teacher class and predicted class are the same, the prediction is considered correct. The perfect correct rate is calculated as the ratio of the number of cases where the teacher and predicted classes are the same as the total number of teacher classes. The perfect correct rate metric ranges from 0–1, where 1 represents the best performance, and 0 indicates the worst performance.

No-misclassified Correct Rate The no-misclassified correct rate indicates the proportion of correctly classified devices with no misclassifications. In addition to the perfect correct rate, this metric considers cases where the teacher class is divided into multiple predicted classes, but all of the predicted classes do not contain access logs from other teachers, i.e., logs from other devices. As shown in Fig. 5a and 5c, even when there are multiple predicted classes, we consider it as a no-misclassified correct result.

No-missed Correct Rate The no-missed correct rate indicates the proportion of correctly classified devices with no-missed logs. As shown in Fig. 5a and 5d, in addition to the perfect correct rate, this metric considers cases where the predicted class contains access logs from other teacher classes, i.e., logs from other devices, but the predicted class also includes the teacher class. Such cases are considered correct prediction results.

Partial Correct Rate The partial correct rate metric represents the proportion of cases that satisfy the criteria for the perfect correct rate, no-misclassified correct rate, or no-missed correct rate among the total number of teacher classes.

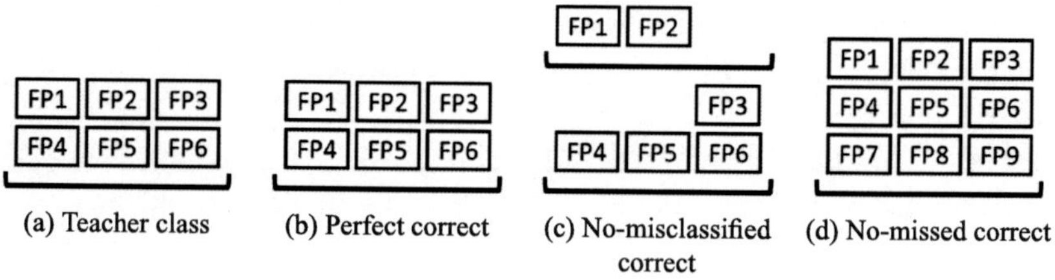

Fig. 5. Example of teacher class and custom evaluation metrics

5.2 Experiment 1

Experiment 1 was conducted to investigate the accuracy and processing time of the proposed method. Here, the proposed method was applied to the test dataset, which comprised the data from the final day of the collection period (Sect. 4). In addition, the goal was to compare the accuracy and processing time of the proposed method based on the number of access logs. To achieve this, three different test datasets were created by randomly extracting 10,000, 50,002, and 100,002 access logs from the available data.

The corresponding results are shown in Table 4. As can be seen, the accuracy increased as the number of access logs used for prediction decreased. Note that we expected the processing time to increase as the number of access logs increased. However, the processing time for the experiment using 100,002 access logs that created more pairs was 2,058 s, while the experiment using 50,002 access logs resulted in a shorter duration of 2,088 s, which is 30 s shorter.

5.3 Experiment 2

Experiment 2 was conducted to investigate the impact of the pair reduction process on the proposed method's accuracy and processing time. To achieve this, the pair reduction process was omitted, and only the grouping process was applied to the same datasets used in Experiment 1.

The results are shown in Table 5. As can be seen, the processing time increased significantly as the number of access log entries for prediction increased. In addition, the adjusted Rand index for the cases with 50,002 and 100,002 access log entries was 6% less than the accuracy obtained in the case with 10,000 entries. However, when comparing with our custom metrics, there is no significant difference from the adjusted Rand index.

5.4 Experiment 3

Experiment 3 was conducted to compare and examine the optimal clustering algorithms used in the pair reduction process. The optimal result in the pair reduction process refers to a state where there is a balanced distribution of access logs in each cluster, and the logs from the same device exist in the same cluster. Generally, when dividing n elements into m sets and considering all possible combinations within each set, the total number of combinations is minimized when the elements are divided evenly. Similarly, in the grouping process, all combinations of access logs within the same cluster are considered. Thus, when there is a balanced cluster size, the number of combinations created in the grouping process is reduced, which results in reduced processing time in the grouping process.

The dataset used in Experiment 3 is the same as described in Sect. 4, and the number of clusters created was 1,000. In terms of clustering algorithms, we compared the k-means and agglomerative clustering methods, Ward's method, average linkage, single linkage, and complete linkage. For agglomerative clustering, we must specify the method for generating clusters. However, in this experiment, we performed clustering using the default values provided by the scikit-learn Python library. In addition, for the agglomerative clustering using Ward's method, we employed the proposed cluster generation method described in Sect. 3.

Here, we considered the adjusted Rand index, the standard deviation of the cluster sizes, and the distribution of the cluster sizes. The results are shown in Table 6. The results of the clustering methods can be divided into two types of groups: those with relatively high the adjust rand index and those without. The former includes k-means, Ward's method, and the proposed method, while the latter includes the average linkage method, the single linkage method, and the complete linkage method. The group with a high the adjust rand index has a low standard deviation of cluster sizes and minimal bias within clusters, indicating balanced clusters. It is observed that the proposed method has the smallest maximum cluster size of 195, which is about half compared to the second smallest, k-means, with 423.

In addition, we found that the standard deviation of the cluster sizes was large when using the single linkage method. This highlights that only one cluster contained nearly all logs.

6 Discussion and Limitations

As shown in Tables 4 and 5, we found that the pair reduction process of the proposed method effectively improved the adjusted Rand index, perfect correct rate, and no-misclassified correct rate while reducing the no-missed correct rate. The reason for the decrease in the no-missed correct rate is believed to be due to access logs being separated in grouping, thereby resulting in accesses from the same device to be allocated to different clusters. However, from an overall comparison of the evaluation metrics, we consider that the pair reduction process contributed to improved accuracy. Furthermore, the processing time of the proposed method with Pair Reduction reduced the dataset of 10,000 entries to approximately 11% compared to the case without it. For datasets of 50,002 and 100,002 entries, the reduction was in about 1%. This indicates that Pair Reduction is effective in reducing processing time.

Table 4. Results of Experiment 1

# of access logs	# of Pairs	Adjusted Rand index	Perfect correct rate	No-misclassified correct rate	No-missed correct rate	Partial correct	Processing time [s]
10,000	187,393	0.3783	0.8303	0.9092	0.9076	0.9864	1,738
50,002	2,855,972	0.3142	0.7969	0.8613	0.9175	0.9820	2,088
100,002	9,841,949	0.2978	0.7750	0.8377	0.9183	0.9810	2,058

Table 5. Results of Experiment 2

# of access logs	# of Pairs	Adjusted Rand index	Perfect correct rate	No- misclassified correct rate	No-missed correct rate	Partial correct rate	Processing time [s]
10,000	49,995,000	0.1618	0.6683	0.7179	0.9349	0.9846	15,410
50,002	1,250,075,001	0.0097	0.6158	0.6563	0.9465	0.9871	187,297
100,002	5,000,150,001	0.0005	0.5807	0.6207	0.9488	0.9888	223,670

Table 6. Adjusted Rand index and distribution of cluster size

Algorithm	Adjusted Rand index	Distribution of cluster size					
		Standard deviation	min	25%	50%	75%	max
Proposed Method	0.0003247	61.00	1	63	116	146	195
k-means	0.0003503	48.14	1	73	94	122.25	423
Ward's method	0.0003419	54.56	1	66	91	124	547
Average Linkage	0.0000113	652.90	1	1	4	14	12,262
Single Linkage	−0.0000082	2,972.12	1	1	1	1	95,638
Complete Linkage	0.0001263	154.27	1	9	49	115	1,404

Based on the results of Experiment 3 (Table 6), clusters created using the proposed method reduced the maximum cluster size to approximately half compared to the existing ones. Assuming a similar distribution of access logs during prediction, it is expected that the maximum cluster size will also be reduced during prediction. In the grouping process, all possible pair creations in the generated clusters are considered; however, given that the combinations are $O(n^2)$, if the number of access logs in the cluster is reduced by half, the number of pairs will be reduced to approximately one-fourth. Thus, the reduction in terms of the maximum cluster size has a significant impact on processing time. In addition, the adjusted Rand index achieved an accuracy of 92.6% compared to the best-performing method that does not utilized the proposed cluster creation method. Even when compared to the best of the relatively low-performing methods, i.e., the average linkage, single linkage, and complete linkage methods, the proposed method surpasses the average linkage method by 257.1%, and the same trend was observed in terms of the cluster size distribution.

Considering the reduction in the maximum cluster size and the relative accuracy, it can be concluded that the proposed cluster creation method outperformed the compared methods.

However, in terms of limitations, in this study, we only used a single deep learning model for the grouping process. Thus, to enhance accuracy, using separate deep learning model for each cluster can be explored. The clusters obtained by the clustering algorithm contain similar fingerprints; thus, creating separate deep learning models for each cluster may facilitate more precise learning of the differences between devices.

7 Conclusion

This paper has proposed a method to aggregate website access logs based on individual devices. The proposed method implements a technique to reduce the number of considered pairs using clustering algorithms, which was found to yield positive results. Compared to methods that do not employ the pair reduction technique, the proposed method achieved predictions in approximately 1% of the processing time on a dataset of 100,002 access logs. In addition, the results in terms of the adjusted Rand index, perfect correct rate, and no-misclassified correct rate exhibited improvements due to the implemented pair reduction process.

8 Ethics

Our experiments were conducted in accordance with the principles outlined in the Menlo report [10]. During the experimentation process, we ensured sufficient privacy protection by avoiding any personal identification. In this study, the original data were subjected to statistical processing to prevent inferences from being made about the individuals represented in the experimental data.

In addition, the dataset used in this study was strictly maintained under our control and was used solely for academic purposes (with no provision for its distribution to others).

References

1. Acar, G., et al.: FPDetective: dusting the web for fingerprinters. In: Proceedings of the 2013 ACM SIGSAC Conference on Computer & Communications Security, CCS 2013, pp. 1129-1140, New York, NY, USA. Association for Computing Machinery (2013)
2. Microsoft Corporation. LightGBM. https://lightgbm.readthedocs.io
3. Eckersley, P.: How unique is your web browser? In: Atallah, M.J., Hopper, N.J. (eds.) PETS 2010. LNCS, vol. 6205, pp. 1–18. Springer, Heidelberg (2010). https://doi.org/10.1007/978-3-642-14527-8_1
4. Fujii, T., Tonaki, M., Toshimitsu, Y., Shibata, S., Hojyo, Y., Saito, T.: Proposal and implementation of an id estimation using deep learning for pcs and mobile devices. Comput. Secur. Symp. **2020**, 10 (2020)
5. Galler, B.A., Fisher, M.J.: An improved equivalence algorithm. Commun. ACM **7**(5), 301–303 (1964)
6. Gómez-Boix, A., Laperdrix, P., Baudry, B.: Hiding in the crowd: an analysis of the effectiveness of browser fingerprinting at large scale. In: Proceedings of the 2018 World Wide Web Conference, WWW 2018, Republic and Canton of Geneva, CHE, 2018, pp. 309-318. International World Wide Web Conferences Steering Committee (2018)
7. Houjou, Y., Saito, Y., Takamichi, S.: Passive fingerprinting enforced with deep neural network. Comput. Secur. Symp. 2019, **2019**, 252–259 (2019)
8. Iqbal, U., Englehardt, S., Shafiq, Z.: Fingerprinting the fingerprinters: learning to detect browser fingerprinting behaviors. In: 2021 IEEE Symposium on Security and Privacy (SP), pp. 1143–1161 (2021)
9. Ward, J.H., Jr.: Hierarchical grouping to optimize an objective function. J. Am. Stat. Assoc. **58**(301), 236–244 (1963)
10. Kenneally, E., Dittrich, D.: The menlo report: ethical principles guiding information and communication technology research. Available at SSRN 2445102 (2012)
11. Masuda, N., et al.: A proposal for identity prediction using passive fingerprints by multi-class classification model. In: 2022 Symposium on Cryptography and Information Security (2022)
12. Takahashi, K., et al.: Consideration of browser fingerprinting only with http headers. In: 2018 Symposium on Cryptography and Information Security (2018)
13. Tonaki, M., Jin, A., Toshimitsu, Y., Takayama, M., Saito, T.: A proposal for identify devices over private relay. In: 2022 Symposium on Cryptography and Information Security (2022)
14. Vastel, A., Laperdrix, P., Rudametkin, W., Rouvoy, R.: Fp-stalker: tracking browser fingerprint evolutions. In: 2018 IEEE Symposium on Security and Privacy (SP), pp. 728–741 (2018)

A Scheme for Source Location Privacy of Multiple Sources in Wireless Sensor Networks

Sain Saginbekov(✉)

Department of Computer Science, School of Engineering and Digital Sciences, Nazarbayev University, Kabanbay Batyr Ave. 53, Astana 010000, Kazakhstan
sain.saginbekov@nu.edu.kz

Abstract. In Wireless Sensor Network applications involving the monitoring of valuable objects, it is crucial to keep the locations of these objects private in order to protect them from potential adversaries. A number of schemes exist to address this problem, of which only a few of them focus on networks where multiple source nodes can report simultaneously. While those schemes provide a reasonable privacy level, they often suffer from inefficiencies in terms of packet overhead. In this paper, we present a scheme called Multi-Source Location Privacy (MSLP). In MSLP, a node uses two-hop neighbourhood information and broadcast nature of wireless communication to route packets to the sink. We perform simulation experiments of MSLP and compare it with other existing schemes and show that MSLP not only provides better location privacy but also offers better performance in terms of energy and delivery delay by reducing packet overhead.

1 Introduction

Wireless Sensor Networks (WSNs) are networks composed of numerous battery-powered sensor nodes that can detect data from the physical environment, and transmitting it to a designated node, called the sink. The sensor nodes, in general, have limited processing power, energy supply, and memory; they can process data and communicate with each other wirelessly to collect sensed data to the sink. The sensor nodes may be equipped with different sensing capabilities, such as light, motion, temperature, humidity, etc., which enable WSNs to be used in various applications, including healthcare, agriculture, environmental monitoring, military, industrial automation, smart cities, etc. [1,11,13]. However, the widespread deployment of WSNs raises concerns about content and context privacy.

The content privacy refers to the preservation of the actual content of information such as confidentiality and message integrity. Whereas the context privacy refers to the preservation of the contextual information about the WSN such as sender and recipient locations, traffic flow, transmission time, etc., that could disclose sensitive information to the adversary [6]. The content privacy

L. Barolli (Ed.): BWCCA 2023, LNDECT 186, pp. 135–148, 2024.
https://doi.org/10.1007/978-3-031-46784-4_12

can be preserved using different cryptographic algorithms. In this paper, we are interested in the context privacy, in particular, privacy regarding the location information of the source nodes which is known as *Source Location Privacy* (SLP).

In WSNs, SLP refers to the preservation of source nodes' location information from adversaries. And it is important in applications where the network collects data from invaluable objects. For example, in applications where valuable entities are being monitored, such as friendly soldiers in a battlefield or endangered animals, it is crucial to insure the privacy of their locations. The disclosure of such information can lead to severe consequences. Examples include capturing animals by an adversary.

The characteristics of the WSNs make the protection of source location privacy a challenging task. Several approaches that consider the characteristics of WSNs have been proposed to address SLP in WSNs. However, to our knowledge, only a few of them consider WSNs with more than one sources [3]. That is, more than one sensor nodes can be reporting to the sink at the same time. But those schemes are not efficient in terms of packet overhead. To this end, in this paper, we propose a scheme called Multi-source location privacy (MSLP), which considers more than one source, provides better location privacy, and efficient in terms of packet overhead. MSLP uses information about two-hop neighbours to route packets so as to obfuscate the adversary and hide the location of source nodes. So, in this paper, our main contribution is a scheme that addresses the SLP problem in a WSN with multiple sources with lower packet overhead. We simulate MSLP and show the efficiency of our protocol compared to existing protocols.

The paper is organized as follows. Section 2 gives an overview of related work. Section 3 defines the network and attacker models we consider in our work, and defines the problem we tackle. Section 4 defines our proposed scheme. Section 5 defines the setup and presents the results of the simulation. Finally, Sect. 6 concludes this work with a summary.

2 Related Work

A number of approaches exist in the literature to solve the SLP problem. They can be classified mainly based on the techniques they use [6].

There exist solutions that use fake sources to solve the SLP problem [2,15–17]. This technique chooses some node as fake sources and let them transmit fake packets to pull the attacker towards themselves. The schemes that use fake sources can provide high SLP level for some type of attackers [8]. However, this approach consumes more energy as fake sources should transmit redundant packets. Therefore, to reduce the number of fake transmissions, in [17], the authors propose to transmit real and fake packets every interval. However, this approach may introduce packet delivery delay. In [4], proposes a scheme where every node computes its own time interval based on an equation and transmits packets in that time interval.

The solution proposed in [14] includes two phases. In the first phase, the source node routes a packet to a random node, called a phantom node, and then in the second phase, that random node floods the packet to the sink. The second phase of this approach is not energy efficient as there are many redundant packet transmissions. Therefore, in [9], after having sent the packet to a random node, the authors propose to route the packet directly to the sink via a shortest path. In [5], the authors propose to choose phantom nodes further away from the source. Whereas, in [10], the authors propose to choose phantom nodes from a special annular region around the sink. In [12], the authors propose a scheme which routes the source packets backwards as well as horizontally and vertically, which necessitates to have special nodes in the network to identify these directions. A scheme proposed in [3], tries to solve the SLP problem in a network with multiple sources. The scheme tries to route packets via different paths as to delay the attacker to reach a source. The above schemes may use longer paths to deliver source packets, and therefore they may incur larger packet overhead and take longer delivery delays. Moreover, their SLP level may depend on a value called random walk, or *h-walk* [9], which is the number of random hops a packet should take before it is directed towards the sink via a shortest path or by flooding.

3 System Models and Problem Statement

In this section, we start with defining the network model used in this paper, then we present the attacker model, and then we proceed to describe the problem we aim to address.

3.1 Network Model

We start by assuming that all nodes in the network are identical, except the *sink node*. The sink node is a special node in the network which collects data from other nodes in the network and then may process and/or forward the data further to the end user. The nodes have the same capabilities such as transmission range, memory, and computation power. The nodes and sink are stationary. We call a node *source node* if it generates data to be sent to the sink. The network may contain multiple sources at the same time. Two nodes are *neighbours* if they are in the transmission range of each other. As the transmission range of a node is usually small, the node may not send the generated data to the sink directly. Instead, the node forwards the data towards the sink by relaying to its neighbour. That is, a packet generated by a source node may reach the sink only after being forwarded by several nodes in between. We call a set of nodes, between a source and the sink, that relayed the packet a *routing path*.

More formally, we consider a WSN as an undirected graph $G = (V, E)$, where V and E are the set of nodes and the set of links that connect the nodes, respectively. We say $n \in V$ and $m \in V$ are neighbours, or 1-hop neighbours, if $(n, m) \in E$. We say $k \in V$ is a 2-hop neighbour of $n \in V$, if $(n, k) \notin E$, but

$(n, m) \in E$ and $(m, k) \in E$. The set of 1-hop and 2- hop neighbours of n are denoted by $N(n)$ and $N_2(n)$, respectively. A path $P = n_1, n_2, ..., n_k$ can be a routing path if $\forall n_{1 \le i < k} \in V$, $(n_i, n_{i+1}) \in E$, and $n_1 \in E$ is a source and $n_k \in E$ is the sink. The hop number of n, denoted by $h(n)$, is the smallest number of links used to connect n and the sink. For example, all neighbours of the sink have the hop number of 1. If $(n, m) \in E$ and $h(n) < h(m)$, then we say that n is a parent node of m, and m is a child of node n. If $h(n) = h(m)$, then we say that n and m are *siblings.* The set of parents of n is denoted by $P(n)$, whereas the set of siblings is denoted by $S(n)$.

3.2 Attacker Model

The attackers can be classified based on their presence and actions. Depending on their presence, the attackers can be grouped mainly into two classes: *local* and *global.* A local attacker can eavesdrop messages transmitted only within its neighbourhood. A global attacker is stronger in terms of resources such as power, memory, and transmission range. Therefore, we assume that a global attacker can eavesdrop any message transmission in the network.

The attackers can also be grouped depending on actions such as *internal* or *external* and *active* or *passive* [6]. An internal attacker has the ability to gain unauthorized access to sensor nodes, but an external attacker lacks that capability. An active attacker possesses the ability to modify the message traffic or manipulate the functioning of sensor nodes, while a passive attacker is incapable of doing so.

In this work, we assume a local, passive, and external attacker. We assume the patient attacker defined in [9]. The *patient attacker* starts at the sink node, and at every node, it waits at that node until it eavesdrops a message from its neighbouring node. We assume also that the attacker has enough resources, such as directional antennas, to identify the source of received message. Whenever the attacker eavesdrops a message from a node, it moves to that node unless the node has not been visited before. That is, let's denote the current location of an attacker located at a node n by A_n, and the set of all visited nodes by $A = \{A_n, A_m, ...\}$. Now, if the attacker is located at node n and it eavesdrops a message from a node k, then the attacker changes its location to A_k only if $A_k \notin A$. Note that, if the attacker owns a GPS-enabled device and can compute the distance between the locations, then the attacker may move from n to k only if $A_k \notin A$ and $h(k) < h(n)$, that is, if k is located closer than n to the sink. However, this kind of movement of attacker does not guarantee that the attacker will locate the source faster. Conversely, as simulation results show, depending on the schemes, this may complicate the attacker's job to reach a source node.

3.3 Problem Statement

As the attacker we assume is the patient and local attacker, and initially it is located at the sink node, it waits until it receives the first packet from one of the sink's neighbour. Then it moves to that neighbour and waits there for another packet. This way, by moving from one node to another node, the attacker eventually may reach the source and find a monitored object. To make the job of the attacker difficult, nodes in the network should route packets in a way to obfuscate the attacker. So, the problem we try to address in this paper is to develop a routing scheme that does not let the attacker to find a source or that requires the attacker to spend a long time to find a source.

There exist several metrics that can be used to access the SLP level of a scheme [9]. The first metric is called *capture likelihood*, which is the probability of an attacker finding the source node in a given time interval. The lower the capture likelihood, the higher the scheme's SLP level. The *safety period* of a scheme is another metric, which is the number of new messages sent by the source until it is found by the attacker. Another metric, called *capture ratio*, is the proportion of total runs to runs in which the source is caught during the safety period. [7].

Let's denote the kth transmitted packet of a source node s by p_k^s. Then we say that the safety period of the scheme is k if s transmits p_k^s when $A_n \in A$ such that $s \in N(n)$, where A_n is the location of the attacker and $N(n)$ is the set of neighbours of node n as mentioned in Sect. 3.2. So, if there is only one source in the network, then a routing scheme should maximize k to increase the SLP level. For a network with multiple sources, we can define the safety period as follows. Given a set of sources $S = \{s_1, s_2, ...\}$, $S \subset V$, sets of transmitted packets $p^{s_1}, p^{s_2}, ...$ by the sources $s_1, s_2, ...$, respectively, until the attacker finds a source $s_i \in S$, then the safety period of a scheme is the sum of all transmitted packets by all sources, that is, safety period $= \sum |p^{s_i}|, \forall s_i \in S$.

To find the safety period of a scheme, it may take a long time in experiments, therefore, in our experiments, we will use the capture ratio as a metric to assess schemes' SLP level. The capture ratio can be defined as follows. Let k be the number of packets to be transmitted by each source and T be the total number of experiment runs. Let R_i be 1 if in the ith experiment run $|p_i^s| \leq k$, there exists a node $n \in V$, a source $s_i \in S$, and an attacker location A_n such that $n \in N(s_i)$ and $A_n \in A$ (see Sect. 3.2); let R_i be 0 otherwise. Then the capture ratio is $\sum_{i<T} R_i/T$.

4 Proposed Scheme

In this section, we present our proposed scheme called Multi-Source Location Privacy (MSLP).

In [3], the proposed protocol tries to address the SLP problem in networks with multiple sources by routing source packets using different paths. After a node n sends a packet to a node m, n adds m to its queue. Then n tries to send another packet to a node k which does not exist in that queue. However, k could have already sent another node's packet, therefore, according to the protocol, k sends a control packet to n to tell that it has already sent and n should choose another node to forward its packet. This step may continue until n finds a node which has not send a packet yet. When there is no more neighbouring node left out of its queue, n chooses a node, say l, from its queue and sends it to l. This process may increase the delivery delay and incur high packet overhead. To reduce the delivery delay and packet overhead, our proposed scheme MSLP uses two-hop neighbourhood information and the broadcast feature of wireless communication - not only the recipient hears the message, but all neighbours of the sender can hear the message. Information about two-hop neighbours can be obtained by having each node in the network broadcast information about its one-hop neighbours.

The MSLP scheme consists of two sequential phases. The first phase, called *Setup phase*, prepares the nodes for forwarding packets towards the sink. The second phase, called *Routing phase*, delivers source packets to the sink. We detail these phases in the following sections.

4.1 Setup Phase

In this phase, each of the nodes learns its hop number, neighbours, and two-hop neighbours. This phase works as follows (see Algorithm 1). First, the sink broadcasts a control packet called *hello*, which includes the hop number of sink (which is set to 0) and *id*. Every neighbouring node n, that is $n \in N(sink)$, which receives the hello packet, updates its hop number $h(n)$ to $h(sink)+1$ and adds the sender *id* to their set of one-hop neighbours. Then n broadcasts hello packet which includes $h(n)$ and its *id*. When node m receives the hello packet from n, m updates its hop number $h(m) = h(n) + 1$ and adds n in its set of one-hop neighbours. If n is closer to the sink than m, that is if $h(n) < h(m)$, then m also adds n to its set of *parents* P. If $h(n)$ is equal to $h(m)$, then m adds n to its set of *siblings* S. Eventually all nodes will learn their hop numbers and neighbours. Then every node n in the networks broadcasts its set of one-hop neighbours. When m receives the neighbours of n from n, m adds the neighbours of n to its two-hop neighbours, to $N_2(n)$.

Algorithm 1: Setup Phase

```
/* N - set of neighbours                                   */
/* N2 - set of 2-hop neighbors                             */
/* P - set of parents                                      */
/* S - set of siblings                                     */
Sink
  my_hop ← 0
  packet hello ←< my_hop, my_id >
  broadcast(hello)
Sensor nodes
  my_hop ← ∞
  event receive(N, id)
    N2(id).add(N)
  event receive(hello)
    N.add(hello.my_id)
    if my_hop > hello.my_hop + 1
      my_hop ← hello.my_hop + 1
      P.clear()
      S.clear()
      P.add(hello.my_id)
    else if my_hop = hello.my_hop + 1
      P.add(hello.my_id)
    else if my_hop = hello.my_hop
      S.add(hello.my_id)
  packet hello ←< my_hop, my_id >
  broadcast(hello)
  broadcast(N, my_id)
```

4.2 Routing Phase

The routing phase begins with a source node sending its packets to the sink. However, as mentioned in Sect. 3.1, the sink may not be in the communication range of the source node. Therefore, the packet may pass through several nodes to reach the sink. The source node, or any other relay node on the routing path, chooses the next node among its neighbours according to the algorithm given in Algorithm 2 until the packet reaches the sink.

The algorithm works as follows. A source node, or any other relay node, n, first tries to find a parent node r, from its parents set P, whose neighbours have not transmitted a packet yet. If there is no such a parent, then n tries to find such a node from its siblings set S. Otherwise, n chooses a relay node r among its parents and siblings $(P \cup S)$ such that the number of neighbours of r that have transmitted packets is the smallest.

Now the question is, how a node knows that its neighbour has transmitted a packet? As the wireless communication is broadcast in nature, a node can snoop its neighbours' transmissions. A node in MSLP takes the advantage of this feature to record all its transmitted neighbours to its queue. For example,

Algorithm 2: Routing Phase

```
/* N(n) - set of neighbours of node n                                   */
/* P - set of parents                                                   */
/* S - set of siblings                                                  */
/* Q - queue of neighbours that have already sent a packet.             */
event senseData()
  relayNode ← chooseNext()
  packet normal ←< data >
  send(normal, relayNode)
event receive(normal, node)
  relayNode ← chooseNext()
  forward(normal, relayNode)
event snoop(normal, node)
  Q ← Q ∪ node
chooseNext()
  /* first try to find a parent                                         */
  for each p ∈ P
    for each r ∈ N(p)
      if r ∉ Q
        return r
  /* else try to find a sibling                                         */
  for each s ∈ S
    for each r ∈ N(s)
      if r ∉ Q
        return r
  /* else try to find a neighbour that has min intersection with Q      */
  min ← ∞
  for each m ∈ (P ∪ S)
    count ← 0
    for each r ∈ N(m)
      if r ∈ Q
        count ← count+1
    if min > count
      min ← count
      returnNode ← m
  return returnNode
```

assume that nodes k and m are neighbours. See Fig. 1a for illustration. If m sends a packet to r, then as k is in the communication range of m, k can also eavesdrop that transmission. Therefore, k can add the ids of m and r in its queue. Now assume that after some time k becomes a relay node. k tries to find the next relay node first among its parents, and then among its siblings, that has no transmitted neighbours. In Fig. 1b, node s is a good relay node for k, as s has no transmitted neighbours (m and r are not neighbours of s). Note that this idea does not allow the attacker to reach the nodes k and s if the attacker is at node r.

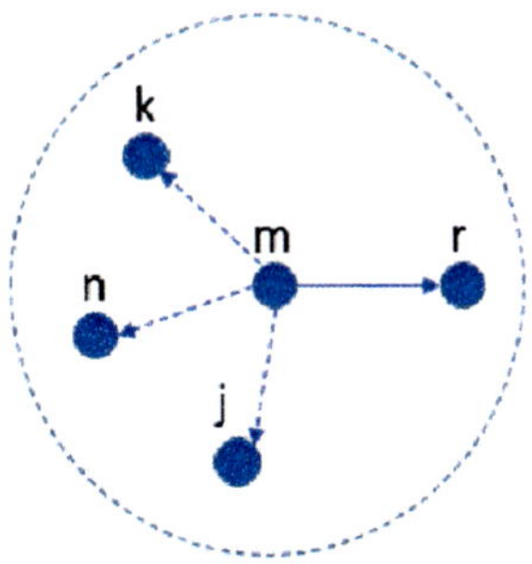

(a) When node m transmits to r, other three neighbours can also eavesdrop that transmission.

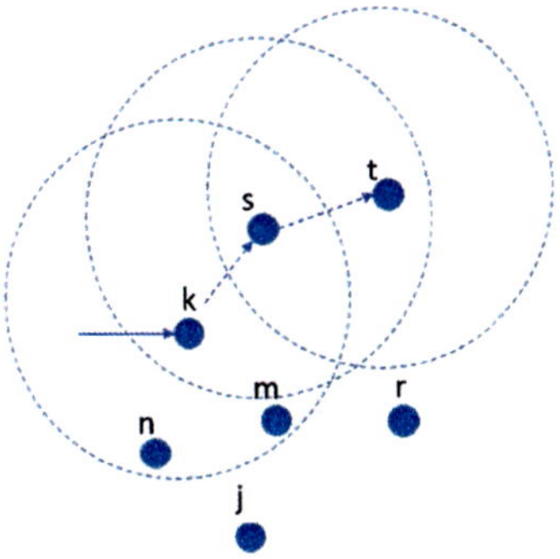

(b) Node k chooses s as its relay node as s is not a neighbour of r.

Fig. 1. Dashed circle is the communication range of the node at the center.

Now, as the number of nodes in the neighbourhood is limited, eventually a node may not find a relay node among its parents and siblings that has not transmitted yet. In that case, the node could choose a random neighbour as a relay node. But in the MSLP scheme, the node tries to find a node that has the smallest number of neighbours that have transmitted. The idea is to route a packet through a region that has less number of packet transmissions. It is because we believe that the attacker is most likely situated at the region with higher message traffic. These ideas are implemented and the results are presented in the next section.

5 Simulation Setup and Results

In this section, we provide details on the simulation configuration and results. In order to show the efficiency of MSLP, we have also implemented two existing schemes and compared the results.

5.1 Simulation Setup

To access the effectiveness of MSLP and compare with other existing schemes, we have developed a discrete event-based simulator using Java programming language. We have compared MSLP with two other schemes, namely, Redpaths [3] and Phantom routing [9], on a network of size 1000 nodes which were located in an area of 100 by 100 units. The nodes were located randomly and had a communication range of 10 units. We have compared MSLP with the other two schemes in networks with 1, 2, 3, and 4 sources, which were located at different corners of the area. See Figs. 2a, 2b, 2c, and 2d, respectively, for illustration.

In order to evaluate the privacy level of the schemes, in our simulations, we have used the capture ratio metric (see Sect. 3.3). We have also evaluated the schemes in terms of packet overhead. Because a node's energy consumption is

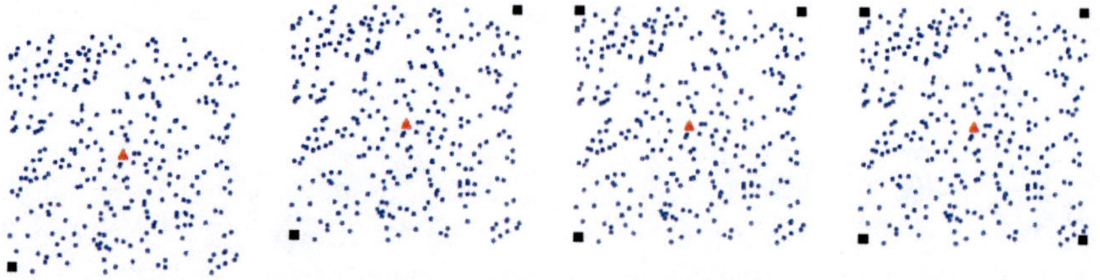

(a) Network with one source.
(b) Network with two sources located at opposite corners.
(c) Network with three sources located at three different corners.
(d) Network with four sources located at different corners.

Fig. 2. Networks with different number of sources (squares) and sink at the center (triangle).

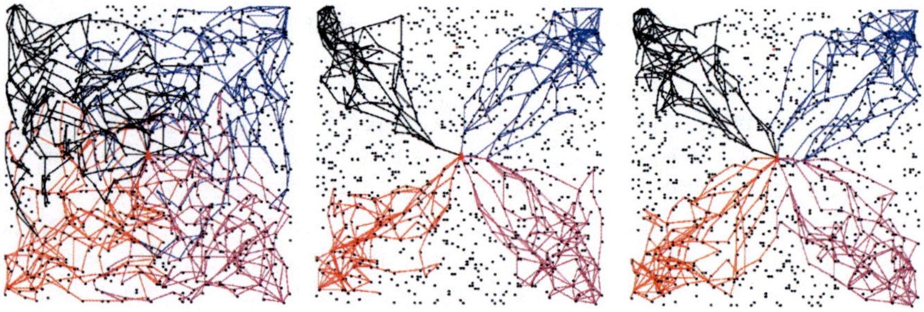

(a) MSLP
(b) Phantom routing
(c) Redpaths

Fig. 3. 15 routing paths from 4 sources to the sink, h-walk is 10, the average hop number between a source and the sink is 8.

directly correlated with the number of packets it transmits, this metric can be used to assess the energy efficiency of MSLP. The lower the packet overhead, the more energy efficient is the scheme. Moreover, the later metric may also affect the packet delivery latency. Increasing the packet overhead may lead to a proportional increase in packet delivery latency.

We have run 1000 simulations and have taken the average of the total number of packet transmissions and the capture ratio. To find the capture ratio, we have used 100 as the safety period. That is, in each simulation run, each source node(s) in the network sends 100 packets. As the other two schemes' SLP level and packet overhead depend on the number of random hops, or h-walk [9], we have also varied the value of h-walk between 6 and 16. Figure 3 illustrates 15 routing paths from 4 sources to the sink (15 packets sent by 4 sources towards the sink) when h-walk is 10.

5.2 Simulation Results

As mentioned in Sect. 2, the SLP level of the Phantom routing and Redpaths schemes depend on the value of h-walk. This can be observed from Fig. 4, as

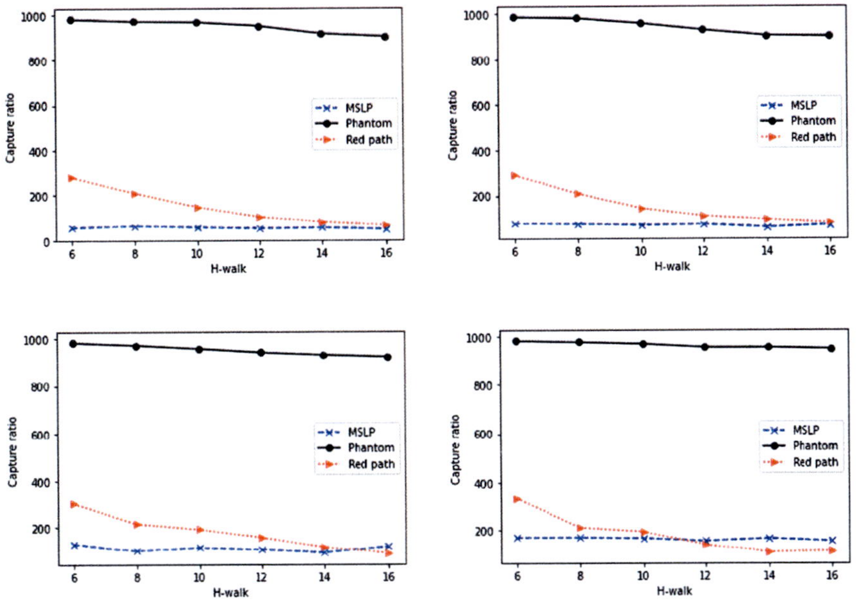

(c) Network with 3 sources.

(d) Network with 4 sources.

Fig. 4. Capture ratio over h-walk, different number of sources.

the h-walk value increases, the capture ratio decreases. Out of 1000 runs, as we increase the value of h-walk, the capture ratio of the Phantom routing scheme decreases from about 98% to about 90% when there is one source, and decreases to 95% when there are four sources. While the capture ratio of Red-paths decreases from about 33% to about 12% when there is one source, and decreases to 7% when there are four sources. As can be observed from the figures, MSLP shows the smallest capture ratio and it does not change based on the h-walk value. It is about 5% irrespective of h-walk value when there is one source, and about 17% when there are four sources in the network. For the h-walk values of 14 and 16, the Redpaths scheme shows slightly better SLP level than MSLP, but as we will see in Fig. 5, the Redpaths scheme obtains this advantage by transmitting more than 3 times more packets.

The number of packet transmissions of the three schemes are shown in Fig. 5. Figure 5a shows the average number of transmitted packets to deliver one source packet to the sink. If in MSLP, about 9 packets are enough to deliver a packet to the sink, in Phantom routing and Redpaths, nodes need to transmit 11 and 16 packets when h-walk is 6; and 18 and 30, when h-walk is 16, respectively. In Fig. 5b, as expected, we can see that the number of packet transmissions increases proportionally as the number of sources increases.

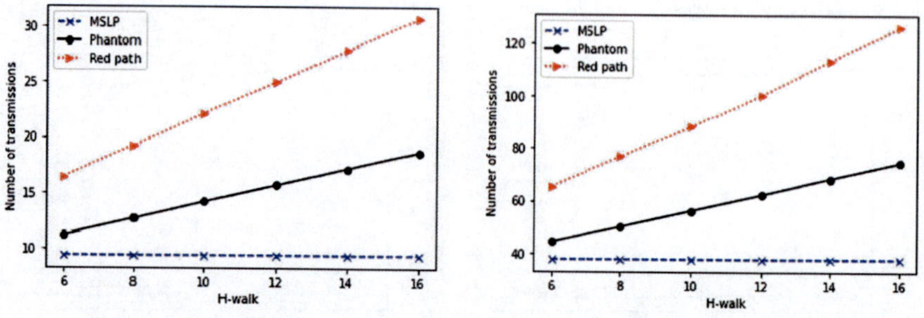

(a) Network with 1 source. (b) Network with 4 sources.

Fig. 5. Number of packet transmissions to deliver a new packet to the sink, different number of sources.

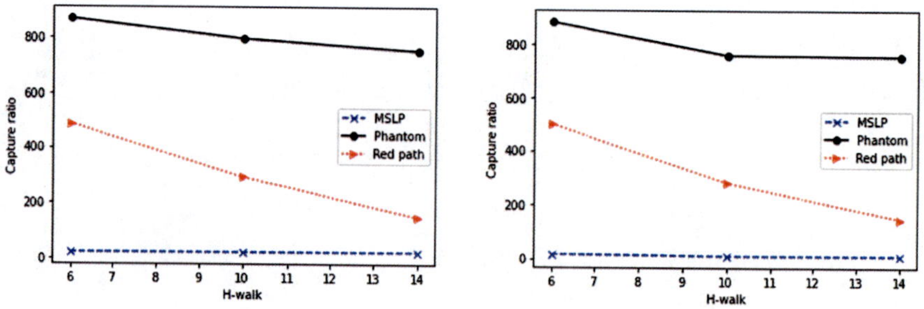

(a) Network with 2 sources. (b) Network with 4 sources.

Fig. 6. Capture ratio, different number of sources. Attacker moves to the farther node to the sink.

As can be observed from the figures, among the three schemes, the number of packet transmissions of Redpaths scheme is the largest for the same h-walk value. We also observe that while the MSLP scheme shows a constant number of packet transmissions, the Redpaths scheme has larger increase than the Phantom routing scheme as the h-walk value increases.

Figure 6 shows the capture ratio of the three schemes when the attacker has a GPS-enabled device and moves only to the farther to the sink node. As can be seen, MPLS shows a constant capture ratio of around 2%, while the capture ratio of Redpaths and Phantom routing decreases from 50% and 88% to 15% and 75%, respectively.

6 Conclusion

In this paper, we have presented a scheme called MSLP that solves the Source Location Problem in a WSN with multiple sources. In MSLP, a node uses two-hop neighbourhood information and broadcast nature of wireless transmission to route packets in order to obfuscate the attacker. We have conducted a series of simulation experiments, and the results show that MSLP can provide high source location privacy level while significantly reducing the packet overhead.

Acknowledgements. This work is supported by the School of Engineering and Digital Sciences, Nazarbayev University, Astana, Kazakhstan.

References

1. Akyildiz, I., Su, W., Sankarasubramaniam, Y., Cayirci, E.: Wireless sensor networks: a survey. Comput. Netw. **38**, 393–422 (2002)
2. Adilbekov, U., Adilova, A., Saginbekov, S.: Providing location privacy using fake sources in wireless sensor networks. In: IEEE 12th International Conference on Application of Information and Communication Technologies (AICT) (2018)
3. Berdibek, A., Saginbekov, S.: A routing protocol for source location privacy in wireless sensor networks with multiple sources. In: Q2SWinet 2019 (2019)
4. Bushnag, A., Abuzneid, A., Mahmood, A.: An efficient source anonymity technique based on exponential distribution against a global adversary model using fake injections. In: Q2SWinet 2017 (2017)
5. Chen, J., Fang, B., Yin, L., Su, S.: A source-location privacy preservation protocol in wireless sensor networks using source-based restricted flooding. Jisuanji Xuebao (Chin. J. Comput.) **33**, 1736–1747 (2010)
6. Conti, M., Willemsen, J., Crispo, B.: Providing source location privacy in wireless sensor networks: a survey. IEEE Commun. Surv. Tutorials **15**, 1238–1280 (2013)
7. Jhumka, A., Bradbury, M., Leeke, M.: Towards understanding source location privacy in wireless sensor networks through fake sources. In: Proceedings of the 2012 IEEE 11th International Conference on Trust, Security and Privacy in Computing and Communications (2012)
8. Jhumka, A., Leeke, M., Shrestha, S.: On the use of fake sources for source location privacy. Comput. J. **54**, 860–874 (2011)
9. Kamat, P., Zhang, Y., Trappe, W., Ozturk, C.: Enhancing source-location privacy in sensor network routing. In: 25th IEEE International Conference on Distributed Computing Systems (ICDCS 2005) (2005)
10. Lightfoot, L., Li, Y., Ren, J.: STaR: design and quantitative measurement of source-location privacy for wireless sensor networks. Secur. Commun. Netw. **9**, 220–228 (2016)
11. Mainwaring, A., Culler, D., Polastre, J., Szewczyk, R., Andferson, J.: Wireless sensor networks for habitat monitoring. In: Proceedings of the 1st ACM International Workshop on Wireless Sensor Networks and Applications, NY, USA, ACM (2002)
12. Manjula, R., Koduru, T., Datta, R.: Protecting source location privacy in IoT-enabled wireless sensor networks: the case of multiple assets. IEEE Internet Things J. **9**, 10807–10820 (2022)
13. Milenkovic, A., Otto, C., Jovanov, E.: Wireless sensor networks for personal health monitoring: issues and an implementation. Comput. Commun. **29**, 2521–2533 (2006)
14. Ozturk, C., Zhang, Y., Trappe, W.: Source-location privacy in energy-constrained sensor network routing. In: Proceedings of the 2Nd ACM Workshop on Security of Ad Hoc and Sensor Networks, ACM (2004)
15. Ouyang, Y., Le, Z., Liu, D., Ford, J., Makedon, F.: Source location privacy against laptop-class attacks in sensor networks. In: SecureComm 2008 (2008)

16. Saginbekov, S., Oryspayev, D.: Snowflake: an adaptive energy and delay efficient scheme for source location privacy in wireless sensor networks. In: Proceedings of the 11th International Conference on Sensor Networks - SENSORNETS (2022)
17. Yang, Y., Shao, M., Zhu, S., Cao, G.: Towards statistically strong source anonymity for sensor networks. ACM Trans. Sens. Netw. **9**, 1–23 (2013)

Blockchain Application for Fish Origin Certification

Riccardo Petracci(✉), Rosario Culmone, and Leonardo Mostarda

School of Science and Technology, Computer Science Division,
University of Camerino, 62032 Camerino, Italy
{riccardo.petracci,rosario.culmone,leonardo.mostarda}@unicam.it

Abstract. Seafood traceability is essential in order to ensure the required level of quality control and management. However, over recent years different food scandals have damaged customer trust. Seafood is particularly affected by different types of frauds that take place at global scale. Often complex and unclear supply chains lead to misleading labels. These can falsely allow the selling of low value species as high valuable ones. Labels can also contain wrong location information. For instance, false location can be used in order to sell farmed fish as wild one. The problem of label misleading appears to be widely widespread in restaurants. A study over 23 different countries showed that restaurant mislabelling can be as high as 40%. While in general many traceability solutions have failed to meet the needs of food chain stakeholders, blockchain technology seems to be a promising solution. The decentralised and self-regulating blockchain nature can enable secure traceability in complex supply chains without the need of a centralised trusted party. Traceability data can be stored inside the blockchain which ensures high integrity, reliability and immutability. In this paper we describe a novel fish traceability system (CERTFish) that integrates in a novel way secure digital tags, blockchain technology, IoT tamper proof devices, location and time information. CERTFish ensures wild fish origin authentication and certifies the fish from the catch, throughout its conservation and transportation till the final customers at the restaurant. CERTFish has been validated in a real case study scenarios in order to certify the origin of a special type of anchovies from a specific Mediterranean area.

1 Introduction

Food traceability is essential in order to provide customer trust and the required level of quality control and management. Various studies on fish products from restaurants and markets found out that 40% of the labels are misleading [1,2,9]. In particular, various studies based on DNA techniques [9] found out that mislabelling is used to sell of low value species as high value ones. For instance,

Work supported by the Italian National Group for Scientific Computation GNCS-INdAM.

L. Barolli (Ed.): BWCCA 2023, LNDECT 186, pp. 149–158, 2024.
https://doi.org/10.1007/978-3-031-46784-4_13

in a study performed in supermarkets and restaurants in Canada, the US, the UK, Singapore, Australia and New Zealand discovered fish with a misleading "snapper" in 40% of the cases. Labels can also contain wrong location information. For instance, false location can be used in order to sell farmed fish as wild one. In Germany, 48% of samples labelled as king scallops were the less valuable Japanese scallop. The problem of label misleading appears to be widely widespread in restaurants. A study over 23 different countries showed that mislabelling can be as high as 40%. Current traceability solutions have failed to meet the needs of fish chain stakeholders. Labels can be easily counterfeit and their location does not denote a specific geographical area since they only define large areas, i.e., twelve ocean and sea location zones specified by the FAO.

While in general many traceability solutions have failed to meet the needs of fish chain stakeholders, blockchain technology seems to be a promising solution. The decentralised and self-regulating blockchain nature can enable secure traceability without the need of a centralised trusted party. Traceability data can be stored inside the blockchain which ensures high integrity, reliability and immutability. In this paper we describe a novel fish traceability system (CERTFish) that integrates secure digital tags, blockchain technology, IoT tamper proof devices, location and time information. CERTFish ensures wild fish origin authentication and certifies the fish from the catch, throughout its conservation and transportation till the final customers at the restaurant. CERTFish has been validated in a real case study scenarios in order to certify the origin of a special type of anchovies from a specific Mediterranean area.

The rest of the article is organised as follows. Section 2 introduces the fish case study, potential attacks, and the system requirements; Sect. 3 describes the CERTFish system; Sect. 4 describes the implementation of our system; Sect. 5 discusses the security of our system, the experimental setup that has been used and presents the results of our validation; Sect. 6 reviews the related work; finally, Sect. 7 concludes the paper and outlines future work.

2 Tyrrhenian Anchovy Case Study

In the Tyrrhenian Sea, at Campania coast, a particular variety of anchovies is fished. This is used to be eaten fresh or to produce a particular type of bottled food that is called "la colatura di alici di Catara" [8]. This bottled food has a particular method of production that dates back the to the ancient Romans. In this paper we focus on the fish-to-fork case study that is the stages that go from the catch to the restaurant or market where fresh anchovies are consumed by customers.

The following entities are involved in the fish-to-fork case study: (i) fishermen; (ii) truck drivers (logistic companies); (iii) restaurants; (iv) markets; (v) customers; (vi) government agency. Fishing light attractors are used by the fishermen during the night in order to attract and catch anchovies. A fleet of boats (usually tens of boats) leaves the harbour at night to fish ancovies and come back during the morning. The fish labels usually include information about the commercial designation and scientific name of the fish (i.e., name and scientific name), catch area, product method (i.e., aquaculture or wild-caught fish), food operator, port of landing and date of landing. The fish is usually put inside a polystyrene box (which are hundreds per boat) and unloaded at the port where truck drivers can take them to the markets or to the restaurants. The fish can go through intermediate sorting centres before reaching the markets or the restaurants. Government agency are in charge of checking the quality and the formal claims that are related to the fish.

Among all the entities involved, (vi) and (i) can usually be trusted. The government agency that provides checks of fish is unlikely to cheat because this will damage its reputation and it can be held liable if found providing false results. The fisherman is normally the victim of counterfeit fish and in need of protection. The logistic service and the intermediate dealers are not trusted. Restaurants and markets are trusted since their reputation would be damaged if they introduce counterfeit fish.

2.1 Potential Attacks on the Fish-to-Fork Case Study

In the scenarios described above, we highlight the following attacks to the traditional fish label system:

- Origin forgery. An adversary can forge a fish label selling anchovies that are not coming from the Cetara origin;
- Specie forgery. An adversary can introduce into the market counterfeit Cetara anchovies that have a different specie;
- Label modification. An adversary can modify various fields from the label such as date in order to have a fresher fish.
- Label reuse. An adversary who has a valid label can reuse it on counterfeit fish.
- Label duplication. An adversary duplicates a label for its use on counterfeit fish.

2.2 System Requirements

The goal of our CERTFish system is to stop the illegal profits made from counterfeit Cetara anchovies. We propose the use of secure digital tags, blockchain technology, IoT tamper proof devices, location and time information in order to prevent counterfeit Cetara anchovies circulation in the markets and restaurants. The high level requirements of the digital tags are:

- Non-forgeable. No one can forge a tag and claim it was generated by a trusted entity.
- Integrity. After being generated, no one can modify the contents of the tag.
- Non-reusable. The tag can be used only once.
- Anti-duplication. It should be hard to duplicate the tag or use the duplicated tags without being detected.

– Tag uniqueness. For each fish box there must be exactly one valid tag at the same time.
– No centralised trusted party should be introduced. Effectively the complex structure of the fish supply chain makes difficult to introduce a centralised trusted authority.

It is worth mentioning that conventional digital certificates or credential can provide integrity or are non-forgeable, but they can be easily reused and duplicated. Therefore, as shown in the following sections, our CERTFish addresses the aforementioned problems.

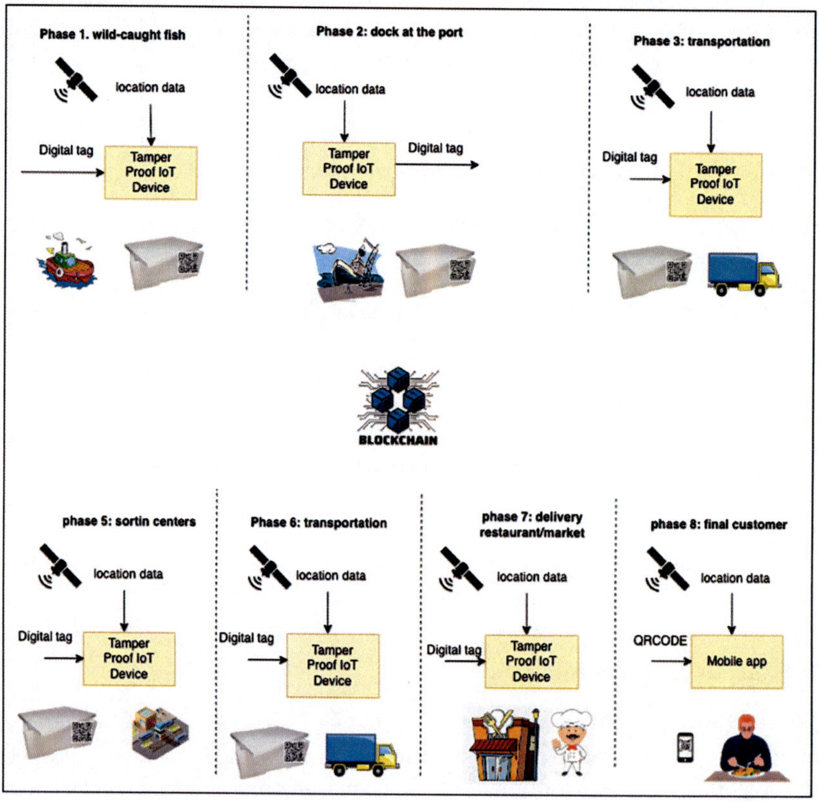

Fig. 1. All phases of the CERTFish system.

3 CERTFish System

Figure 1 shows all phases of the CERTFish system. At the beginning (phase 1), the fishing boat is at open sea for fishing anchovies. Immediately after fishing, the fishermen place the anchovies inside boxes which are equipped with two digital tag that are qrcode and rfid. The boxes are sealed with a cover in order to prevent tampering of its content during transportation. An IoT tamper proof device is used in order to scan the digital tag. The IoT device is tamper proof, water resistant and is completely sealed (it has only a power button). Once the IoT device scans a box it stores a tuple $(id, location, time, fish\_data, fisherman, boat\_id, lifetime)$ inside its internal memory where id is the unique identifier of the box, $location$ and $time$ are the GPS coordinates and its time (i.e., the time and date at which the fish was caught), $fisherman$ and $boat\_id$ are the fisherman identifier and the name of the boat, $fish\_data$ the name and specie of the fish and $lifetime$ the duration of the qrcode. This is usually set to 48 h for fresh fish. The fish tuples cannot be immediately sent to the blockchain since the boat may have not connection while fishing offshore. Each After fishing all boats dock at the port and phase 2 begins. At the port, the IoT device of each boat detects the LTE connection and uploads the data directly into a blockchain. Boats can potentially send data at the same time. This can generate a burst of transactions that must handled by the blockchain. At phase 3, the truck drivers will load all fish boxes inside their lorries. Once the boxes are loaded each driver uses his own IoT device in order to add tracking information into the blockchain. More precisely, a driver IoT device stores a tuple $(id, driverId, truckId, location, time)$ inside its internal memory where id is the unique identifier of the box, $location$ and $time$ are the GPS coordinates and its time (i.e., the time and date at which the fish is loaded into the lorry), $driverId$ and $truckId$ are the driver identifier and the name of the lorry. The driver IoT device can potentially scan a large amount of boxes together thus phase 3 can generate a burst of transactions towards the blockchain. The boxes may be delivered directly to the restaurants/markets or be further elaborated by a sorting centre that dispatches the boxes to the final markets (this is phase 5). Sorting centres add their tracking information directly into the blockchain. For the sake of information we omit the details of the tuples that are provided by the sorting centres. These can generate a quite high amount of transactions when compared to all other phases. At this point the last phase of the supply chain is 7. In this phase the markets or the restaurants gets the fish box, checks that no cover seal has been broken and adds their tracking data by using his IoT device. More precisely, a market/restaurant IoT device stores a tuple $(id, destinationId, location, time)$ inside its internal memory where id is the unique identifier of the box,$destinationId$ is the identifier of the market, $location$ and $time$ are the address of the market and the current time (i.e., the time and date at which the fish was received). The device sends the tuples into the blockchain when there is connection. From the box digital tag the market/restaurant can generate a unique qrcode. This can be read by the customer (phase 8) in order to verify the authenticity of the fish. More precisely, the customer uses a mobile app that retrieves the blockchain tracking data from the qrcode and performs the following checks:

1. the origin is from Cetara (the fish origin is also shown in a visual map);
2. the qrcode was generated by an authorised fisherman;
3. the integrity of the tracing data;
4. the freshness of the food (i.e., no more than 48 h has passed since the fish was caught);
5. the current position is the same as the address of the restaurant/market were the fish was received.

4 Implementation

Figure 2 shows the CERTFish IoT device that has been installed inside fishing boats. This is a tamper-proof and water resistant hardware that, with minimum human involvement, stores the tuples that have been described in Sect. 2. For instance, the IoT boat device reads the digital tag of the box, the time and location where the fish is caught and adds the fisherman and boat information. The IoT device has only the power button and has an internal battery. The hardware has been implemented by using the LILYGO T-SIM7000G module [5]. This device has the micro-controller of the ESP32 family, a modem SIM7000G for the LTE connection, a GPS receiver and a lithium battery. The IoT device has an RFID reader [11] and a qrcode reader [10] to read the digital tag of the box that contains anchovies. Boxes are equipped with both devices that is qrcode and RFID. When a polystyrene box is used a new code is generated

Fig. 2. The CERTFish IoT device that have been installed inside fishing boats.

for each new box. When a reusable plastic box is used the same code is reused thought the time. As soon as the IoT device has LTE connection, it prepares a set of raw blockchain transactions that contain all the box data. The device is equipped with an encryption hardware that is able to perform the Elliptic Curve Digital Signature Algorithm (ECDSA) by using the secp256k1 curve. This is the algorithm that is used by the ethereum platform. The IoT device uploads all transactions to a Hyperledger BESU blockchain (i.e., we have chosen a permissioned blockchain solution). More precisely, the blockchain network is composed of 7 validators which are run by 7 fishing companies. We have chosen the Istanbul Byzantine Fault Tolerant (IBFT) 2.0 consensus protocol. This allows the blockchain to tolerate up to 3 Byzantine validators. We have developed smart contracts in solidity in order to implement the fish traceability system. A smart contract is used to store information about all authorised entities such as boats, lorries, sorting centres, restaurants, and markets. Smart contracts are used to store all digital tags that are generated by the various entities. Mobile applications have been developed in order to read tracking information from the blockchain.

5 Discussion and Results

Our FISHCert system allows the fulfilment of all system requirements that have been defined in Sect. 2. Non-forgeability is ensured by means of the smart contracts that are described in Sect. 4. They ensures that only authorised devices that provide a valid digital signature can write tracking data into the blockchain. For instance, only authorised fishermen can write data that are related to fishing information. Integrity and immutability is ensured by the use of blockchain. Anti-duplication is ensured by the use of tamper proof RFID devices. In particular the unique EPC represents a hidden digital DNA that counterfeiters cannot reproduce. Qrcode that are generated for the markets and the restaurants cannot be duplicated and reused as well. More precisely, time and location checks are performed by the customer application in order to ensure anti-duplication and non-reusability. In fact, duplicating and reusing a digital tag in other locations will trigger an alert in the customer app since the current location does not correspond to the one stored in the blockchain (i.e., the valid location stored by an authenticated restaurant/market). Moreover, a digital tag information lasts for a certain amount of time (i.e., 48 h) since its generation (i.e., when the fisherman writes the fishing information at the port). Each tag has unique id and time information. More precisely when the tag is not reused its id can be used as primary key. When the tag is reused over time the couple $(id, time)$ where $time$ is the time recorded during the dock of the boat are unique.

Constraints can be inserted in the chain creation phase to protect against possible fraud. For example, by imposing the time constraint derived from the geographical distance, obtained with the GPS of the devices, between the place of loading of the fish and the cold storage room. For example, if the places are 100 km away, the registration upon arrival in the cold storage room must be carried out within a time interval between 1 and 3 h.

The reading of the RFId code is tamper-proof (if the reading device is tamper-proof) while the QRCode does not guarantee this feature. However, the alteration of the QRCode relating to the box can be easily discovered by comparing the RFId code with the assigned QRCode. A device that has both acquisition methods (and tamper-proof) is a guarantee of the authenticity of the tracking.

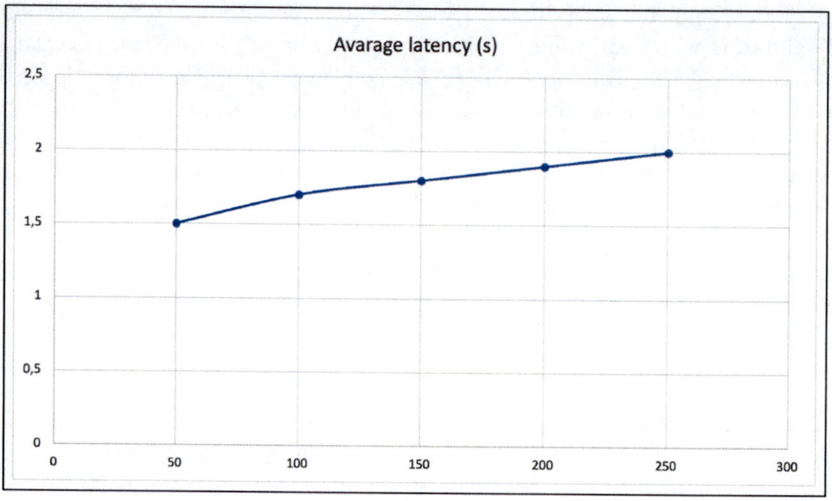

Fig. 3. Transaction per seconds are on the X-axis while the average latency in seconds is on the Y-axis.

Figure 3 shows the results of the experiments we have performed. We have simulated the IoT devices of each entity (i.e., boats, drivers, intermediate centres and restaurants/markets). All IoT devices send an equal amount of transactions per seconds. For instance, when 100 transactions per seconds are considered we assume a total of 25 transactions per seconds are sent by each entity (i.e., boats, drivers, intermediate centres and restaurants/markets). The Hyperldger BESU blockchain is able to accept transactions with a very small delay. When we have more than 500 transactions per seconds the blockchain can have high delay or lose some transactions.

6 Related Work

As the authors in [13] describe, the problem of food traceability is particularly important and affect both producers and consumers. Italian products such as "Grana padano" (i.e., one of the best-selling cheese worldwide with 130,000 tons of product) or "prosecco" (i.e., one of the best-selling wine in the world with 627.5 million bottles) are Italian products that are particularly affected by fraud and counterfeit. Generally speaking, the European union has recognised hundreds of typical Italian products that suffer from the lack of a reliable, safe, scalable systems that allow the certification of origin and traceability. As described in [6] many users are willing to pay more money when a product shows clear, reliable and secure information. Therefore, the EU encourages the creation of organisations that certify the [12] food supply chain.

The blockchain technology [14] can be used successfully if two main features are considered in its application: transparency and simplicity. Many blockchain-based approaches integrate IoT as [4] information providers. The data that are provided by IoT devices must be certified and validated as trusted otherwise the system cannot provide any useful information. More precisely, the IoT data must be certified before any subsequent use. For example, the security of tamper-proof biometric sensors must be certified before their use for people's identity certification [7]. The data certification problem is present in any food supply chain where reliable data must be delivered during the food processing. This problem is particularly critical for the fishing supply chain where the moment of fishing is very important for the certification of the quality of the catch. Currently only [3] briefly introduces a methodology for certifying fish without describing its security and its the implementation details. More precisely, in [3] entities can provide false data on the fish product for their own advantage. An insecure tagging process [2] can indeed allow fraud. Some tracking solutions using blockchain involve hybrid approaches through the use of images and artificial intelligence technologies [15]. These approaches are complex and difficult to manage as well as expensive. To the best of our knowledge CERTFish is the only blockchain based approach that combine IoT tamper proof devices, location and time information, and blockchain in order to certify the origin of the fish.

7 Conclusions and Future Work

In this paper we have presented CERTFish a system that can be used in order to certify the fish from catch-to-fork. More precisely, CERTIFish combines tamper proof IoT devices, the blockchain, location and time information to ensure the origin and the authenticity of the fresh fish that is sold in the market and consumed in the restaurants. We have implemented CERTIFish in all its components. We have developed and programmed IoT devices in order to send tracking information directly to a Hyperldger BESU permissioned blockchain. Smart contracts have been written in order to specify all valid entities (i.e., boats, drivers, sorting centres, restaurants and markets) and to record all digital tag information. Boxes equipped with both RFID and qrcode have been used in order to securely transport the fish from its catch to the restaurants/markets. We discussed the security of the system and tested the approach by simulating hundreds of IoT sensor sending transactions. As future work we plan to extend the approach to other types of fresh food such as truffle and fruits. We aim at using AI to substitute tags by visually identifying and authenticating the product.

References

1. Revealed: seafood fraud happening on a vast global scale (2021). https://www.theguardian.com/environment/2021/mar/15/revealed-seafood-happening-on-a-vast-global-scale
2. Donlan, C.J., Luque, G.M.: Exploring the causes of seafood fraud: a meta-analysis on mislabeling and price. Mar. Policy **100**, 258–264 (2019). https://doi.org/10.1016/j.marpol.2018.11.022. https://www.sciencedirect.com/science/article/pii/S0308597X18304731
3. Larissa, S., Parung, J.: Designing supply chain models with blockchain technology in the fishing industry in Indonesia. IOP Conf. Ser.: Mater. Sci. Eng. **1072**(1), 012020 (2021). https://doi.org/10.1088/1757-899X/1072/1/012020
4. Li, J., Maiti, A., Springer, M., Gray, T.: Blockchain for supply chain quality management: challenges and opportunities in context of open manufacturing and industrial internet of things. Int. J. Comput. Integr. Manuf. **33**(12), 1321–1355 (2020). https://doi.org/10.1080/0951192X.2020.1815853
5. Lilygo: Ttgo t-sim7000g (2022). http://www.lilygo.cn/claprod_view.aspx?TypeId=21&Id=1297&FId=t28:21:28. Accessed 1 Dec 2022
6. Pierucci, A., Columbu, S., Kell, L.T.: A global review of MSC certification: why fisheries withdraw? Mar. Policy **143**, 105124 (2022). https://doi.org/10.1016/j.marpol.2022.105124. https://www.sciencedirect.com/science/article/pii/S0308597X22001713
7. Poh, N., Wong, R., Marcialis, G.L.: Toward an attack-sensitive tamper-resistant biometric recognition with a symmetric matcher: a fingerprint case study. In: 2014 IEEE Symposium on Computational Intelligence in Biometrics and Identity Management (CIBIM), pp. 175–180 (2014). https://doi.org/10.1109/CIBIM.2014.7015460
8. Mercogliano, R., D'Ambrosio, R., De Felice, A., Chirollo, C., Citro, A., Anastasio, A.: "COLATURA DI ALICI" CETARESE: EVALUATION OF AMINIC PROFILE. Ital. J. Food Safety. **1**(2), 45–48 (2008)
9. Rasmussen, R.S., Morrissey, M.T.: DNA-based methods for the identification of commercial fish and seafood species. Compr. Rev. Food Sci. Food Safety **7**(3), 280–295 (2008). https://doi.org/10.1111/j.1541-4337.2008.00046.x
10. SparkFun: SparkFun 2D barcode scanner breakout (2022). https://www.sparkfun.com/products/18088. Accessed 1 Dec 2022
11. SparkFun: SparkFun simultaneous RFID reader - m6e nano (2022). https://www.sparkfun.com/products/14066. Accessed 1 Dec 2022
12. Sun, S., Wang, X.: Promoting traceability for food supply chain with certification. J. Cleaner Prod. **217**, 658–665 (2019). https://doi.org/10.1016/j.jclepro.2019.01.296. https://www.sciencedirect.com/science/article/pii/S0959652619303233
13. Tian, F.: A supply chain traceability system for food safety based on HACCP, blockchain & internet of things. In: 2017 International Conference on Service Systems and Service Management, pp. 1–6 (2017). https://doi.org/10.1109/ICSSSM.2017.7996119
14. Viriyasitavat, W., Hoonsopon, D.: Blockchain characteristics and consensus in modern business processes. J. Ind. Inf. Integr. **13**, 32–39 (2019). https://doi.org/10.1016/j.jii.2018.07.004. https://www.sciencedirect.com/science/article/pii/S2452414X18300815
15. Wang, X., et al.: Blockchain-enabled fish provenance and quality tracking system. IEEE Internet Things J. **9**(11), 8130–8142 (2022). https://doi.org/10.1109/JIOT.2021.3109313

A Fuzzy-Based System for Assessment of Recognition Error in VANETs

Ermioni Qafzezi[1(✉)], Kevin Bylykbashi[1], Shunya Higashi[2], Phudit Ampririt[2], Keita Matsuo[1], and Leonard Barolli[1]

[1] Department of Information and Communication Engineering, Fukuoka Institute of Technology (FIT), 3-30-1 Wajiro-Higashi, Higashi-Ku, Fukuoka 811-0295, Japan
qafzezi@bene.fit.ac.jp, {kt-matsuo,barolli}@fit.ac.jp

[2] Graduate School of Engineering, Fukuoka Institute of Technology (FIT), 3-30-1 Wajiro-Higashi, Higashi-Ku, Fukuoka 811–0295, Japan
bd21201@bene.fit.ac.jp

Abstract. This paper introduces a novel system for recognition error detection in Vehicular Ad Hoc Networks (VANETs) using Fuzzy Logic (FL). The proposed system leverages a comprehensive set of input parameters, including Internal and External Distraction, Driver's Inattention, and Inadequate Surveillance, to effectively evaluate and mitigate potential errors in vehicle recognition and response. In order to recognize the critical role of driver behavior and external factors in shaping road safety, we employ FL to model the intricate relationships and uncertainties inherent in such contexts. By incorporating linguistic variables and a rule-based inference mechanism, the system transforms the multidimensional input parameters into actionable insights regarding the likelihood of recognition errors. The distinctive contribution of this research lies in its holistic consideration of both driver-related and external variables, encompassing a wide spectrum of influences on recognition accuracy. Through simulatio validation, our proposed system demonstrates its efficacy in capturing subtle variations in driver attention and environmental conditions. Ultimately, the FL-based recognition error system holds significant promise in advancing the capabilities of VANETs, paving the way for more adaptive and responsive vehicular communication systems that prioritize safety in dynamic road environments.

1 Introduction

Car accidents continue to be a significant global concern, resulting in substantial human and economic costs. Among the various factors contributing to these accidents, the actions of drivers remain a primary cause. The National Highway Traffic Safety Administration reports that drivers are responsible for 94% of car accidents due to poor decision-making. The decisions and behaviors of drivers play a pivotal role in shaping road safety outcomes, making the study of driver-related factors crucial for devising effective accident prevention strategies. Understanding the underlying causes of drivers' actions can lead to targeted interventions aimed at reducing accident rates and their associated societal impacts.

L. Barolli (Ed.): BWCCA 2023, LNDECT 186, pp. 159–167, 2024.
https://doi.org/10.1007/978-3-031-46784-4_14

Recognition errors represent a pivotal subset of drivers' actions that contribute significantly to road accidents. These errors arise when drivers fail to accurately perceive and interpret their surroundings, leading to incorrect decisions and maneuvers. Factors such as Internal and External Distraction, Driver's Inattention, and Inadequate Surveillance have been identified as key parameters to recognition errors. Internal distractions, such as cognitive load and multitasking, divert drivers' attention away from the road. External distractions, like electronic devices or roadside attractions, further compound the risk. Driver inattention, often linked to fatigue or preoccupation, can result in delayed reactions to critical situations. Inadequate surveillance, where drivers fail to thoroughly scan their environment, can lead to collisions with unexpected obstacles. Recognizing the intricate interplay of these factors is essential for developing targeted interventions that mitigate the occurrence of recognition errors and subsequently reduce accident rates.

In [12], we propose a FL-based system that takes into account internal factors such as recognition error, decision error, and performance error as the primary reasons for driver mistakes. This work aims to delve into the role of recognition errors as a significant driver-related cause of car accidents. By analyzing the complex relationship between Internal and External Distraction, Driver's Inattention, Inadequate Surveillance, and their impact on recognition errors, we aim to contribute to the understanding of accident causation and prevention. Through the utilization of FL-based recognition error system in Vehicular Ad Hoc Networks (VANETs), the potential for real-time detection and mitigation of such errors emerges can improve the road safety.

The structure of the paper is as follows. In Sect. 2, we present an overview of VANETs and its evolution from traditional to cloud-fog-edge SDN-VANETs. In Sect. 3, we describe in detail the emerging technologies in VANETs such are: SDN, Network Slicing and their advantages. In Sect. 4, we introduce the proposed FL-based system and its implementation. In Sect. 5, we discuss the simulation results. Finally, conclusions and future work are given in Sect. 6.

2 Overview of VANETs

VANETs are specialized subsets of Mobile Ad Hoc Networks (MANETs) that enable communication among vehicles and between vehicles and roadside infrastructure. VANETs leverage wireless communication technologies to create an interconnected network within the vehicular environment. Vehicles act as mobile nodes, capable of exchanging real-time information with each other and with fixed infrastructure elements such as traffic lights and road signs. This communication enables a wide range of applications that can significantly impact road safety, traffic efficiency and overall driving experience. VANETs enable vehicles to share real-time information, such as traffic conditions, road hazards, and vehicle trajectories, leading to improved situational awareness and informed decision-making for both drivers and traffic management systems [3,4,11,14].

Applications of VANETs span in multiple domains. One of the most critical areas is road safety. Cooperative collision warning systems leverage VANETs to alert drivers about nearby vehicles, reducing the risk of accidents. Additionally, VANETs contribute to traffic management, enabling dynamic traffic signal control and congestion avoidance through real-time data exchange. Enhanced traffic flow leads to reduced travel times and fuel consumption. VANETs also support infotainment and navigation applications by providing drivers with up-to-date information about points of interest, parking availability, and optimal routes.

However, the adoption of VANETs is not without challenges. High vehicle mobility and rapidly changing network topologies can result in intermittent connectivity, impacting the reliability of communication. Ensuring secure and privacy-preserving communication in VANETs is crucial, as malicious attacks can compromise safety and data integrity. Additionally, the deployment of infrastructure for VANETs, such as roadside units, presents logistical and cost challenges. Handling the massive amount of data generated by vehicles in real time requires efficient data dissemination strategies and scalable communication protocols.

Despite these challenges, advancements in VANET technology continue to be made. The integration of VANETs with emerging technologies like autonomous vehicles holds a great potential. Autonomous vehicles can leverage VANET data to enhance their decision-making processes, leading to safer navigation and coordination. Moreover, the introduction of 5G and beyond communication technologies provides opportunities for improved VANET connectivity, reduced latency, and enhanced data rates. Research efforts are also directed towards developing machine learning and Artificial Intelligence (AI) algorithms to analyze and predict traffic patterns, enabling more proactive traffic management strategies.

3 Network Technological Advancement in VANETs

In addition to the integration of autonomous vehicles and the utilization of machine learning and AI-based algorithms, VANETs have embraced an array of advanced technologies to enhance their capabilities. Software-Defined Networking (SDN) has emerged as a game-changing paradigm, allowing centralized control over network resources and enabling dynamic reconfiguration of communication paths. This is particularly valuable in VANETs, where seamless connectivity and adaptable communication routes are crucial to maintaining consistent data exchange.

Moreover, the concept of Network Slicing (NS) has gained traction in VANETs. The NS involves dividing a single physical network infrastructure into multiple virtual networks, each tailored to specific requirements. This approach allows VANETs to allocate resources and prioritize communication according to the diverse needs of various applications, from safety-critical communications to infotainment services [1,5,13].

Furthermore, the integration of cloud-fog-edge computing into VANETs has introduced powerful computational paradigms. Cloud computing enables resource-intensive processing and data storage in remote data centers, while fog and edge computing push computing closer to the network edge, reducing latency and enhancing real-time data processing capabilities. In the context of VANETs, this trio technology enables efficient data analysis, immediate decision-making, and timely response to critical events, resulting in enhanced road safety and traffic management.

By integrating SDN, NS and cloud-fog-edge computing into the VANET ecosystem, researchers and practitioners are equipping these networks with the tools to efficiently manage data, optimize communication, and adapt to changing traffic conditions. These technological advancements align with the broader goal of creating smarter and more resilient transportation systems that leverage the full potential of connectivity, computation, and communication within the vehicular environment.

4 Proposed Fuzzy-Based System

Our proposed system, known as the Fuzzy System for Assessment of Recognition Errors (FS-ARE), represents a novelty in the field of intelligent driving support systems. With a primary focus on enhancing road safety, FS-ARE introduces a new approach that assesses the level of recognition errors in real-time by considering a comprehensive set of input parameters. By analyzing Internal and External Distraction, Driver's Inattention, and Inadequate Surveillance, FS-ARE aims to provide an accurate and dynamic assessment of driver behavior and potential risks.

Table 1. FS-ARE parameters and their term sets.

Parameters	Term Sets
Internal and External Distraction (IED)	Low (Lo), Medium (Me), High (Hi)
Driver Inattention (DI)	Low (L), Medium (M), High (H)
Inadequate Surveillance (IS)	Small (GS), Medium (MS), Big (PS)
Recognition Error (RE)	RE1, RE2, RE3, RE4, RE5 RE6, RE7, RE8, RE9

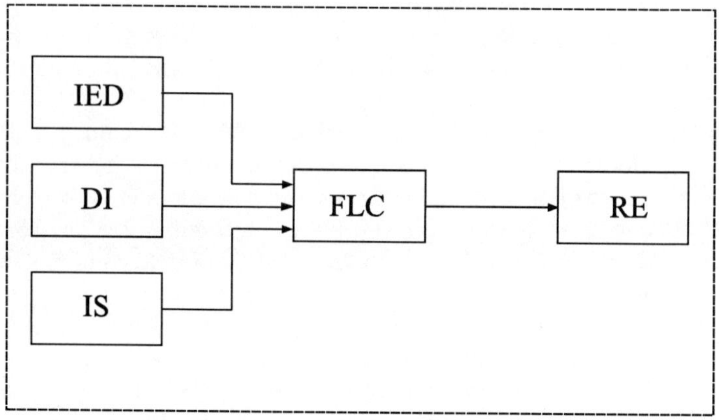

Fig. 1. Proposed system structure.

We explain the input parameters in following.

Internal and External Distraction (IED): Internal distractions refer to thoughts, emotions, or activities that divert the driver's attention from the primary task of driving. External distractions, on the other hand, are stimuli from the environment that compete for the driver's attention. These distractions can include activities like talking to passengers, using a smartphone, adjusting the radio, or eating while driving. Both internal and external distractions lead to recognition errors by causing cognitive overload, diverting attention from the road, and impairing a driver's ability to recognize and process relevant information, such as road signs, pedestrians, or other vehicles.

Driver Inattention (DI): Driver inattention refers to a lack of focus and engagement with the driving task. It can result from various factors, such as fatigue, drowsiness, daydreaming, or divided attention. Inattention reduces a driver's capacity to detect and recognize critical information in the driving environment, leading to recognition errors. For instance, a fatigued driver might miss important road signs or fail to notice sudden changes in traffic conditions, contributing to errors in recognition.

Inadequate Surveillance (IS): Inadequate surveillance occurs when a driver fails to adequately monitor the road and surroundings. This can happen due to overconfidence, complacency, or distraction. When drivers do not sufficiently

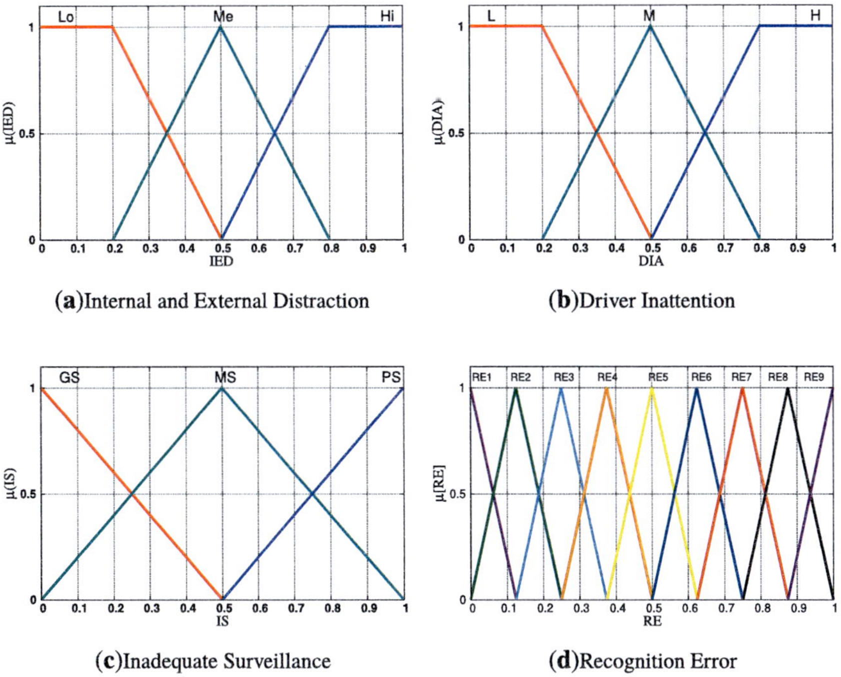

(**a**)Internal and External Distraction

(**b**)Driver Inattention

(**c**)Inadequate Surveillance

(**d**)Recognition Error

Fig. 2. Membership functions of FS-ARE.

scan their environment, they may miss important cues, fail to notice potential hazards, or misinterpret traffic situations. Inadequate surveillance leads to errors in recognizing road conditions, traffic signals, and unexpected events, increasing the likelihood of recognition errors.

FS-ARE leverages the power of FL to model the complex and uncertain relationships between the input parameters and recognition errors. The FL ability to handle imprecise and vague data makes it a suitable choice for capturing the nuanced interactions among variables that influence driver performance [2,6–10,15–17]. By incorporating linguistic variables and a rule-based inference mechanism, the system transforms raw sensor data and contextual information into meaningful insights about the driver's state of attention and the likelihood of recognition errors.

Table 2. FRB of FS-ARE.

No	IED	DI	IS	RE
1	Lo	L	GS	RE1
2	Lo	L	MS	RE2
3	Lo	L	PS	RE3
4	Lo	M	GS	RE3
5	Lo	M	MS	RE4
6	Lo	M	PS	RE5
7	Lo	H	GS	RE5
8	Lo	H	MS	RE6
9	Lo	H	PS	RE7
10	Me	L	GS	RE3
11	Me	L	MS	RE4
12	Me	L	PS	RE5
13	Me	M	GS	RE5
14	Me	M	MS	RE6
15	Me	M	PS	RE7
16	Me	H	GS	RE6
17	Me	H	MS	RE8
18	Me	H	PS	RE9
19	Hi	L	GS	RE3
20	Hi	L	MS	RE4
21	Hi	L	PS	RE5
22	Hi	M	GS	RE5
23	Hi	M	MS	RE6
24	Hi	M	PS	RE7
25	Hi	H	GS	RE7
26	Hi	H	MS	RE8
27	Hi	H	PS	RE9

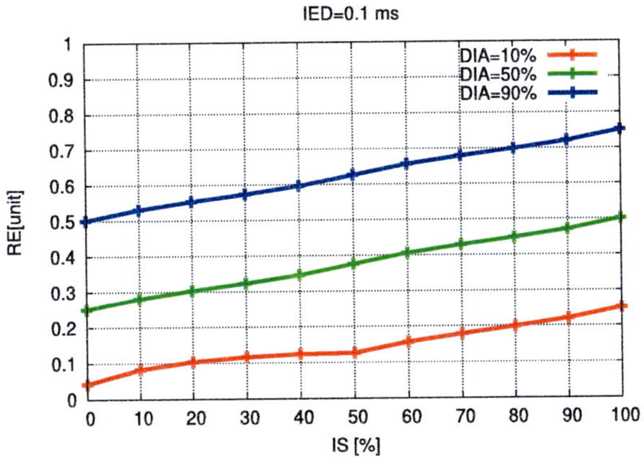

(**a**)IED = 0,1

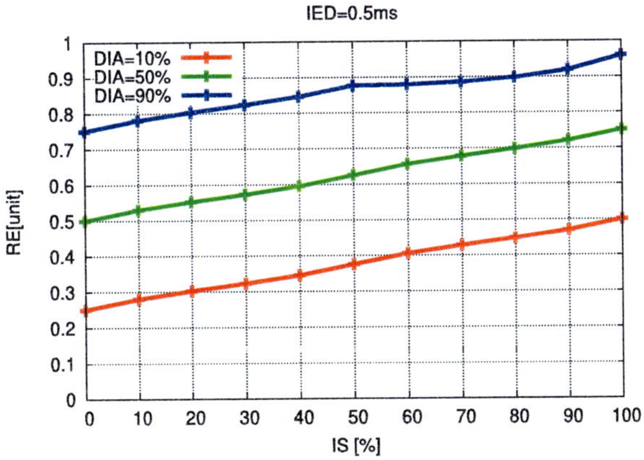

(**b**)IED = 0.5

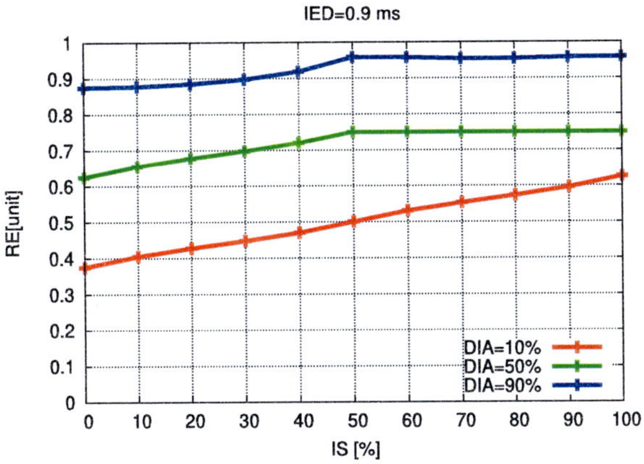

(**c**)IED = 0.9

Fig. 3. Simulation results for FS-ARE.

The structure of the proposed system is shown in Fig. 1, whereas the term sets of the input and output parameters are given in Table 1. The membership functions of input and output parameters are shown in Fig. 2 and the Fuzzy Rule Base (FRB) is given in Table 2.

5 Simulation Results

The simulations were conducted using FuzzyC, and the outcomes are presented across three distinct scenarios. Figure 3a, Fig. 3b, and Fig. 3c depict the results corresponding to low, medium, and high levels of Internal and External Distraction respectively. The figures illustrate the relationship between Recognition Error and Inadequate Surveillance for different levels of Driver Inattention, while maintaining constant IED values.

Specifically, Fig. 3a shows the scenario with minimal IED errors, leading to lower RE values in comparison to the other scenarios. In terms of the driving experience and accident occurrence, the observed pattern in Fig. 3a suggests that when drivers engage in fewer internal and external distractions (low IED), the recognition errors are lower. This implies that reduced distractions result in better awareness and attention to the road, leading to a more accurate perception of the driving environment. Consequently, the probability of accidents caused by recognition errors is diminished in such scenarios.

When IED levels increase as depicted in Fig. 3b and Fig. 3c, for DIA = 90% we can see that RE values are more than 50%. The likelihood of recognition errors rises, potentially leading to compromised driving experiences and an increased risk of accidents due to inadequate surveillance and inattention. Therefore, managing internal and external distractions through strategies such as minimizing device usage, staying focused, and practicing attentive driving can contribute significantly to safer driving conditions and accident prevention.

6 Conclusions

In the realm of vehicular safety, where driver errors persist as a major contributor to road accidents, VANETs present a potential avenue for mitigating such risks and enhancing overall road safety.

This study has demonstrated the outcomes of the proposed FS-ARE system, which operates in real-time to predict and prevent recognition errors, thereby contributing to accident prevention.

By examining diverse scenarios centered around three parameters IED, DI and IS, we have gained insights into the intricate relationship between these variables. For high values of DIA, IED and IS, RE values also deteriorates. This underscores the importance of strategic distraction management, encompassing techniques like minimizing device use and practicing attentive driving, to foster safer road conditions.

In the future work, we will consider other parameters which affect the recognition errors.

References

1. Al-Heety, O.S., Zakaria, Z., Ismail, M., Shakir, M.M., Alani, S., Alsariera, H.: A comprehensive survey: benefits, services, recent works, challenges, security, and use cases for sdn-vanet. IEEE Access **8**, 91028–91047 (2020)
2. Bojadziev, G., Bojadziev, M., Zadeh, L.A.: Fuzzy Logic for Business, Finance, and Management, Advances in Fuzzy Systems - Applications and Theory, vol. 12. World Scientific (1997)
3. Hartenstein, H., Laberteaux, K.P. (eds.): VANET: Vehicular Applications and Inter-Networking Technologies. Intelligent transportation systems. Wiley, Hoboken (2010). https://doi.org/10.1002/9780470740637
4. Hartenstein, H., Laberteaux, L.: A tutorial survey on vehicular ad hoc networks. IEEE Commun. Mag. **46**(6), 164–171 (2008)
5. Hussein, A., Elhajj, I.H., Chehab, A., Kayssi, A.: Sdn vanets in 5G: an architecture for resilient security services. In: 2017 Fourth International Conference on Software Defined Systems (SDS), pp. 67–74 (2017)
6. Kandel, A.: Fuzzy Expert Systems. CRC Press Inc, Boca Raton (1992)
7. Klir, G.J., Folger, T.A.: Fuzzy sets, uncertainty, and information. Prentice Hall, Upper Saddle River (1988)
8. Klir, G.J., Yuan, B.: Fuzzy Sets and Fuzzy Logic - Theory and Applications. Prentice Hall, Upper Saddle River (1995)
9. McNeill, F.M., Thro, E.: Fuzzy Logic: A Practical Approach. Academic Press Professional Inc, San Diego (1994)
10. Munakata, T., Jani, Y.: Fuzzy systems: an overview. Commun. ACM **37**(3), 69–77 (1994)
11. Peixoto, M.L.M., et al.: Fogjam: a fog service for detecting traffic congestion in a continuous data stream VANET. Ad Hoc Netw. **140**, 103046 (2023). https://doi.org/10.1016/j.adhoc.2022.103046
12. Qafzezi, E., Bylykbashi, K., Higashi, S., Ampririt, P., Matsuo, K., Barolli, L.: A fuzzy-based error driving system for improving driving performance in VANETs. In: Barolli, L. (ed.) CISIS. LNDECT, vol. 176, pp. 1–9. Springer, Cham (2023)
13. Sadio, O., Ngom, I., Lishou, C.: SDN architecture for intelligent vehicular sensors networks. In: 2018 UKSim-AMSS 20th International Conference on Computer Modelling and Simulation (UKSim), pp. 139–144 (2018)
14. Schünemann, B., Massow, K., Radusch, I.: Realistic simulation of vehicular communication and vehicle-2-x applications. In: Proceedings of the 1st International Conference on Simulation Tools and Techniques for Communications, Networks and Systems & Workshops, SimuTools 2008, Marseille, France, March 3-7, 2008, p. 62. ICST/ACM (2008). https://doi.org/10.4108/ICST.SIMUTOOLS2008.2949
15. Zadeh, L.A., Kacprzyk, J.: Fuzzy Logic for the Management of Uncertainty. Wiley, New York (1992)
16. Zadeh, L.A., Klir, G.J., Yuan, B.: Fuzzy Sets, Fuzzy Logic, and Fuzzy Systems - Selected Papers by Lotfi A Zadeh, *Advances in Fuzzy Systems - Applications and Theory*, vol. 6. World Scientific (1996). https://doi.org/10.1142/2895
17. Zimmermann, H.J.: Fuzzy control. In: Fuzzy Set Theory and Its Applications, pp. 203–240. Springer (1996). https://doi.org/10.1007/978-94-010-0646-0

An Intelligent System Based on Cuckoo Search for Node Placement Problem in WMNs: Tuning of Scale and Host Bird Recognition Rate Hyperparameters

Shinji Sakamoto[1(✉)], Kaho Asakura[1], Leonard Barolli[2], and Makoto Takizawa[3]

[1] Department of Information and Computer Science, Kanazawa Institute of Technology, 7-1 Ohgigaoka, Nonoichi, Ishikawa 921-8501, Japan
shinji.sakamoto@ieee.org, b1906534@planet.kanazawa-it.ac.jp

[2] Department of Information and Communication Engineering, Fukuoka Institute of Technology, 3-30-1 Wajiro-Higashi, Higashi-Ku, Fukuoka 811-0295, Japan
barolli@fit.ac.jp

[3] Department of Advanced Sciences, Faculty of Science and Engineering, Hosei University, Kajino-Machi, Koganei-Shi, Tokyo 184-8584, Japan
makoto.takizawa@computer.org

Abstract. Wireless Mesh Networks (WMNs) have become popular and they have many advantages. However, they have inherent wireless communication issues. To solve the problem, optimization of locations of mesh routers is a good approach. But optimizing the locations of mesh routers is an NP-hard problem. In this research work, we present an intelligent simulation system using Cuckoo Search (CS) algorithm called WMN-CS. We tune the hyperparameters and carry out computer simulations. The simulation results show that a good performance is achieved when the scale parameter (γ) values range from 0.07 to 0.09 and the host bird recognition rate (p_a) values from 0.900 to 0.925.

1 Introduction

Wireless Mesh Networks (WMNs) are cost-effective, easily deployable and have high robustness. However, when the location of mesh routers is not optimized, they have different issues such as interference, congestion, reduced throughput, increased packet loss, and extended delays [11].

Optimizing the location of mesh routers can solve the issues. However, this problem is NP-hard [2,4,7]. In order to solve this problem, we propose meta-heuristics-based intelligent systems.

A lot of research works focus on resource allocation, node selection, and routing in WMNs [1,13]. In terms of node placement problems, Mixed-Integer Linear Programming (MILP)-based optimization models, as well as meta-heuristic approaches such as Hill Climbing (HC), Simulated Annealing (SA), Genetic Algorithm (GA), and Particle Swarm Optimization (PSO) are proposed and evaluated [5,6,8–10,12].

L. Barolli (Ed.): BWCCA 2023, LNDECT 186, pp. 168–177, 2024.
https://doi.org/10.1007/978-3-031-46784-4_15

Especially, heuristic approaches are frequently used for solving NP-hard problems because they offer efficiency compared to MILP-based techniques [14]. Heuristic-based methods are promising approaches for solving the node placement problem in WMNs and other related issues [15].

In this work, we focus on the node placement problem in WMNs. We present an intelligent system called WMN-CS, which is based on Cuckoo Search (CS) [3]. In this paper, we tune the hyperparameters and carry out computer simulations. The simulation results show that a good performance is achieved when the scale parameter (γ) values range from 0.07 to 0.09 and the host bird recognition rate (p_a) values from 0.900 to 0.925.

The rest of this paper is organised as follows. The mesh router nodes placement problem is defined in Sect. 2. In Sect. 3, we present CS algorithm. The proposed simulation system is discussed in Sect. 4. Simulation results are presented in Sect. 5, followed by our conclusions and suggestions for future work in Sect. 6.

2 Node Placement Problem in WMNs

To solve the node placement problem in WMNs, we consider a continuous 2D area. Inside the considered area, we distribute mesh clients and mesh routers at arbitrary points. The main objective is to find suitable locations of mesh routers that maximize network connectivity and user coverage.

To evaluate network connectivity, we consider a graph of the mesh router nodes and we use the Size of Giant Component (SGC) as the network connectivity metric. The Number of Covered Mesh Clients (NCMC) is used to evaluate user coverage. The NCMC is determined by a number of mesh clients within the radio range of at least one mesh router node. It should be noted that the network connectivity and user coverage directly affect the network's performance. Hence, the SGC and NCMC are very important metrics for assessment of WMNs.

To provide a formal framework for the problem, we introduce the adjacency matrix of the WMN graph. The individual nodes correspond to the router and client nodes, with edges representing the interconnections of nodes within the mesh network. Each node has a triple $\boldsymbol{v} = <x, y, r>$, where the values are 2D positional coordinates and the communication distance. Here, a connection exists linking two nodes, $\boldsymbol{u}$ and $\boldsymbol{v}$, if $\boldsymbol{v}$ is in the communication distance of $\boldsymbol{u}$.

3 CS Algorithm

3.1 CS Overview

The CS is a meta-heuristic which is inspired by the phenomenon of brood parasitism observed in certain cuckoo species. To apply CS as an optimization tool, there are three rules [16].

1. Each cuckoo can lay one egg at a time and randomly place it in a chosen nest.
2. The nests with high-quality eggs are considered the best and are carried over to the next generations.
3. The number of available host nests remains fixed and the host bird recognizes the egg laid by a cuckoo with a probability of $p_a \in (0, 1)$. If the host bird recognizes the egg, it can either get rid of it or abandon the nest and build a new one elsewhere.

Algorithm 1. Pseudo Code of CS algorithm.

Initialize Parameters:
    Computation Time $t = 0$ and T_{max}
    Number of Nests $n(n > 0)$
    Host Bird Recognition Rate $p_a(0 < p_a < 1)$
    Lévy Distribution Scale Parameter $\gamma(\gamma > 0)$
Fitness Function to get Fitness Value as f
Generate Initial n Solutions S_0
while $t < T_{max}$ **do**
  while $i < n$ **do**
    $j := i \ \% \ \mathrm{len}(S_t)$ // j is the remainder of dividing i by number of solutions.
    Generate a new solution S^i_{t+1} from S^j_t by Lévy Flights.
    if $(f(S^i_{t+1}) < f(S^j_t))$ and (rand() $< p_a$) **then**
      Discard Solution S^i_{t+1}
    end if
    $i = i + 1$
  end while
  $t = t + 1$
end while
return Best solution

We show the pseudo code of CS in Algorithm 1. In CS algorithm, there are three hyperparameters: number of nests, host bird recognition rate and scale parameter of Lévy distribution.

The CS algorithm starts with initialising its parameters. Then, the fitness value of each nest is calculated and the CS algorithm identfies the most optimal nest within the entire set. After that, the CS try to find suitable solutions by iterating the process. This iterative process terminates when a specified stopping criterion is satisfied, typically expressed as $(t \geq T_{max})$.

Within each iteration, CS generates new solutions (finds nests to lay its egg) using Lévy flight, which is a type of random walk. Then, the fitness value of a new solution is calculated. When the new solution is better than the previous solution, the current solution is supplanted with this improved new solution. When the new solution is worse than the previous solution, the present solution may be dropped by using a stochastic process.

After completing the calculation of all fitness values, the best solutions are retained, while the remaining nests are dropped. New solutions are generated to take the place of the dropped solutions. Then, the fitness value of each nest is evaluated again. This process is repeated until the stopping criterion is satisfied.

3.2 Lévy Flight

In CS algorithm, the Lévy flight is used to generate new solutions using a stochastic process. The Lévy flight is based on the Lévy distribution and is characterized by its proclivity for long-tailed behaviour. This characteristic causes them to move to long distances occasionally. This feature makes CS capable of delving deep into the search space, potentially uncovering superior solutions.

The Lévy flight in general moves short, but occasionally moves in long distances with a low probability. This pattern of movement is observed in nature. For instance, honeybees use the Lévy flight to find flower gardens. They usually try for short distances to find a suitable flower. When they can not find a good flower nearby, they move in long distances to discover a flower garden. This phenomenon is not exclusive to insects. The humans also exhibit behaviours aligned with the principles of the Lévy flight.

The Lévy distribution is a continuous probability distribution widely used in probability theory and statistics to model non-negative random variables. It is a specific case of the inverse-gamma distribution and is classified as stable.

The probability density function of Lévy distribution is:

$$P(x;\mu,\gamma) = \sqrt{\frac{\gamma}{2\pi}}\frac{e^{-\gamma/2(x-\mu)}}{(x-\mu)^{3/2}},$$

where μ is the local parameter ($\mu \leq x$) and γ is the scale parameter.

While, the cumulative distribution function of Lévy distribution is:

$$F(x;\mu,\gamma) = \text{erfc}\left(\sqrt{\frac{\gamma}{2(x-\mu)}}\right),$$

where erfc is the complementary error function.

The Lévy distribution can be generated by using the inverse transformation method.

$$F^{-1}(x;\mu,\gamma) = \frac{\gamma}{2(\text{erfc}^{-1}(x))^2} + \mu$$

4 Overview of WMN-CS Simulation System

We show the flowchart of WMN-CS in Fig. 1. The system generates randomly the initial solution. The solutions are considered nests and selected solutions are considered eggs. Then, it finds nests and evaluates nests in order to lay the egg. When a suitable nest to lay the egg is found, it replaces the solution using a value that follows the Lévy distribution. The WMN-CS evaluates the new solutions using the fitness function by comparing nests found during CS. It should be noted that sometimes the egg may be swapped, resulting in the worse solution. The process is terminated until a predetermined condition is satisfied.

We define the fitness function as:

$$\text{Fitness} = \alpha \times \text{SGC}(\boldsymbol{x},\boldsymbol{y},\boldsymbol{r}) + \beta \times \text{NCMC}(\boldsymbol{x},\boldsymbol{y},\boldsymbol{r}).$$

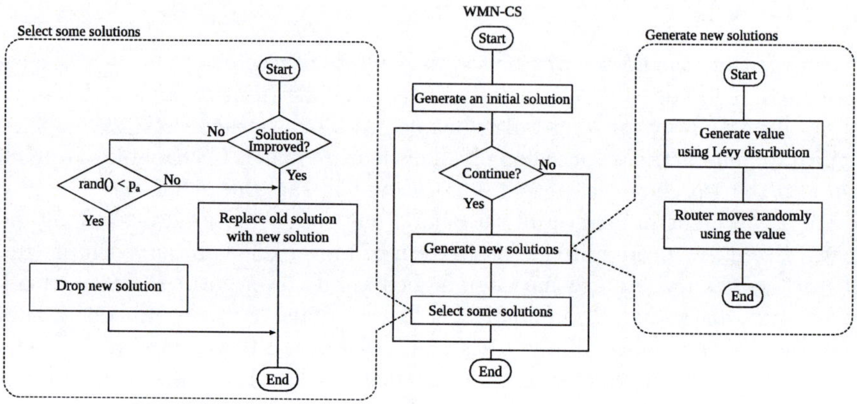

Fig. 1. WMN-CS flowchart.

The fitness function is the combination of SGC and NCMC metrics, which are calculated from WMN graph using vectors $\boldsymbol{x}$, $\boldsymbol{y}$, and $\boldsymbol{r}$ representing the positions and communication distances of mesh nodes. The SGC is used for measuring the connectivity of mesh routers, while NCMC for the user coverage. The α and β are the weight coefficients for SGC and NCMC, respectively.

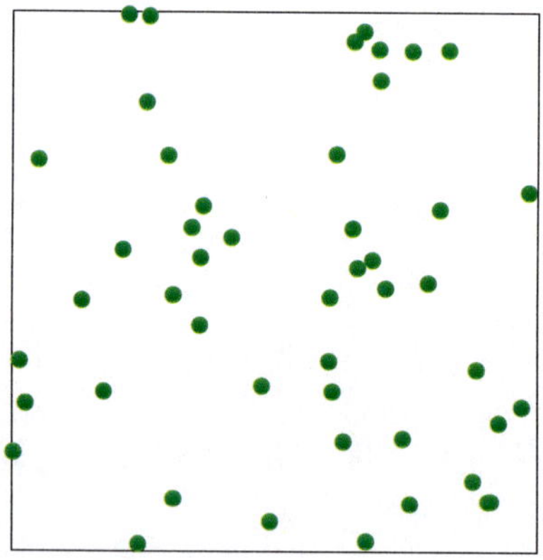

Fig. 2. Uniform distribution.

WMN-CS can generate different mesh clients distributions. We consider Uniform distribution of mesh clients as shown in Fig. 2, where mesh clients are spread out in the considered area.

Table 1. Parameter settings.

Parameters	Values
Clients Distribution	Uniform
Area Size	32×32
Number of Mesh Routers	16
Number of Mesh Clients	48
Total Iterations	2000
Iteration per Phase	10
Number of Nests	$10, 30, 50, 70, 90$
Radius of a Mesh Router	From 2.0 to 3.0
Fitness Function Weight-coefficients (α, β)	0.7, 0.3
Scale Parameter (γ)	From 0.01 to 100.00
Host Bird Recognition Rate (p_a)	From 0.01 to 0.99

5 Simulation Results

In this section, we show simulation results using WMN-CS simulation system. The area size is considered 32×32 and the simulation parameters are shown in Table 1. We set the number of nests 10, 30, 50, 70, 90. Then, we calculate the average of the results of above five patterns of the number of nests. Also, we conducted simulations 10 times for the same parameters.

Table 2. Simulation results for first step parameter tuning.

		γ				
		0.01	0.10	1.00	10.00	100.00
p_a	0.01	67.08	69.12	68.26	60.15	49.93
	0.25	81.83	82.23	80.76	76.23	65.48
	0.50	84.52	84.42	82.28	78.26	70.91
	0.75	85.18	**85.70**	83.45	80.13	73.62
	0.99	84.38	84.88	84.32	83.32	82.17

We present the simulation results for the first step parameter tuning in Table 2. While, the heatmap of simulation results is shown in Fig. 3. The good performance is archved when γ and p_a are 0.10 and 0.75, respectively.

We show the simulation results for the second step parameter tuning and heatmap in Table 3 and Fig. 4, respectively. The good performance is archved when γ is 0.07 and the p_a is 0.90.

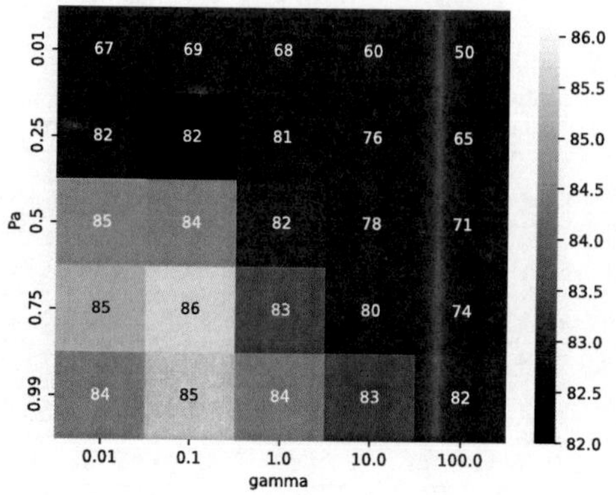

Fig. 3. Heatmap of simulation results for first step parameter tuning.

Table 3. Simulation results for second step parameter tuning.

		γ								
		0.01	0.02	0.05	0.07	0.10	0.20	0.50	0.70	1.00
p_a	0.50	84.52	84.67	83.26	84.46	84.42	84.07	82.83	82.85	82.28
	0.55	84.77	84.92	83.20	84.83	84.80	84.32	83.51	82.85	82.57
	0.60	85.15	85.06	83.41	85.25	84.93	84.77	83.60	83.27	82.75
	0.65	85.03	85.57	83.60	85.41	85.23	84.83	83.85	83.51	82.96
	0.70	85.80	85.62	83.76	85.73	85.42	85.22	83.95	83.72	83.17
	0.75	85.18	85.75	83.81	85.67	85.70	85.31	84.50	83.80	83.45
	0.80	85.37	85.88	84.00	86.08	85.77	85.58	84.55	84.18	83.52
	0.85	85.60	86.06	84.10	86.02	86.02	85.78	84.72	84.46	83.83
	0.90	85.71	85.65	84.38	**86.26**	86.22	85.67	85.15	84.66	84.11
	0.95	85.38	85.48	84.58	85.68	85.97	85.78	85.25	85.12	84.51
	0.99	84.38	84.66	84.21	84.95	84.88	84.82	84.37	84.45	84.32

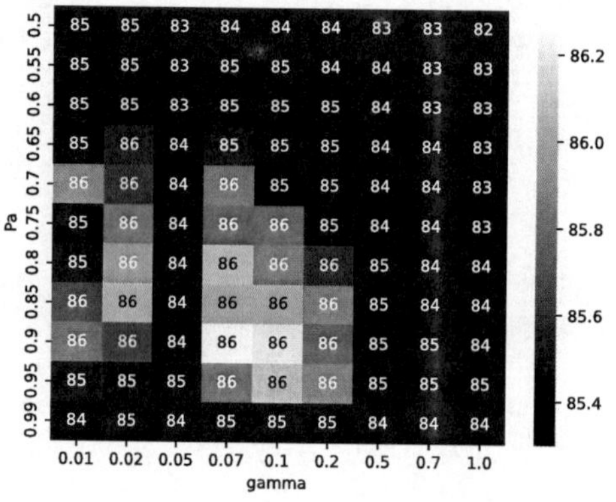

Fig. 4. Heatmap of simulation results for second step parameter tuning.

Table 4. Simulation results for third step parameter tuning.

		γ					
		0.05	0.06	0.07	0.08	0.09	0.10
p_a	0.850	84.10	86.16	86.02	86.13	86.15	86.02
	0.875	84.41	86.17	85.95	85.96	86.01	86.06
	0.900	84.38	85.93	86.26	86.10	86.13	86.22
	0.925	84.58	86.06	86.11	86.11	**86.31**	86.00
	0.950	84.58	86.00	85.68	85.85	85.70	85.97
	0.975	84.48	85.51	85.58	85.32	85.52	85.62

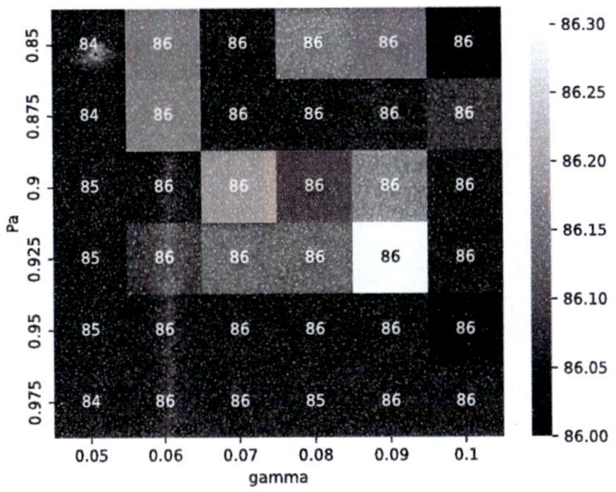

Fig. 5. Heatmap of simulation results for third step parameter tuning.

The simulation results for the third step parameter tuning and heatmap are shown in Table 4 and Fig. 5, respectively. The good performance is archved when γ is 0.09 and the p_a is 0.925.

As shown in Table 4, the fitness value is more than 86.00 when γ is from 0.07 to 0.09, and p_a is from 0.900 to 0.925. Thus, we conclude that a good performance is achieved when γ values range from 0.07 to 0.09 and p_a values from 0.900 to 0.925.

6 Conclusions

In this work, we proposed and implemented a CS-based system for node placement problem in WMNs (called WMN-CS). We conducted simulations for tuning of scale and host bird recognition rate hyperparameters. The simulation results show that a good performance is archived when γ values range from 0.07 to 0.09 and p_a values from 0.900 to 0.925.

In our future work, we would like to tune the number of nest parameters.

References

1. Ahmed, A.M., Hashim, A.H.A.: Metaheuristic approaches for gateway placement optimization in wireless mesh networks: a survey. Int. J. Comput. Sci. Netw. Secur. (IJCSNS) **14**(12), 1 (2014)
2. Amaldi, E., Capone, A., Cesana, M., Filippini, I., Malucelli, F.: Optimization models and methods for planning wireless mesh networks. Comput. Netw. **52**(11), 2159–2171 (2008)
3. Asakura, K., Sakamoto, S.: A cuckoo search based simulation system for node placement problem in wireless mesh networks. In: Barolli, L. (ed.) CISIS 2023. LNDECT, vol. 176, pp. 179–187. Springer, Cham (2023). https://doi.org/10.1007/978-3-031-35734-3_18
4. Basirati, M., Akbari Jokar, M.R., Hassannayebi, E.: Bi-objective optimization approaches to many-to-many hub location routing with distance balancing and hard time window. Neural Comput. Appl. **32**, 13,267–13,288 (2020)
5. Coelho, P.H.G., do Amaral, J.F., Guimaraes, K., Bentes, M.C.: Layout of routers in mesh networks with evolutionary techniques. In: The 21st International Conference on Enterprise Information System (ICEIS-2019), pp. 438–445 (2019)
6. Elmazi, D., Oda, T., Sakamoto, S., Spaho, E., Barolli, L., Xhafa, F.: Friedman test for analysing WMNs: a comparison study for genetic algorithms and simulated annealing. In: 2015 9th International Conference on Innovative Mobile and Internet Services in Ubiquitous Computing, pp. 171–178. IEEE (2015)
7. Gharehchopogh, F.S., Shayanfar, H., Gholizadeh, H.: A comprehensive survey on symbiotic organisms search algorithms. Artif. Intell. Rev. **53**, 2265–2312 (2020)
8. Lee, S.C., Tan, S.W., Wong, E., Lee, K.L., Lim, C.: Survivability evaluation of optimum network node placement in a hybrid fiber-wireless access network. In: IEEE Photonic Society 24th Annual Meeting, pp. 298–299. IEEE (2011)
9. Lin, C.C.: Dynamic router node placement in wireless mesh networks: a PSO approach with constriction coefficient and its convergence analysis. Inf. Sci. **232**, 294–308 (2013)
10. Oda, T., Elmazi, D., Barolli, A., Sakamoto, S., Barolli, L., Xhafa, F.: A genetic algorithm-based system for wireless mesh networks: analysis of system data considering different routing protocols and architectures. Soft. Comput. **20**, 2627–2640 (2016)
11. Qiu, L., Bahl, P., Rao, A., Zhou, L.: Troubleshooting wireless mesh networks. ACM SIGCOMM Comput. Commun. Rev. **36**(5), 17–28 (2006)
12. Sakamoto, S., Oda, T., Ikeda, M., Barolli, L.: Design and implementation of a simulation system based on particle swarm optimization for node placement problem in wireless mesh networks. In: 2015 International Conference on Intelligent Networking and Collaborative Systems, pp. 164–168. IEEE (2015)
13. Sanni, M.L., Hashim, A.H.A., Anwar, F., Naji, A.W., Ahmed, G.S.: Gateway placement optimisation problem for mobile multicast design in wireless mesh networks. In: 2012 International Conference on Computer and Communication Engineering (ICCCE), pp. 446–451. IEEE (2012)

14. Seetha, S., Anand John Francis, S., Grace Mary Kanaga, E.: Optimal placement techniques of mesh router nodes in wireless mesh networks. In: Haldorai, A., Ramu, A., Mohanram, S., Chen, M.-Y. (eds.) 2nd EAI International Conference on Big Data Innovation for Sustainable Cognitive Computing. EICC, pp. 217–226. Springer, Cham (2021). https://doi.org/10.1007/978-3-030-47560-4_17
15. Taleb, S.M., Meraihi, Y., Gabis, A.B., Mirjalili, S., Ramdane-Cherif, A.: Nodes placement in wireless mesh networks using optimization approaches: a survey. Neural Comput. Appl. **34**(7), 5283–5319 (2022)
16. Yang, X.S.: Nature-Inspired Metaheuristic Algorithms. Luniver Press (2010)

Dynamic Spatiotemporal Graph Convolution Network for Cellular Communication Traffic Prediction

Pan Ruifeng(✉)

Nanchang Vocational University, No. 8, Qianjin East Road, Anyi Fenghuang Development Zone, Nanchang, Jiangxi, China

prf@sohu.com

Abstract. With the increasing complexity of the spatial topology of cellular communication networks and dynamic temporal characteristics of mobile network services, accurate prediction of cellular services is challenging. This article proposes the use of Dynamic Spatiotemporal Graph Convolutional Network (DSGCN) for mobile traffic prediction. This network modelling the dynamic characteristics of nodes in the cellular network traffic graph, and that capturing the dynamic spatiotemporal characteristics of border nodes by converting the cellular network traffic graph into hypergraph. In addition, for all nodes on different timestamps, complex spatiotemporal correlations are gathered through dynamic graph convolutional networks. By performing collaborative convolution on traffic flow maps and their hypergraphs, mobile traffic prediction is enhanced. Compared with conventional methods, the results of experiments show that this model has better predictive performance and higher training efficiency.

1 Introduction

With the rapid explosion of cellular network devices and 5th generation communication network, predicting cellular traffic is becoming increasingly complex and a key module in communication network management. In this situation, the major task of cellular traffic prediction is to model the dynamic spatiotemporal characteristics of large-scale cellular traffic information. For decades, People have studied numerous prediction approaches, such as linear method historical average (HA) [1], autoregressive comprehensive moving average (ARIMA) [2] method based on clustering model, support vector regression (SVR) [3] method with excellent abstraction capability and minor error, etc. However, from the current perspective, the performance of these methods is not very good.

Recently, some prediction methods based on deep neural networks (DNN) have been proposed, such as recursive neural networks (RNN) method [4], Bayesian networks (BN) method [5], etc. Compared with the traditional methods mentioned above, DNN has a good representation of the nonlinear relationship between traffic information. In order to get hold of the temporal characteristics of cellular traffic information, time neural networks are usually used, such as Long short-term memory networks (LSTM) [6], but it is difficult to train sequence discovering models with high computational load. In order

L. Barolli (Ed.): BWCCA 2023, LNDECT 186, pp. 178–186, 2024.
https://doi.org/10.1007/978-3-031-46784-4_16

to cognizance the spatial characteristics of cellular traffic information, some methods based on Convolutional Neural Network (CNN) [7] are introduced, these approaches describe the network as an image of a convolution operator, which will lose critical information of mobile networks. In addition, due to the complex network structure, these methods often take longer to train the predictive algorithm.

In this paper, we propose the use of Dynamic Spatiotemporal Graph Convolutional Network (DSGCN) for mobile cellular service prediction. The main conception of DSGCN is to build a cellular traffic graph to reveal the complex spatiotemporal characteristics of different timestamp nodes. We implemented Graph Convolution Network (GCN) on the traffic graph sequence. We use the hypergraph of the cellular traffic graph to mine the dynamic correlation of the border nodes of the cellular traffic graph, which helps to explore the complex potential relationship of the dynamic cellular traffic system. Due to the aforementioned characteristics represented by edge correlation and collaborative GCN on graphs with conflated spatiotemporal characteristics, our model has achieved significant improvements compared to current GCN based methods and was experimentally verified.

2 DSGCN Predictive Model

Figure 1 shows the proposed DSGCN framework is composes of four parts. The input cellular traffic information handled by a linear transform component in the beginning and then input into a stacked Spatiotemporal Module (STM), which is the core component for implementing dynamic graph convolution on cellular traffic graphs and their hypergraphs. Then, the known characteristics are mixed through hop joins. At the end, the final traffic prediction result was obtained through Leaky ReLU and linear transformation layer.

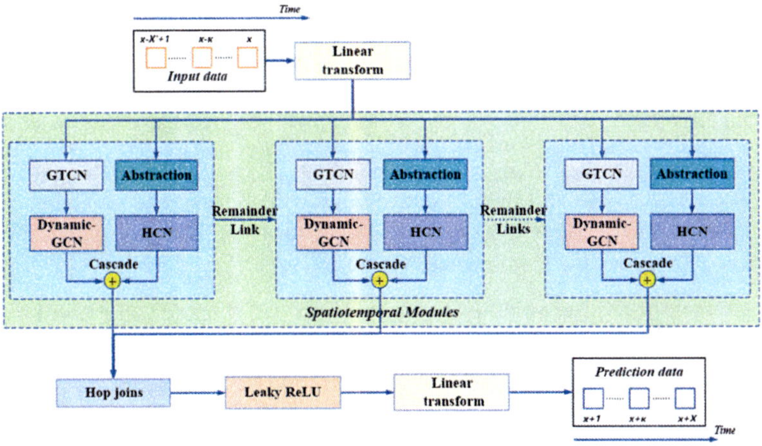

Fig. 1. The DSGCN framework for cellular service prediction. It consists of multiple stacked spatiotemporal modules (STMs), input layers, and output layers. Each STM has a remainder link and adds the outcome of the module by hopping join and inputting it to the outcome layer.

2.1 Cellular Traffic Prediction Problem

The indirect graph $\boldsymbol{T} = (\nu, \varepsilon, \alpha)$ is the digitized representation of the mobile traffic, $\boldsymbol{\nu}$ represents a collection of network regions, called $\boldsymbol{M}$ nodes, which means $\boldsymbol{\nu}$ is the number $\boldsymbol{M}$, $\boldsymbol{\varepsilon}$ represents a collection of border nodes and adjacent matrix $\boldsymbol{\alpha} \in \{0, 1\}_{M \times M}$ is the connectedness of network regions.

The mobile cellular traffic information of $\boldsymbol{M}$ nodes at time $\boldsymbol{x}$ can be represented as $\mathbf{D}_{\boldsymbol{x}} = (\mathbf{d}_x^1, \mathbf{d}_x^2, ..., \mathbf{d}_x^{\mathrm{M}})^{\mathsf{T}}$, where $\mathbf{d}_{\mathrm{x}}^{\mathrm{i}}$ is the measured value of the $\boldsymbol{i}$th node at time x. Next for all nodes τ time, the traffic sequence of a time slice can be represented as $\boldsymbol{S}^{1:\kappa} = (\mathbf{D}_1, \mathbf{D}_2, \ldots, \mathbf{D}_\kappa)^{\mathsf{T}}$. Therefore, the problem is to predict $\boldsymbol{S}^{1:\kappa}$. It can be explained as the historical cellular traffic information of the given cellular traffic graph $\boldsymbol{T}$ and $\boldsymbol{X'}$ time slot before time $\boldsymbol{x}$, the discovering function $\mathcal{L}$ is used to predict the cellular traffic information of the X time slot after time $\boldsymbol{x}$. The formula can be as follows,

$$\mathcal{L}(\mathcal{S}^{\left(x-X'+1\right):x}, T) = \mathcal{S}^{(x+1):(x+X)} \tag{1}$$

2.2 Spatiotemporal Module

The architecture of STM dynamically grabs the spatiotemporal characteristics of nodes and border nodes in mobile cellular traffic graphs, where edge characteristics are represented as hypergraphs. The remainder links are made between different stacked STMs. Each STM consists of three components: GTCN, Dynamic-GCN, and HCN.

Gate Temporal Convolutional Network (GTCN). We use GTCN to discover the temporal dependence of cellular traffic information by expanding convolutions in the temporal dimension. For the input characteristic sequence $\mathcal{S}$ of cellular traffic graph nodes. The GTCN formula is as follows,

$$TCN_\omega(\mathcal{S}) = Conv_h(\mathcal{S})$$

$$GTCN(\mathcal{S}) = th(TCN_{\omega 1}(\mathcal{S})) \odot sigmoid(TCN_{\omega 2}(\mathcal{S})) \tag{2}$$

among them, $Conv_h(\cdot)$ is an extended convolution with the convolution kernel h in the time dimension, $TCN_{\omega 1}(\cdot)$ and $TCN_{\omega 2}(\cdot)$ are two models of $TCN_\omega(\cdot)$, with different $\boldsymbol{\omega}$ parameters $\omega 1$ and $\omega 2$. $th(\cdot)$ and $sigmoid(\cdot)$ are activation functions.

Graph Convolutional Network (GCN). In current GCN based communication flow prediction framework, the adjacent matrices $\boldsymbol{W}$ is used to gather the spatial relationship between separate nodes at a specific time. Kipf et al. [8] proposed the formula of GCN as follows,

$$GCN(\mathbf{D}) = \widehat{W}\mathbf{D}\Upsilon \tag{3}$$

where $\widehat{\boldsymbol{W}}$ is the normalized adjacency matrix, $\mathbf{D}$ represents input characteristic matrix, and $\boldsymbol{\Upsilon}$ represents discovering parameter.

We consider Hypergraph Convolutional Network (HCN) [9] as an upgradation of GCN. Feng et al. [10] proposed the following expression of HCN,

$$HCN(\boldsymbol{D}_k) = M_{ks}^{-\frac{1}{2}} EHM_{kt}^{-\frac{1}{2}} E^X M_{ks}^{-\frac{1}{2}} \boldsymbol{D}_k \Upsilon \quad (4)$$

among them, $\boldsymbol{M}_{ks}$ and $\boldsymbol{M}_{kn}$ represents the diagonal matrix of super edge degree and super node degree respectively. Here, we refer to the nodes of a hypergraph as hypernodes to distinguish them from the nodes of a general graph. $\boldsymbol{E}$ represents incidence matrix, $\boldsymbol{H}$ represents weighted diagonal matrix of the hyperedge. $\boldsymbol{D}_k$ represents input characteristic matrix, $\boldsymbol{\Upsilon}$ represents discoverable parameter.

In addition, Temporal Convolutional Network (TCN) not only has a lower computing cost, but has much more time dimension receptive field.

Dynamic Graph Convolution Network. We propose the dynamic GCN to gather the spatiotemporal characteristics of cellular traffic information, it's expression is shown as below, according to the ideas of Graph WaveNet [12],

$$G_e = Reshape(HCN^n_{\Upsilon''_e}(\boldsymbol{D}^n_k))$$

$$G_c = Reshape(HCN^n_{\Upsilon''_c}(\boldsymbol{D}^n_k))$$

$$DynamicGCN(S) = \sum\nolimits_{i=0}^{I-1} (R_j \odot G_e)^i S\Upsilon'_{i,j} + (R_k \odot G_c)^i S\Upsilon'_{i,k} + R^i_f S\Upsilon'_{i,f} \quad (5)$$

where $\boldsymbol{R}_f = Softmax(Relu(L_1 L_2^X))$ is an adaptive adjacency matrix, where $\boldsymbol{L}_1$, $\boldsymbol{L}_2$ are discoverable parameters. $\boldsymbol{\Upsilon}^{'}_{j}$, $\boldsymbol{\Upsilon}^{'}_{b}$ and $\boldsymbol{\Upsilon}^{'}_{f}$ are discoverable parameters. $Reshape(RV)$ represents calculating reshaping vector $\boldsymbol{RV}$ as sparse matrix $\boldsymbol{G}$, and the result is $\boldsymbol{G}[e_s, e_d]$ equals to $\boldsymbol{RV}$.

Hypergraph Convolution Network (HCN). We use the HCN derived from (3) to acquire the dynamic border node characteristics as below,

$$Z_h = G_{ks}^{-\frac{1}{2}} HZ_f G_{kt}^{-\frac{1}{2}} H^\top G_{ks}^{-\frac{1}{2}}, (Z_f = diag(R_f))$$

$$HCN^n_\Upsilon(\boldsymbol{D}^n_k) = \sum\nolimits_{i=0}^{I-1} Z^i_k \boldsymbol{D}^n_k \Upsilon_i \quad (6)$$

among them, $\boldsymbol{\Upsilon} = [\Upsilon_0, \ldots, \Upsilon_{I-1}]$ represents the argument to be discovered, and $\boldsymbol{L}_f$ represents the weighted vectors of the hyperedge of self-adaption discovering. $HCN^n_\Upsilon(\boldsymbol{D}^n_k)$ is the input for dynamic GCNs in the cellular traffic graph.

2.3 Input and Output

Figure 1 shows that the input part consists of a linear transform that primarily converts the input cellular traffic information into a AmI-Space to enhance the network's expressiveness. The output part runs hop-join operations at the outputs on all STMs, then mixes the spatiotemporal characteristics and inputs to Leaky ReLU and linear transform parts to derive the eventual prediction result $\widehat{\mathcal{S}}^{(x+1):(x+X)} = (\widehat{\mathbf{D}}_{x+1}, \widehat{\mathbf{D}}_{x+2}, ..., \widehat{\mathbf{D}}_{x+X})^{\mathsf{T}}$. . In this paper, we use Mean Absolute Error (MAE) to form the loss function, that is, for ground truth $\mathcal{S}^{(x+1):(x+X)} = (\mathbf{D}_{x+1}, \mathbf{D}_{x+2}, ..., \mathbf{D}_{x+X})^{\mathsf{T}}$, the depletion is defined as follows,

$$depl = MAE(\widehat{\mathcal{S}}^{(x+1):(x+X)}, \mathcal{S}^{(x+1):(x+X)}) = \frac{\sum_{i=1}^{I}\sum_{j=x+1}^{x+X}\left|\widehat{\mathcal{S}}_j^i - \widehat{\mathcal{S}}_j^i\right|}{I \times X} \quad (7)$$

3 Performance Experiments

This paper considers the ground truth mobile traffic dataset publish by Telecom Italia [13]. The polymerized cellular traffic information from 20 × 20 regions in Milan is hourly recorded. The cellular traffic information is divided into training set, validation set, and test set in ratio of 7:1:2. Before using the forecast 7 days and 3-h data is used as former cellular traffic information. In the DSGCN method, we assume that the number of STM K is 3, the Lot size is 8. The optimizer is selected with a discovering rate of 0.001. Table 1 shows more experimental parameter settings. All experiments were conducted on computers equipped with Intel (R) Core (TM) i5-1240P CPUs and 16GB of memory.

Table 1. The parameters used in the experiment.

Category	Hyper-Parameter	Setup
Train	Lot size b	8
	Epochs	100
Optimizer	Primary discovering rate	0.001
	Discovering rate attenuation	0.97
Model	STMs S	3
	mid-filters M	60
	Top-k	4

Performance of DSGCN Method. We compared our DSGCN method with several traffic prediction approaches, Historical Average (HA), Auto Regressive Integrated Moving Average (ARIM) [14], Support Vector Regression (SVR) [15]. We measure the performance of each approach through three indexes: Mean Absolute Errors (MAE), Root Mean Squared Errors (RMSE) and Accuracy [16].

As show in Table 2, the best result is bold. The results show that the DSGCN performs better than other approaches in all indexes. Based on the data comparison, the traffic prediction performance of DSGCN reduces over 61% and 37% respectively in terms of the MAE and RMSE, and the Accuracy improves over 13%.

Table 2. Traffic prediction performance results using different methods.

Methods	MAE	RMSE	Accuracy
ARIMA	0.0362	0.0498	0.305
HA	0.0363	0.0506	0.3
SVR	0.0381	0.042	0.313
DSGCN	**0.0139**	**0.0264**	**0.356**

Parameter Adjustment. The primary hyperparameters of the DSGCN method include lot size b and the amount of mid-filters M. We investigated the sensitiveness of mobile cellular traffic information from different settings in Milan through traffic prediction experiments. Here, we use MAE as an evaluating indicator. Figure 2 shows the experiment and adjust results as below,

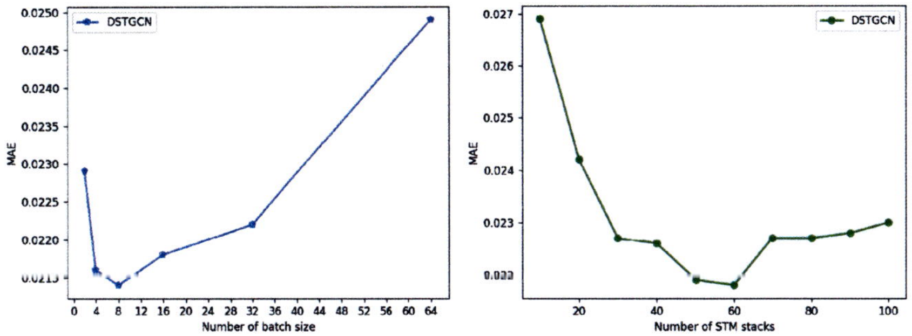

(a) The number of lot size b.

(b) The number of mid-filters M.

Fig. 2. Research on hyperparameters of cellular cellular traffic information.

Lot size b. As shown in Fig. 2(a), change the number of lot size to adjust the amount of data processed by each iteration of the model, thereby adjusting the model's expressiveness and performance. However, as the number increases, performance gradually decreases. Based on the results, we set b = 8 lot size in the model.

Mid-filters M. The amount of mid-filters is the dimensionality of the node characteristic channels in the graph. Figure 2(b) shows that as M increases, performance first decreases and then stabilizes. However, excessive numbers can make the pattern too complex, causes over-fitting and efficiency degradation. Therefore, we set $M = 60$ in the model.

Training Loss. To evaluate the training effectiveness of the model and the stability of the optimization process, we record the training loss value for each time, which convergence is an important indicator in GCN training methods, as the number of iterations increased. The training loss values for each epoch after running 100 iterations are shown in Fig. 3.

Figure 3 shows that as the number of iterations increases, the training loss of our DSGCN method gradually decreases and tends to stabilize, showing convergence. This indicates that the model is gradually optimizing and the difference between the predicted results of the model and the ground truth cellular traffic information is decreasing. In other words, the model is discovering the key characteristics from the data and is able to fit the training data well.

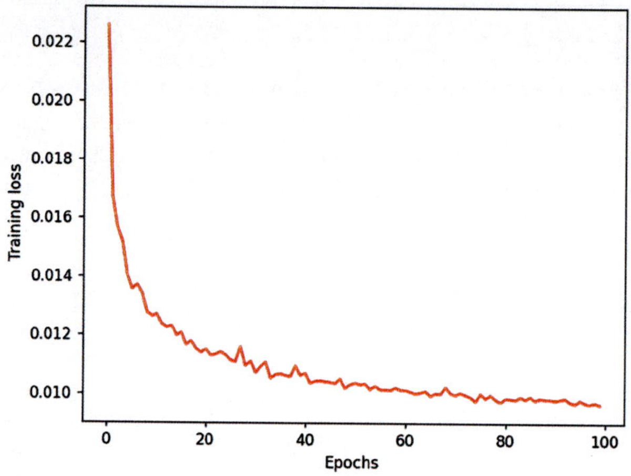

Fig. 3. The training loss of each epoch. The results show that the training loss has convergence.

Visualization of Result. In order to achieve comparison, the experiment is based on default hyperparameters with the same number of training periods in the corresponding literature. We visualized the traffic prediction results of DSGCN method for 120 h on a real-world dataset. Figure 4 shows the predicted results in contrast with the ground truth. The results indicate that our DSGCN method has good performance throughout the entire time period.

Figure 4 shows the predicted results of DSGCN method in a specific area of cellular services. The model can predict the peaks and valleys of cellular services. That is to say, GCN based methods can achieve good performance, demonstrating the excellent performance of GCN in capturing spatial correlations and dependencies. In addition, compared with the ground truth, the predicted curve of our proposed model DSGCN shows that except for the significant difference in the peak flow rate, the curves of other time periods tend to be consistent, indicating that the DSGCN method can effectively gather complex spatiotemporal characteristics between different nodes.

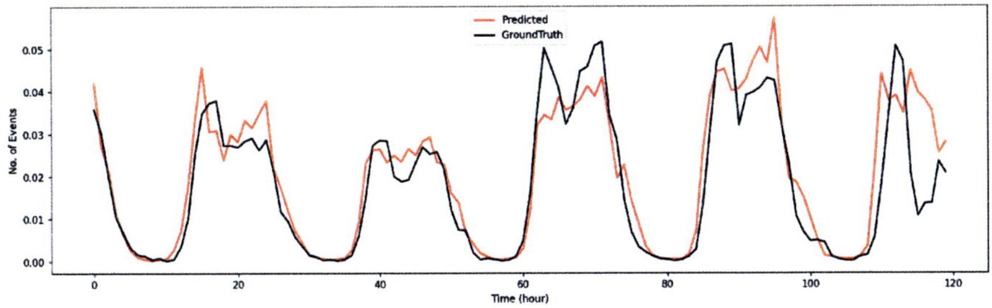

Fig. 4. The predicted results of DSGCN prediction method.

4 Conclusion

This paper proposes a dynamic GCN traffic prediction method to model the short-range spatiotemporal correlation of cellular traffic. Especially, we proposed an uncomplicated algorithm to abstract the former patterns and current daily patterns of cellular traffic information. Our method converts the border nodes of the cellular traffic graph into hypergraph and utilizes the GCN on the hypergraph to utilize their spatiotemporal characteristics. The proposed model was appraised on ground truth cellular traffic information and demonstrated to have advanced effects. However, in the prediction process, only the relationship between time series and mobile traffic was considered, and the impact of weather, holidays, and other conditions on mobile traffic is needed further research.

References

1. Liu, J., Guan, W.: A summary of traffic flow forecasting methods. J. Highway Transp. Res. Develop. **21**(3), 82–85 (2004)
2. Xu, F., et al.: Big data driven mobile traffic understanding and forecasting: a time series approach. Services Comput. **9**(5), 796–805 (2016)
3. Drucker, H., Burges, C.J., Kaufman, L., Smola, A., Vapnik, V.: Support vector regression machines. Proc. Adv. Neural Inf. Process. Syst. **9**, pp. 1–7 (1996)
4. Zhou, C., Nelson, P.C.: Predicting traffic congestion using recurrent neural networks. Proc. World Congr. Intell. Transp. Syst. (2002)
5. Fusco, G., Colombaroni, C., Isaenko, N.: Comparative analysis of implicit models for real-time short-term traffic predictions. IET Intell. Transp. Syst. **10**(4), 270–278 (2016)
6. Zhao, Z., et al.: LSTM network: a deep discovering approach for shortterm traffic forecast. IET Intell. Transp. Syst. **11**(2), 68–75 (2017)
7. Yu, H., Wu, Z., Wang, S., Wang, Y., Ma, X.: Spatiotemporal recurrent convolutional networks for traffic prediction in transportation networks. Sensors **17**(7), 1501 (2017)
8. Kipf, T.N., Welling, M.: Semi-supervised classification with graph convolutional networks. Proc. Int. Conf. Discover. Represent. pp. 1–14 (2017)
9. Yadati, N., Nimishakavi, M., Yadav, P., Nitin, V., Louis, A., Talukdar, P.: HyperGCN: a new method for training graph convolutional networks on hypergraphs. Proc. Adv. Neural Inf. Process. Syst. 1511–1522 (2019)
10. Feng, Y., You, H., Zhang, Z., Ji, R., Gao, Y.: Hypergraph neural networks. In: Proceedings of the AAAI Conference on Artificial Intelligence, vol. 33, pp. 3558–3565 (2019)

11. Dauphin, Y.N., Fan, A., Auli, M., Grangier, D.: Language modeling with gated convolutional networks. In International Conference on Machine Learning, pp. 933–941 (2017)
12. Wu, Z., Pan, S., Long, G., Jiang, J., Zhang, C.: Graph WaveNet for deep spatial–temporal graph modeling. In: Proceedings 28th International Joint Conference Artificial Intelligent, pp. 1907–1913 (2019)
13. Barlacchi, G., et al.: A multi-source dataset of urban life in the city of Milan and the province of Trentino. Sci. Data **2** (2015). Art. no. 150055
14. Makridakis, S., Hibon, M.: ARMA models and the box-jenkins methodology. J. Forecasting **16**(3), 147–163 (1997)
15. Drucker, H., et al.: Support vector regression machines. Proc. Adv. Neural Inf. Process. Syst. **9**, 155–161 (1997)
16. Zhao, L., et al.: T-GCN: a temporal graph convolutional network for traffic prediction. IEEE Trans. Intell. Transp. Syst. **21**(9), 3848–3858 (2019)

Caching Strategy Utilizing Social Network Analysis Algorithm for Effective Cache Storage Allocation in Named Data Networking

Ridha Muldina Negara[1,2(✉)], Nana Rachmana Syambas[1], Eueung Mulyana[1], and Rashid Muhammad Fajri[2]

[1] School of Electrical Engineering and Informatics, Institut Teknologi Bandung, Bandung, Indonesia
ridhanegara@telkomuniversity.ac.id
[2] School of Electrical Engineering, Telkom University, Bandung, Indonesia

Abstract. Named Data Networking (NDN) utilizes caching strategies to manage data caching on network nodes, making it suitable for high-traffic future networks. In-network caching brings desired data packets closer to users topologically, enhancing network performance and satisfying users. However, a critical challenge in NDN network design is determining the Content Store (CS) size allocation for each router. Previous studies overlooked router position, assuming uniform CS sizes. To address this, we proposed a caching strategy with the Social Network Analysis (SNA) algorithm was developed to calculate the influence of an NDN router on the topology. The SNA algorithm includes degree centrality, betweenness centrality, and consolidated centrality. Validated using the ICARUS caching simulator, the study's results propose an efficient caching strategy for allocating cache storage on all NDN routers. The SNA-based approach outperforms alternatives in cache hit rate, latency, link load, and path stretch, contributing to effective and efficient network planning.

1 Introduction

Rapid technological developments have significantly increased the amount of content and data traffic. According to the Cisco Annual Internet Report, the number of Internet users will increase by 15% in 2023 from 2018 [1, 2]. So, to meet the needs of the network, which will experience an increase in traffic, a paradigm is proposed, namely Information-Centric Networking (ICN). This paradigm uses the concept of content-centric or information-centric, namely, data requests are made by the user based on the desired content no matter where the content is located [3, 4]. One of the ICN technologies is Named Data Networking (NDN) [5, 6]. Named Data Networking (NDN) has 3 data structures, namely Pending Interest Table (PIT), Content Store (CS), and Forwarding Information Base (FIB).

Content Store (CS) is a temporary cache storage on each cache router. The mechanism for managing and managing content stores on NDN is called caching. In setting the cache space capacity on each router, there are still deficiencies, which are determined

L. Barolli (Ed.): BWCCA 2023, LNDECT 186, pp. 187–198, 2024.
https://doi.org/10.1007/978-3-031-46784-4_17

based on heuristics or assumptions [7]. Previous research determines the size of cache capacity based on this heuristic or assumption; this method cannot be used in an actual topology. In actual conditions, the topology can change over time and has high complexity. Then a technique is needed to determine cache capacity based on the existing topology conditions. The technique that can be used based on the state of the topology is to use the application of network centrality.

Applying a Social Network Analysis (SNA)-based caching strategy to cache nodes or NDN routers can help improve network efficiency and performance by storing the most relevant content in the cache and minimizing network traffic [8]. However, it is essential to note that a more detailed and optimized implementation will require more complex modeling and tweaks tailored to the specific network environment. In this paper, we will evaluate the performance of implementing SNA-based network mentality techniques to determine the capacity of cache space on routers. Performance evaluation using the Icarus simulator by looking for the most efficient according to cache hit ratio, latency, link load, and path stretch.

2 Related Works

Cache placement in the Icarus simulator is related to the placement of cache capacity sizes on nodes in the topology [6]. Some of the cache placements that can be used are as follows:

a. Uniform
This cache placement uses uniform centrality as a technique to place a uniform size of cache capacity across all cache nodes. The total cache capacity space and the number of existing cache nodes are divided.
b. Degrees
The concept of degree centrality gives a centrality value based on the number of degrees at the node [9]. So, the application to cache placement is to place the size of the cache capacity proportionally to the degree level of the node.
c. Betweenness Centrality
The concept of betweenness centrality [10] is to provide a value of centrality using the following formula:

$$c_B(v) = \sum\nolimits_{s,t \in V} \frac{\sigma(s, t|v)}{\sigma(s, t)} \quad (1)$$

σ(s, t) = Sum of possible shortest paths from all pairs of nodes

σ(s, t|v)= The number of shortest paths that pass through the node whose centrality value you want to calculate.

After looking for the value of betweenness centrality using formula 1, its application in cache placement is to place the size of the cache capacity proportionally according to the betweenness centrality value of the cache node in the topology.

d. Consolidated
The consolidated concept combines two centrality techniques. Unlike the other cache placements, which assign a cache capacity size to all nodes, Consolidated places the

same/uniform cache size for the selected group of nodes. In the consolidated, there is a spread factor. This factor determines the selected group of nodes. Where the numbers entered for this factor are in the form of a float from 0 to 1, with each addition, the spread of the placement of the cache size is also getting more significant. If the author fills in the number 1.0, then all candidate nodes will be selected, and if we fill in the number 0.0, then only the one with the highest/weakest centrality will be selected.

3 System Models

3.1 Icarus Simulator

Icarus [11] is a Python programming simulator that evaluates caching strategies without being tied to the ICN architecture. Icarus has the advantages of scalability and extensibility. Regarding scalability, Icarus supports hundreds of thousands of request simulations every minute so that Icarus can complete more than millions of request simulations in a few minutes. Extensibility means Icarus is easy to use and developed for research in caching. In Icarus, a cache policy is a strategy for managing the content exchange within a node. In contrast, the strategy for selecting a node to store content is called a strategy. Icarus has four stages in running the simulation process [10]. These stages are:

1. Scenario Generation
   Setting up the network topology that has been configured in Icarus.
2. Experiment Orchestration
   In the second stage, Icarus will read the configuration file, which has added the parameters desired by the user.
3. Experimental Execution
   At this stage, Icarus starts running the scenarios that the user creates.
4. Results Collection and Analysis
   After the experiment has finished running, the results will be combined in a file, and then the user can carry out the analysis.

3.2 Formulas

To evaluate the performance of the impact of implementing centrality, the topology that must be used is the actual topology. Here the topology that will be used is the Geant topology, which illustrates the European topology [12].

In this topology, there are a total of 53 nodes, including eight receiver nodes, 19 cache router nodes, 13 content source nodes, and the remaining 13 nodes are ordinary routers. Figure 1 is the Geant topology, and Table 1 is the configuration [12].

In the simulation, all cache strategies use leave copy everywhere (LCE), which is a strategy that makes all cache nodes that are passed by content (cache) store that content [13]. The cache policy used is the least recently used (LRU), which is a cache policy that will insert new content into the content store, and if the content store is full, it will delete the most recently requested content [14].

The network cache capacity ratio is 1%, 4%, 10%, and 15% of the total content. For population/workload content using the Zift (Stationary) distribution model. With variations in the parameters, α are 0.8 and 1.2. Content placement uses Uniform for

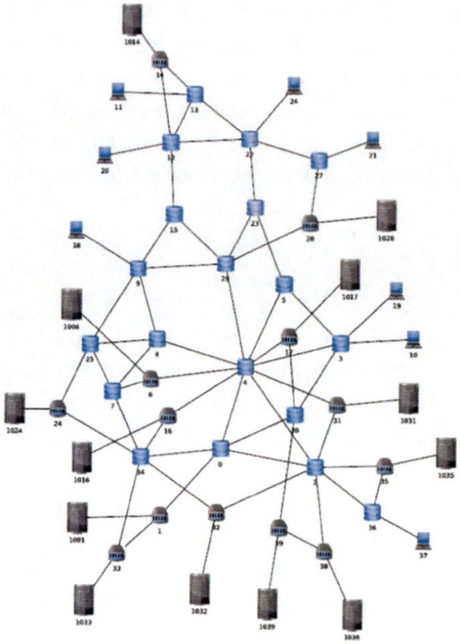

Fig. 1. GEANT Topology

Table 1. GEANT Topology Configuration

Network Characters	Graph Vertex	ID nodes	NUM
Users	Degree = 1	{10, 11, 18, 19, 20, 21, 26}	8
Routers (cache-disabled)	Degrees = 2	{1, 6, 14, 16, 17, 24, 28, 31, 32, 33, 35, 38, 39}	13
Routers (cache-able)	Degrees > 2	(0, 2, 3, 4, 5, 7, 8, 9, 12, 13, 15, 22, 23, 25, 27, 29, 30, 34, 36}	19
Sources	Attached on nodes whose degrees are 2	{1001, 1006, 1014, 1016, 1017, 1024, 1028, 1031, 1032, 1035, 1038, 1039}	13

cache placement or large cache capacities on nodes using several cache placements, namely Uniform, Degree, consolidated, and betweenness centrality. The paper includes a list of crucial abbreviations and symbols in Table 2.

The simulation is carried out with a small content catalog. So that the simulation can be implemented smoothly, the amount of content on the network is set at 100,000. Then the total cache capacity space will be 1000, 4000, 10000, and 15000 with a cache warm-up of 10,000 and a cache measurement of 200,000. The experimental parameters can be seen in Table 3. The experiment was carried out 15 times to reduce the experimental error,

Table 2. (a) Symbols are used to describe the algorithm, (b) The pseudo-code for obtaining the cache placement with consolidated algorithm.

(a)

Symbols	Descripstion
R	The set of all router node
R_c	The set of all cache node
R_{c_n}	The set of all selected target cache node
R_{c_T}	The number of cache node R_c in the topology
R_c^v	Cache node of node v
$R_{c_n}^T$	The number of target cache node R_{c_n} in the topology
C	the percentage value of the cache budget of all content to be retrieved.
Sc	Cache size in each cache node
D	Degree value of all node in topology
D_v	Degree of cache node v
D_T	Number of D in topology
B	Betweenness value of all node in topology
B_v	Betweenness of cache node v
B_T	Number of B in topology
N	Number of all content
Sp	The spread factor to determine the selected group of cache nodes.
T	The subsection of the ranked node on which to the deploy caches.
Md	The centrality metric according to which nodes are selected.

(b)

Algorithm 4 Obtaining the cache placement with Consolidated.

```
Input: Topology, cache budget C, spread factor Sp, metric selected Md and target T
Output: None
if Sp < 0 or Sp > 1 then
    /*checking the spread factor value must be between 0 and 1 */
  return Error;
if T is not "top" or "bottom" then
    /*checking the target argument must be either "top" or "bottom"*/
  return Error;
if Md is None and Sp < 1 then
  /*If Md not specified, betweenness centrality is selected*/
  Md ← B;
Cache node R_c;
if Sp == 0 then
  R_cn ← R_c;
else if Sp > 0 and Sp < 1 then
  nodes ← sorted cache node Rc according to the smallest Md value;
  if T is "top" then
      /*If T not specified, "top" is selected*/
      nodes ← sorted cache node Rc according to the largest Md value;
  Cutoff ← max(1, round (Sp * R_cT));
  R_cn ← get all nodes before cutoff index;
Sc ← (C * N) / R_cT;
if Sc == 0 then
  return None;
for node v in R_cn do
  R_c^v ← Sc;
return None;
```

and the results were averaged. Performance is measured by four parameters, namely, based on cache hit ratio (CHR), latency, link load, and path stretch.

4 Evaluation Performance

This session analyzes the simulation results of the schemes that have been made.

Table 3. Experimental Parameters

Parameter	Value
Topology	GEANT
Cache placements	Uniform, Degree, Consolidated, and Betweenness centrality
Cache Capacity	1%, 4%, 10%, and 15% of the total content
Workloads	Stationary
Number of contents	100000
Warm-ups	10000
Measured	200000
Alpha	0.8 and 1.2
Cache policies	LRU
Strategy	LCEs
Content placement	Uniform

4.1 Cache Hit Rate (CHR)

The Cache Hit Ratio (CHR) value represents the average number of cache hits served when a user sends a request [15]. CHR is a fundamental metric for evaluating performance in NDN. The cache hit ratio is calculated using the following formula [1]:

$$\mathrm{CHR} = \frac{\sum_{k=1}^{k} N_k}{NUM} \tag{2}$$

In formula 2, K stands for the number of users, and N_k represents the count of cache hits or the number of requests that have been successfully served. NUM denotes the total number of requests sent by the user.

The average performance results for the cache hit ratio with a parameter $\alpha = 0.8$ and an increase in the network cache in the Geant topology are illustrated in Fig. 2 (a). For a network cache of 0.01, the Cache placement Betweenness centrality achieves the highest CHR at 11.2%, surpassing the cache placement consolidated (10.5%) by 0.7%. When the network cache is set to 0.15, Cache placement Betweenness centrality still holds the highest CHR at 33%. This is because the node with the highest centrality value in the Geant topology possesses significant cache capacity and is frequently traversed, leading to an increased likelihood of cache hits.

The augmentation of network cache during the simulation expands the cache space across routers/cache nodes, resulting in a more abundant content pool. Consequently, the probability of cache hits becomes more significant, leading to an increase in the CHR value.

Performance with the parameter $\alpha = 1.2$ is depicted in Fig. 2(b). It is evident that for a network cache of 0.01, the Cache Hit Ratio (CHR) in the Betweenness centrality cache placement reaches 67.6%. Even in the simulation with the parameter $\alpha = 1.2$,

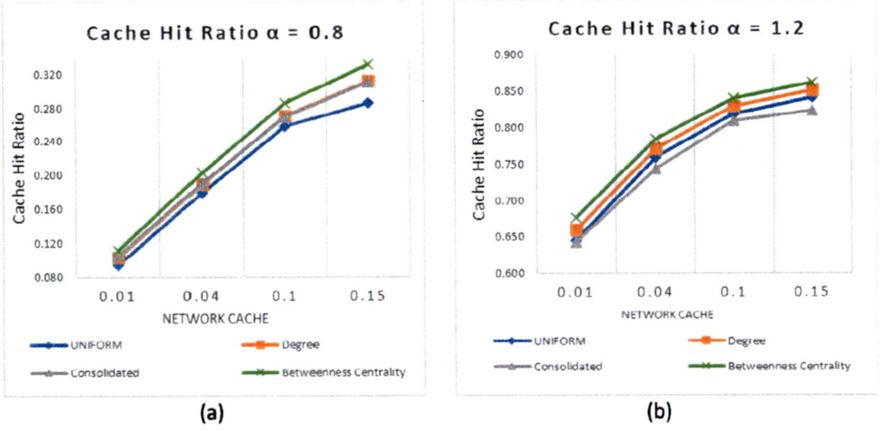

Fig. 2. Cache Hit Ratio: (a) $\alpha = 0.8$ and (b) $\alpha = 1.2$

the Betweenness centrality cache placement maintains the highest CHR compared to other cache placements. In Fig. 2(b), the CHR value experiences a significant increase in comparison to Fig. 2(a); however, the cache placement performance ranking shows only marginal differences. This outcome arises because a higher value of α corresponds to a greater frequency of popular content requests being sent, consequently resulting in an elevated likelihood of cache hits.

4.2 Latency

In addition to the cache hit ratio, latency is another crucial performance metric. Latency refers to the time required from when the user sends the interest packet until the user successfully receives the requested packet data. Lower latency leads to faster content delivery, while higher latency results in slower content delivery. The unit of latency is milliseconds (ms) [16].

The performance of latency results for the four cache placements in the simulation of 15 trials with the parameter $\alpha = 0.8$ and experiencing an increase in the network cache in the Geant topology, is depicted in Fig. 3(a). The results show that the lowest latency in network cache 0.01 is achieved by cache placement betweenness centrality, at 78.5 ms, which is 0.5 ms shorter than cache placement consolidation (79 ms). When the network cache is set to 0.15, the lowest latency is achieved by cache placement betweenness centrality, at 61.2 ms.

An increased network cache leads to a broader allocation of cache space and more content available on cache nodes. This condition reduces the likelihood of cache misses. A cache miss occurs if the cache node/router does not possess the requested content. Consequently, the request is forwarded to another cache node/router until the desired content is located, resulting in even higher latency.

The performance of average latency results with the parameter $\alpha = 1.2$ is presented in Fig. 3(b). All cache placements exhibit lower latency. In network cache 0.01, the lowest latency is recorded for cache placement betweenness centrality, at 32.5 ms, which is 1 ms lower than cache placement degree (33.5 ms). Similarly, when the network cache is set to 0.15, although there is a substantial decrease in the latency values across the four cache placements, the ranking of the lowest latency performance remains relatively consistent. Cache placement betweenness centrality continues to exhibit the lowest latency.

With an increase in the α parameter in the latency value for the four cache placements, there is a notably significant decrease in value. In contrast to CHR (Cache Hit Ratio), an increase in the parameter α results in a more modest reduction in latency. A higher α value leads to more frequent user requests for popular content, subsequently increasing the likelihood of cache hits. This results in faster processing of user interest data and higher chances of obtaining the desired data, as the probability of the cache node containing the content becomes greater.

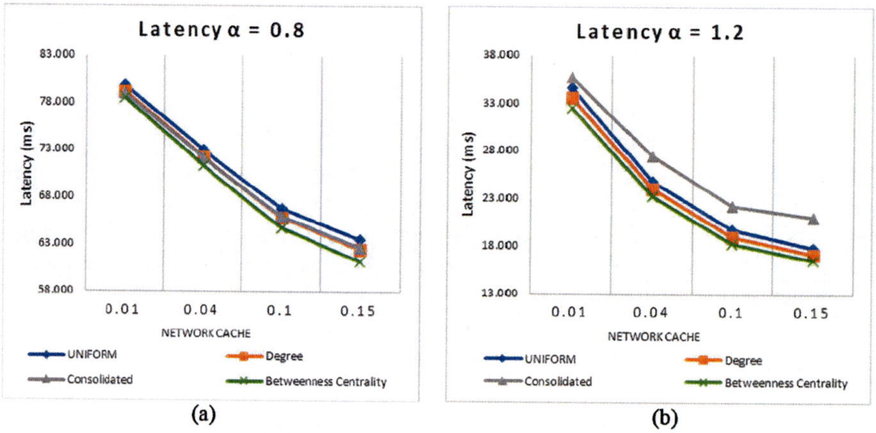

Fig. 3. Latency (a) $\alpha = 0.8$ and (b) $\alpha = 1.2$

4.3 Link Load

Link load refers to the average traffic carried by all links within the network. In the Icarus framework, the link load for each individual link can be computed by quantifying the volume of packet interests and data packets transiting through it over a specific time interval [1]. A higher link load value signifies elevated data traffic on that particular link, while conversely, a lower link load value indicates comparably lower data traffic.

The performance of latency results for the four cache placements is demonstrated in a simulation of 15 trials, utilizing the parameter $\alpha = 0.8$, and reflecting an augmentation in the network cache within the Geant topology, as depicted in Fig. 3(a). The outcomes reveal that cache placement betweenness centrality achieves the lowest latency at a network cache of 0.01, registering 78.5 ms – a marginal 0.5 ms reduction compared to cache placement consolidation (79 ms). For a network cache set at 0.15, the cache placement betweenness centrality retains the shortest latency at 61.2 ms.

Figure 4(a) illustrates the performance of average link load outcomes utilizing the parameter $\alpha = 0.8$. Notably, when the network cache value is 0.01, the lowest link load

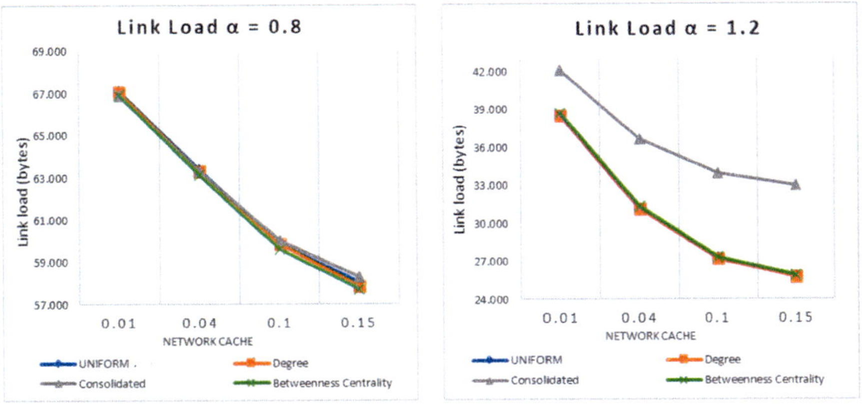

Fig. 4. Link Load: (a) α = 0.8 and (b) α = 1.2

value is observed in the consolidated placement, totaling 66,882 bytes. This is minutely lower by 0.067 bytes compared to betweenness centrality (69,949 bytes). In scenarios where the network cache is set at 0.15, the cache placement betweenness centrality again attains the lowest value, measured at 61,208 bytes.

With each successive increment in the network cache value, the link load value for the four cache placements experiences a decline. This phenomenon is attributed to the expanded cache space capacity within cache nodes, facilitating increased content availability and subsequently enhancing the likelihood of cache hits. Consequently, the need to forward requests to alternative cache nodes diminishes, leading to reduced network traffic.

Figure 4(b) depicts the performance of average link load outcomes using the parameter $\alpha = 1.2$. An increase in the α parameter results in a reduction of the link load value. This trend mirrors that of latency: as the α value increases, the link load value decreases. The α value amplifies the potential for popular content requests, leading to an accumulation of content within cache nodes. Consequently, cache hits become more frequent, reducing overall traffic.

4.4 Path Stretch

Path stretch refers to the ratio between the number of nodes from requests to routers that register cache hits and the number of nodes from requests to origin server providers [17].

The performance of average path stretch results using the parameter $\alpha = 0.8$ is displayed in Fig. 5(a). For a network cache of 0.01, the cache placement betweenness centrality exhibits the lowest path stretch value at 0.794, a marginal 0.001 reduction compared to the consolidated placement (0.795). As the network cache increases to 0.15, the path stretch value diminishes further. In the case of cache placement betweenness centrality, it drops to 0.672. The augmentation of network cache enhances the cache space capacity, allowing cache nodes to accommodate more content. Consequently, the likelihood of finding user-requested content within the nearest cache node increases.

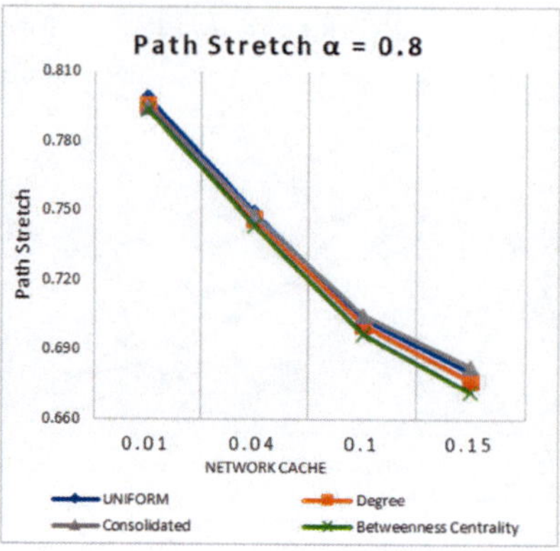

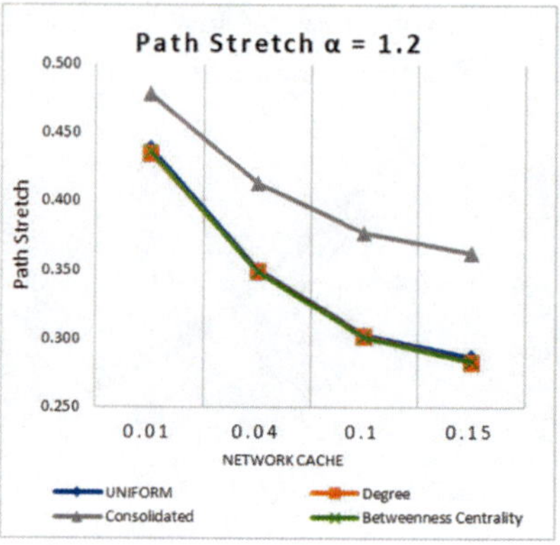

Fig. 5. Path Stretch (a) $\alpha = 0.8$ and (b) $\alpha = 1.2$

Figure 5(b) illustrates the performance of average path stretch outcomes with the parameter $\alpha = 1.2$. An escalation in the α parameter leads to a decreased path stretch value. This shift is due to the heightened chances of popular content requests, thereby increasing the probability of the closest cache node containing the desired content.

An increased network cache results in a more widespread distribution of cache space and an abundance of content available on cache nodes. This circumstance reduces the likelihood of cache misses. A cache miss arises when the cache node/router does not possess the requested content. This scenario prompts the request to be redirected to another cache node/router until the desired content is located, subsequently leading to elevated latency.

5 Conclusion

Caching plays a crucial role in optimizing NDN nodes' cache utilization. Within the Icarus software, cache placement stands as a pivotal caching parameter. This placement pertains to allocating cache space capacity on NDN nodes or cache nodes based on simulation outcomes conducted on the Geant topology. The most effective cache placement, denoted as "Betweenness centrality," distributes cache space capacity proportionally, aligned with the centrality of the respective NDN node. In this approach, nodes boasting higher centrality values possess larger cache capacities to accommodate an expanded content repository. Notably, the Geant topology exhibits a central node configuration, yielding elevated node centrality values and substantial cache space capacities. Consequently, the likelihood of cache hits surges. Furthermore, augmenting the network cache bolsters cache space capacity, thereby elevating the Cache Hit Ratio (CHR) and concurrently reducing latency, link load, and path stretch. The manipulation of the α parameter yields similar effects—increasing it diminishes the CHR, as well as latency, link load, and path stretch. This pattern is attributed to user preferences for popular content, consequently leading to frequent cache hits.

Acknowledgments. This research is part of dissertation research of Ridha Muldina Negara funded by The Indonesia Endowment Funds for Education (LPDP) under Grant Nos.

0001825/TRP/D/PDD-2020. Ridha Muldina Negara is thankful for the LPDP Scholarship supporting her doctoral study at ITB.

References

1. Gui, Y., Chen, Y.: A cache placement strategy based on entropy weighting method and topsis in named data networking. IEEE Access **9**, 56240–56252 (2021). https://doi.org/10.1109/ACCESS.2021.3071427
2. Negara, R.M., Rachmana Syambas, N.: Caching and machine learning integration methods on named data network: a survey (2020). https://doi.org/10.1109/TSSA51342.2020.9310811
3. Nour, B., Khelifi, H., Moungla, H., Hussain, R., Guizani, N.: A distributed cache placement scheme for large-scale information-centric networking. IEEE Netw. **34**(6), 126–132 (2020). https://doi.org/10.1109/MNET.011.2000081
4. Peng, X., Garg, H.: Intuitionistic fuzzy soft decision making method based on CoCoSo and CRITIC for CCN cache placement strategy selection. Artif. Intell. Rev. **55**(2), 1567–1604 (2022). https://doi.org/10.1007/s10462-021-09995-x
5. Zhang, L., et al.: "NDN Project 2010," Relatório Técnico NDN-0001, Xerox Palo Alto Res. Center-PARC, vol. 157, no. October, pp. 1–24 (2010). http://www.named-data.net/techreport/TR001ndn-proj.pdf
6. Saino, L., Psaras, I., Pavlou, G.: Framework and algorithms for operator-managed content caching. IEEE Trans. Netw. Serv. Manag. **17**(1), 562–576 (2020). https://doi.org/10.1109/TNSM.2019.2956525
7. Mahananda, I.G.E., Yovita, L.V., Negara, R.M.: Performance of homogeneous and heterogeneous cache policy for named data network. In: 2022 10th International Conference on Information and Communication Technology ICoICT 2022, pp. 120–123 (2022). https://doi.org/10.1109/ICoICT55009.2022.9914827
8. Muldina Negara, R., Rachmana Syambas, N., Mulyana, E.: C3CPS: CRITIC-CoCoSo-based caching placement strategy using multi-criteria decision method for efficient content distribution in named data networking. J. King Saud Univ. - Comput. Inf. Sci., 101714 (2023). https://doi.org/10.1016/j.jksuci.2023.101714
9. Behera, R.K., Rath, S.K., Misra, S., Damaševičius, R., Maskeliunas, R.: Distributed centrality analysis of social network data using MapReduce. Algorithms **12**(8) (2019). https://doi.org/10.3390/a12080161
10. Chai, W.K., He, D., Psaras, I., Pavlou, G.: Cache 'less for more' in information-centric networks (extended version). Comput. Commun. **36**(7), 758–770 (2013). https://doi.org/10.1016/j.comcom.2013.01.007
11. Saino, L., Psaras, I., Pavlou, G.: Icarus: a caching simulator for Information Centric Networking (ICN) ∗ categories and subject descriptors (2014)
12. Cite, P., Published, T.H.E., Record, R.: Distributed Caching in the Future Generation Networks Doctor of Philosophy Loughborough University, Loughborough University (2020)
13. Dutta, N., Patel, S.K., Faragallah, O.S., Baz, M., Rashed, A.N.Z.: Caching scheme for information-centric networks with balanced content distribution. Int. J. Commun. Syst. **35**(7) (2022). https://doi.org/10.1002/dac.5104
14. Putra, M.A.P., Situmorang, H., Syambas, N. R.: Least recently frequently used replacement policy named data networking approach. In: 2019 International Conference on Electrical Engineering and Informatics (ICEEI), vol. 2019-July, no. September 2021, pp. 423–427, (2019). https://doi.org/10.1109/ICEEI47359.2019.8988828

15. Meng, Y., Naeem, M.A., Ali, R., Kim, B.S.: EHCP: an efficient hybrid content placement strategy in named data network caching. IEEE Access **7**, 155601–155611 (2019). https://doi.org/10.1109/ACCESS.2019.2946184
16. Naeem, M.A., Nguyen, T.N., Ali, R., Cengiz, K., Meng, Y., Khurshaid, T.: Hybrid cache management in IoT-based named data networking. IEEE Internet Things J. **9**(10), 7140–7150 (2022). https://doi.org/10.1109/JIOT.2021.3075317
17. Kumar, S., Tiwari, R.: Optimized content centric networking for future internet: dynamic popularity window based caching scheme. Comput. Networks **179**(June), 107434 (2020). https://doi.org/10.1016/j.comnet.2020.107434

A Model to Estimate the Minimum Data Transmission Time in IoT Networks Using Multiple Fungible Paths

David W. White[1(✉)], Isaac Woungang[2], Felix O. Akinladejo[1], and Sanjay K. Dhurandher[3]

[1] University of Technology, Kingston, Jamaica
{dwwhite,fakinladejo}@utech.edu.jm
[2] Toronto Metropolitan University, Toronto, Canada
iwoungan@torontomu.ca
[3] Netaji Subhas University of Technology, New Delhi, India

Abstract. Reports have emerged of the vast amounts of data being transmitted across the Internet by Internet of Things (IoT) devices. Amidst this, the number of connected IoT devices continues to grow at a very rapid pace but over the last five years peak internet utilization has remained just under 50%. While applications have been processing increasingly massive amounts of data, users have been demanding faster or more acceptable response times and others have touted the economic value of having faster response times.

In this paper we proffer arguments for adding more fungible paths to IoT and other smart devices to obtain theoretical speedup in network data transmission, and we present equations to compute the number of fungible paths needed to complete transmission of a set amount of data in a specific required minimum time frame, and the resulting speedup achieved. Our proposed model can assist IoT and smart device manufacturers, network administrators, and networked applications developers in deciding how many physical and logical layer one communication units to build into the devices they manufacture, and how many of these communication paths should be utilized during a communication session to achieve the desired data transfer/response times.

1 Introduction

Computer networks emerged in the later part of the 1950s followed by the creation of the ARPANET in 1969, which gave rise to the Internet [1–3]. The ARPANET itself benefited tremendously from the mathematical models created by Leornard Kleinrock to represent the performance of packet-switched computer networks [4, 5]. Connecting smart devices to a computer network began in earnest in the 1980s, and by the 1990s the term Internet of Things (IoT) had entered the lexicon to describe such a network [6–8]. Today, the world has seen a phenomenal growth in the IoT network. Figure 1 shows the graph from [9] which portrays the real and projected growth in IoT devices compared to non-IoT devices for the period 2010 to 2025. The years super-scripted with an asterisk are forecasts, while the other years represent actual growth numbers. As

L. Barolli (Ed.): BWCCA 2023, LNDECT 186, pp. 199–208, 2024.
https://doi.org/10.1007/978-3-031-46784-4_18

expected, while the number of connected IoT devices has been growing, so too has the volume of data that is transmitted by these devices across the Internet. In [10], Cisco estimated that volume of data created (but not necessarily stored) by devices in 2021 was 847 zettabytes, and explained that data created by devices is two orders of magnitude higher than the data that is stored. In [11], Cisco estimated that about 500 zettabytes of data was transmitted on the Internet by IoT devices in 2018. This state of affairs implied that more bandwidth will be needed to facilitate acceptable data transfer times, which in turn will contribute to acceptable response times. In [12], the authors listed and explained the economic value and other benefits of having an acceptable rapid response time. These included substantial cost savings and sizable increases in productivity. In [12], the authors further noted that as the response time decreased, the time taken to complete a task also decreased dramatically while improving the quality of service.

Network bandwidth is the maximum carrying capacity of a computer network. If the network connection between two communication peers is considered to be analogous to a road connecting two points, and bandwidth as the maximum amount of cars that can travel from the source point to the destination point in a given time, then it is foreseeable that increasing the number of lanes on the road leading from the source to the destination will theoretically increase the bandwidth. This paper proposes a theoretical model to determine the minimum time it will take for a given set of data to be transmitted from the source to the destination if a set number of fungible paths are used, the number of fungible paths to use if a specific time for the data transmission of a given set of data is required, and the speedup achieved as more fungible paths are added to the computer network. In the rest of the paper, we present a review of the literature, and our methodology, results, conclusions and recommendations.

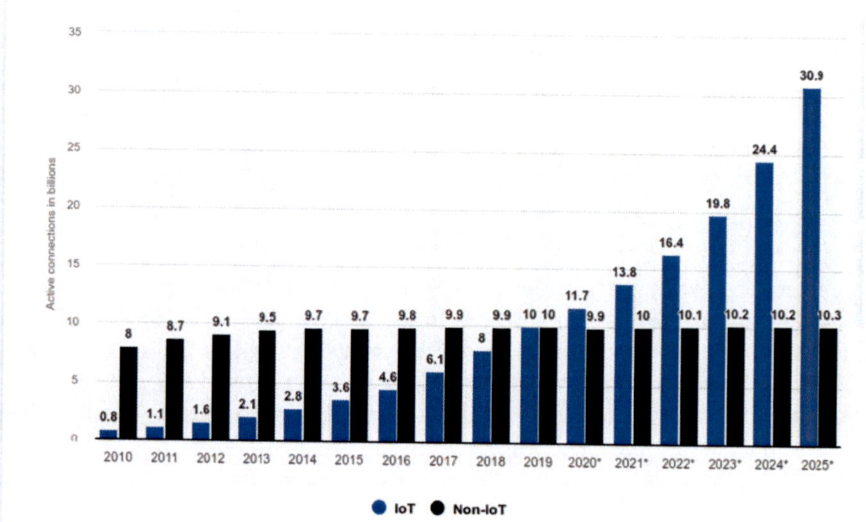

Fig. 1. IoT and non-IoT connections worldwide 2010–2025. Adopted from [9]

2 Related Work

In this section, we present some related literature [13–20]. Both [13, 14] defined bandwidth as the maximum amount of data that can be transmitted across a computer network. In [15], the author explained that just as how adding more lanes to a roadway will increase the number of cars that can enter that roadway per unit of time, then similarly adding more parallel transmission paths to a network will increase the throughput of the network. However, [15] also explained that just as how adding more lanes to the roadway will not decrease the total time required to traverse the road, adding more parallel transmission paths will not decrease the propagation delay of the network.

The increase in performance achieved by adding more resources to a computer system is termed as speedup. In the high performance and parallel computing fields, [16] introduced Amdahl's law as one of the pioneering methods used to compute speedup. In [17] and [18] respectively, the authors explained how Amdahl's law works and gave the formula used to compute speedup. In [17], the author stated that in Amdahl's law, the workload remains fixed as the number of processors is increased, and [18] listed the speedup equation given in Eq. (1) below,

$$Speedup = \frac{s+p}{s+\left(\frac{p}{N}\right)} = \frac{1}{s+\left(\frac{p}{N}\right)} \tag{1}$$

where N is the number of processors, s is the percent of time spent on the serial (or non-parallelizable) parts of the program load, p is the percent of time spent on parts of the program that can be done in parallel, and $(s + p)$ always sum to 1 for simplicity.

In [19] however, the author presented a critique of Amdahl's law, observing that the overall computed speedup was inhibited by the serial/non-parallelizable section of the program load. Instead, [19] stated that the Work-Span model for parallel computation produced more realistic speedup calculations, and presented a discussion of its use. In the Work-Span model, individual sub-tasks of the program workload are represented as vertices in a directed acyclic graph (DAG), an example of which is shown in Fig. 2 below. Work is computed as the total number of vertices in the graph and span is computed as the critical path from the initial task to the final task. For example, in the DAG in Fig. 2, the work is 8 and the span is 5. In [19], the author further explained that a sub-task cannot be run until its predecessor sub-tasks had been completed, for example, sub-task 2 must be completed before sub-task 6 can be run and sub-task 3 must be completed before sub-task 5 can be run and so on.

In [19], the author presented the relation in Eq. (2) as the sanitized formula to compute speedup using the Work-Span model. Using Eq. (2) to compute the speedup for the DAG in Fig. 2 results in $speedup \leq 8 \div 5 \leq 1.6$.

$$speedup \leq \frac{work}{span} \tag{2}$$

Notwithstanding what was stated above by [19, 20] spoke to the merits of using Amdahl's law and expounded how it could be useful, using a lawnmower metaphor to illustrate the concept of speedup using Amdahl's law. The author separated the time slots into serial section time, single path time, per-unit parallel setup time, overhead parallel setup time, potential parallel time, and total time for all sections.

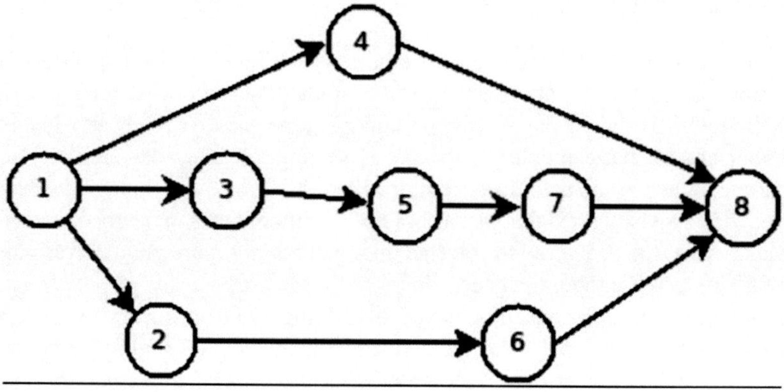

Fig. 2. Work-Span model example

3 Methodology

We employed a quantitative research design, which involved building and testing our proposed mathematical model to compute the ideal number of paths to use to achieve a desired transfer time, and the resulting speedup that would be achieved if that specific number of paths were utilized. We designed the DAGs in Fig. 3, Fig. 4, and Fig. 5 to represent respectively the use of a single fungible path, then two fungible paths, then three fungible paths to transfer a fixed amount of data from the sender to the receiver. In Fig. 3 we show a pair of WiFi radios used to generate each fungible path, but we also used pairs of Bluetooth radios and pairs of Ethernet Network Interface Cards to generate other logical fungible paths as well in our other experiments. The sender consisted of the nodes labeled 1, 2 and 3, while the receiver consisted of the nodes labeled 4, 5, and 6. We then used the pseudo-code given in Algorithm 1 to model the sub-tasks (represented as vertices) in Fig. 3, Fig. 4, and Fig. 5, and allowed the number of available fungible paths along with the data to be transmitted, to be passed in as input. Next we analyzed the algorithms to determine an equation to represent the time T, it would take to transfer a fixed amount of data C, when we used a specific number of fungible paths L, as portrayed in Fig. 3, Fig. 4 and Fig. 5, with $1 \leq L \leq 3$. We graphed the results and noted the characteristics of the resulting curve. We then extrapolated the data in our mathematical model to 10, 100 and finally to 2 million paths, graphed the data and compared the resulting curves with the previously generated curves. We solved our generated equation for T to obtain a new equation for L, which would allow us to predict how much paths L should be used in order to transfer a fixed amount of data C in a specific time frame T. In our experiments we first used a value of 4 MB for the size of the data. This value was the size of full plaintext of the King James version of the Bible obtained from [21].

Next we constructed an equation to determine the theoretical speedup S, as more fungible paths were added to our model portrayed in Fig. 3, Fig. 4 and Fig. 5, using fungible pairs of network radios or network interface cards, while we kept the span constant at 6. We computed the speedup S, both as a function of the number of fungible paths L and also as a function of the number of vertices and edges in our DAGs, based on the formula for enumerating the first Betti number in graph theory. We recorded the results and graphed them. We again computed the speedup S, but this time using the Work-Span model, recorded the results, graphed them and compared them to the results from our constructed equation for S.

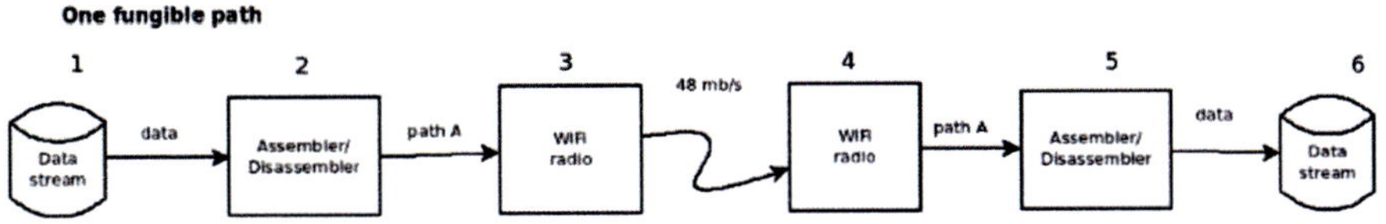

Fig. 3. Directed Acyclic Graph showing the use of one fungible path

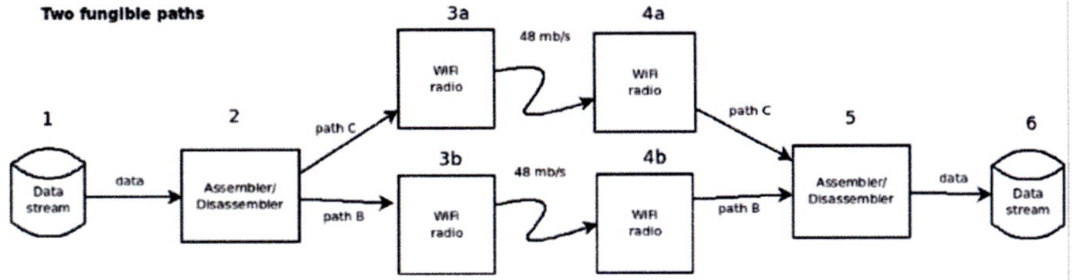

Fig. 4. Directed Acyclic Graph showing the use of two fungible paths

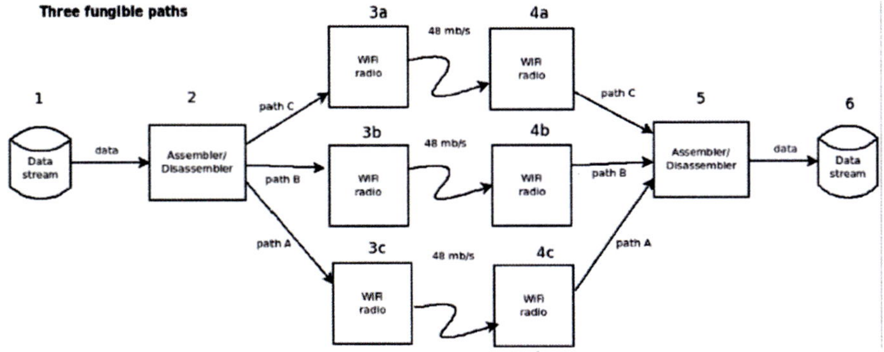

Fig. 5. Directed Acyclic Graph showing the use of three fungible paths

Finally, we borrowed the data from the lawnmower example in [10] which was inspired by Amdahl's law. We extrapolated and then graphed the data to determine the nature of the theoretical speedup as the number of paths were added from one to infinity, i.e. $1 \leq L \leq \infty$. In our setup, we fixed the serial section time at 20, the single path time at 40, and the per-unit parallel setup time at 1. We used these values to compute the overhead parallel setup time, potential parallel time and the total time for all sections, as we varied the number of fungible paths. We recorded and graphed the results.

```
Algorithm 1: Transmit data using fungible Paths
INPUT: inputDataStream, AVAILABLE_PATHS
OUTPUT:
BEGIN
        //Establish network connection for AVAILABLE_PATHS

        pathNum = 1  //initialize the counter to the first available fungible path
        READ inputDataStream  //get the data to be transmitted
        WHILE there is more data to be sent  //loop until all data has been transmitted
                SEND data along path pathNum //simultaneously send the data in parallel along the
                                             //next available fungible path
                INCREMENT pathNum            //go to the next available fungible path
                IF pathNum > AVAILABLE_PATHS THEN //restart from fungible path 1
                        pathNum = 1
                END IF
                READ inputDataStream //get next basket of data to be transmitted
        END WHILE
END
```

4 Results

The equation we derived for the time T, taken to transmit a fixed number of data C, using L number of fungible paths, each operating at a speed of S, is shown in Eq. (3).

$$T = \frac{C}{\sum_{i=1}^{L} S} \tag{3}$$

As C was held constant, we obtained a series of hyperbolic paraboloids as we varied the number of paths from 1 to 2,000,000 and experimented with different fungible path speeds (e.g. path speed $S = 25$, $S = 48$, $S = 100$, etc.). Figure 4 shows the graph we obtained for $C = 4$ MB, $S = 25$ MB/S, and $1 \leq L \leq 10$.

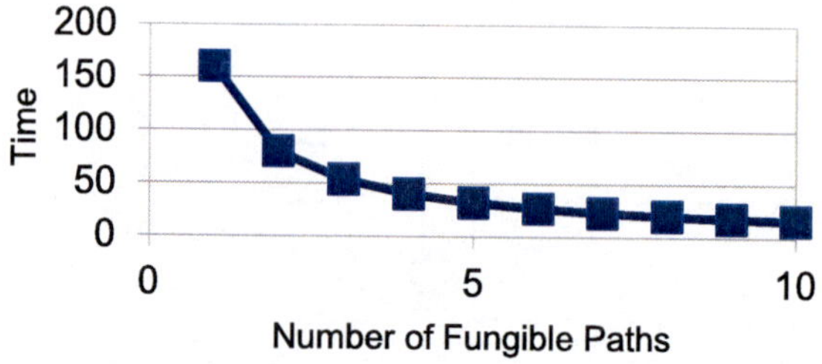

Fig. 6. Decrease in time as number of fungible paths increases

When we solved Eq. (3) to make L the subject of the equation, we get:

$$L = \frac{C}{T \times S} \tag{4}$$

Again, as we held C constant, we obtained a series of hyperbolic paraboloids as we varied the time T and experimented with different fungible path speeds (e.g. path speed

$S = 25$, $S = 48$, $S = 100$, etc.). Figure 5 shows the graph we obtained for $C = 4$ MB, $S = 25$ MB/S, and $20 \leq T \leq 200$. This reveals that if 80 s is acceptable to complete transmission of 4 MB then we only need 2 fungible paths operating at speeds of 25 MB/S each, but if we need to accomplish the transmission in 40 s or 20 s, then 4 paths or 8 paths at these speeds are required respectively. Of course, as the path speed increases then less number of paths will be needed to achieve the desired total transmission time.

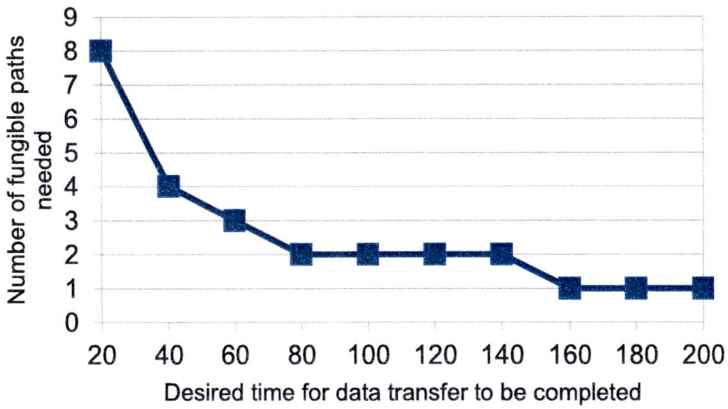

Fig. 7. Number of paths needed to complete data transmission in a specific time

To see what speedup s we would achieve if we used L fungible paths to ensure that our data transmission would be completed in time T, we obtain:

$$s = \frac{L+2}{3} \tag{5}$$

Equation (5) assumes the span is kept constant at 6 as the number of fungible paths are increased. When we sought to restate Eq. (5) in terms of the number of vertices and edges in the DAGs, we noticed that the number of fungible paths L was always equivalent to the first Betti number of the graph plus one, ceteris paribus. This allowed us to obtain Eq. (6) in which s represents the speedup achieved, v represents the number of vertices in the DAG and e represents the number of edges connecting those vertices.

$$s = \frac{1}{3}(e - v + 4) \tag{6}$$

Figure 6 depicts the linear relationship between the number of fungible paths used and the resulting theoretical speedup. For this graph, we increased the paths from 1 to 15 and the work from 6 to 34, while we kept the span constant at 6 and used Eq. (5) to compute the speedup. Alternatively, we obtained the exact relationship when we used Eq. (6) to compute the speedup, whereby we varied the vertices from 4 to 32, and the corresponding edges connecting them from 3 to 45 (Figs. 8 and 9).

Finally, we noted from graphing the extrapolated lawnmower data from [20] based on Amdahl's law, that initially speedups in total time taken to complete data transmission

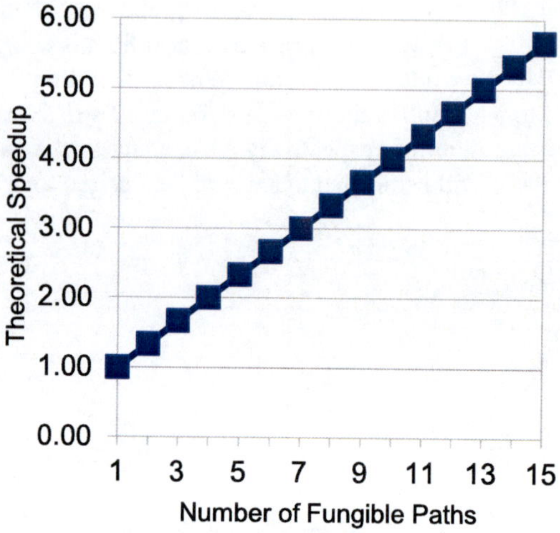

Fig. 8. Theoretical speedup as paths increase

were achieved as more and more fungible paths were added to the DAG. However, at some trigger point, the law of diminishing returns kicked in and adding any more fungible paths after that point was reached resulted in a falloff of speedup. This is portrayed in Fig. 7, where initially as we added more fungible paths to our network (1 to 5 paths) we saw a drop in the total transmission time from 60 s to just over 30 s. However, as we increased the number of paths from 5 onward we noticed that the total transmission time started to increase to the extent where using beyond 40 fungible paths produced a worse transmission time than using only a single path.

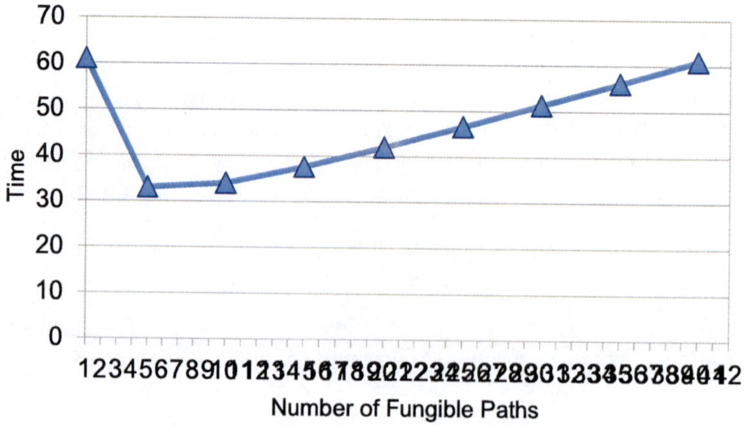

Fig. 9. Time taken to complete data transmission

5 Conclusion

In this paper, we traced the development and growth of the Internet and later the Internet of Things as massive computer networks. We also discussed how mathematical models played a key role in their development. We also discussed how more bandwidth would be required as the IoT continued to grow at a phenomenal rate, and also the positive economic value and improvements in quality of service when there was a short or acceptable response time.

We discussed speedup models for high performance computing such as Amdahl's law and the Work-Span model, and showed how these models could be applied to IoT networks to compute the speedup as more fungible paths were introduced to the network. We then presented our own model, which allowed one to estimate how many fungible paths were needed to fully transmit a fixed amount of data in a specific time. Our model also allowed the transmission time to be estimated for a specific number of fungible paths, and the resulting speedup to be computed. However, we also showed that after a certain cut-off point, the speedups achieved and hence time taken to complete full transmission started to exhibit the law of diminishing returns.

We recommend that IoT and other network application software and hardware could be designed using our model to achieve estimated transmission times and speedups by employing multiple fungible paths. These paths could be leveraged statically or dynamically. We recommend that our model could be used to generate metrics on network performance in a network involving multiple fungible paths.

We further recommend that more research be conducted in this area to augment the work we have presented here on time estimation and speedup with fungible paths.

References

1. Sterling, C.H., (ed.): Military Communications: From Ancient Times to the 21st Century. ABC-Clio, p. 399 (2008)
2. Haigh, T., Ceruzzi, P.E.: A New History of Modern Computing. MIT Press, pp. 87–89 (2021). 14 Sep 2021
3. Tanenbaum, A., Wetherall, D.: Computer Networks. Prentice Hall, 5th Edn (2010)
4. Gillies, J., Cailliau, R.: How the Web was Born: The Story of the World Wide Web. Oxford University Press, p. 25 (2000)
5. Hempstead, C., Worthington, W.: Encyclopedia of 20th-Century Technology. Routledge (2005)
6. "The "Only" Coke Machine on the Internet". Carnegie Mellon University (2023). https://www.cs.cmu.edu/~coke/history_long.txt. Accessed 16 July 2023
7. Raji, R.S.: Smart networks for control. IEEE Spectr. **31**(6), 49–55 (1994)
8. Ashton, K.: That 'Internet of Things' Thing. RFID Journal (2009). Accessed 22 June 2009
9. Vailshery, L.: IoT and non-IoT connections worldwide 2010–2025 6 Sep 2022. https://www.statista.com/statistics/1101442/iot-number-of-connected-devices-worldwide/. Accessed 17 July 2023
10. Global Cloud Index Projects Strong Multicloud Traffic Growth. Cisco Newsroom, 25 Feb 2018. https://newsroom.cisco.com/c/r/newsroom/en/us/a/y2018/m02/global-cloud-index-projects-cloud-traffic-to-represent-95-percent-of-total-data-center-traffic-by-2021.html. Accessed 22 July 2023

11. Networking, C.V.: Cisco global cloud index: Forecast and methodology, 2016–2021. White paper. Cisco Public, San Jose, 1 (2016)
12. Doherty, W.J., Thadhani, A.J.: The economic value of rapid response time. IBM Report, November 1982
13. Bandwidth Noun - Definition, Pictures, Pronunciation and Usage Notes. Oxford Advanced Learner's Dictionary at OxfordLearnersDictionaries.com. Oxford University Press. www.oxfordlearnersdictionaries.com/definition/english/bandwidth. Accessed 23 July 2023
14. Bandwidth. Cambridge Dictionary. Cambridge University Press & Assessment 2023. https://dictionary.cambridge.org/dictionary/english/bandwidth. Accessed 19 July 2023
15. Comer, D.: Computer Networks and Internets, 6th Edn. Pearson (2014)
16. Amdahl, G.M.: Validity of the single-processor approach to achieving large scale computing capabilities, AFIPS conference proceedings, vol. 30 (Atlantic City, N.J., Apr. 18-20), AFIPS Press, Reston, Va., 1967, pp. 483-485 (1967)
17. Rodgers, D.: Improvements in multiprocessor system design. ACM SIGARCH Computer Architecture News, vol. 13, issue 3, June 1985, pp 225–231. New York, NY, USA: ACM (1985)
18. Gustafson, J.L.: Reevaluating Amdahl's law. Commun. ACM. **31**(5), 532–533 (1988)
19. McCool, M.D., Robison, A.D., Reinders, J.: "Performance theory". Structured Parallel Programming: Patterns for Efficient Computation. Elsevier, pp. 61–62 (2012). ISBN 978–0–12–415993–8
20. Eadline, D.: The Lawnmower Law. Linux Magazine. CA, USA: QuinStreet, Inc. (2008). http://www.linux-mag.com/id/6020/. Accessed 13 Nov 2019
21. Powell, M.: The Canterbury Corpus (2001). https://corpus.canterbury.ac.nz/descriptions/. Acceseed 20 Mar 2020

An Effective Technique for Collecting Multi-languages Hotel Reviews: A Case Study of 5 Stars Hotels in Bali Island

Dahlan Nariman(✉)

College of Sustainability and Tourism, Ritsumeikan Asia Pacific University, Beppu, Japan
dahlan@apu.ac.jp

Abstract. In recent years, online platforms have become more popular due to the spread of the Internet and its applications. The Internet provides a variety of human-created facts and data sources. But it consists of a vast array of disjointed structured and unstructured data that are difficult to collect by physical means and problematic to use in mechanical processes. Recently, modules and applications have been developed with various systems to collect data and transform it into organized information. However, those systems have complex menus and difficult for individual users to operate. Collecting multilingual hotels reviews data requires storing each language in a separate database to facilitate data analysis processes base on each natural language analytical method. Therefore, we consider that we need a simple tool with single button click system to collect multilingual reviews and store it in separate data base. In this research project We are developing a tool to collect hotels or attractions reviews data called MULARS (Multi-Languages Reviews Scrapers) for collecting large-scale of multi-language reviews. This paper is started by describing the web scraping process using different kinds of web scraping techniques. Then, the proposed system will be explained, and the results of data collection performances are presented. Finally, the paper discusses the advantages and future possibilities of the system applications on the different areas to exploit these data, such as development of business intelligence data, and development of new applications for data analysis.

1 Introduction

In recent years, online platforms have become more popular due to the spread of the Internet and its applications. In the hospitality industry, Information technology has changed the face of the hospitality industry. The Internet also provides a way for travelers to express their opinions and recommendations. A good example is online reviews. This has become an important consideration for hotel bookings decision. These reviews are the decision-making reference for many people when evaluating places to visit, such as which hotel or restaurant to book. Following this trend, online reviews have taken a giant leap forward, and their importance has grown to peak in the last few years. The hotel industry should take this trend very seriously, they need to focus on managing these reviews in an effective and efficient manner. The tourism industry also has promoted its products and services based on the reviews that people frequently write about their

L. Barolli (Ed.): BWCCA 2023, LNDECT 186, pp. 209–217, 2024.
https://doi.org/10.1007/978-3-031-46784-4_19

trips. For the hotel industry and consumer behavior researcher, the internet presents many interesting opportunities for Understanding consumer choice. Any information posted on travel website represents a potential data source to be analyzed. For instance, a hospitality researcher may collect a list of reviews from TripAdvisor and perform text mining.

Although these possibilities are attractive, however, the fact remains that manually collecting a large amount of Online data is inefficient and practically impossible. A typical hospitality research study may require data from multiple hotels across different markets over a prolonged period of time; collecting this data manually can be incredibly time consuming. Therefore, hospitality researchers need to automate the process of gathering and storing the Information presented on travel websites. This is where web Scraping comes into play. Web scraping is the process of creating a computer program to download, parse, and organize data from the web in an automated manner [1]. Studies show that the Internet in contrast to offline, became the main channel for travel reservations [2]. Therefore, this is important in studying the behaviors of online customers cannot be overestimated.

However, in tourism sector in local region such as in Japan, the utilization of big data such us online platform reviews still low, compared to others sector. It still has not yet taken root and relies on the experience and intuition of local people [3]. For example, according to a survey on inbound tourism of Japan, less than 40% of local governments in Japan are promoting the use of ICT and big data utilization. In addition, only less than 10% of the municipalities have started or are planning to start analyzing various logs such as access logs and reviews. Regarding challenges in big data utilization, more than 60% of the municipalities answered that "the specific situation of utilization is not clear" and 50% answered that "the effects and benefits are not clear." In addition, more than 40% of them answered "we lack finances and human resources" [4]. Therefore, it is important to create models of big data utilization, and to show not only the analysis but also the effects and benefits. In addition, as more data is utilized, the utilization and its methods will become more common, the necessary costs will decrease, and the problems of financial and human resource shortages will be solved.

This research provides a tool called MULARS (Multi-Languages Reviews Scrapers) for collecting large-scale of multi-language reviews. The main features of the developed tool are:

i) Ability to aggregate data obtained from TripAdvisor reviews and ability to store multi-language reviews to separate database
ii) Possibility to collect all reviews with minimum operation which is input an URL and one-click
iii) Possibility to analyze both hotel and reviews comments.
iv) The ability to detect the meaning (positive, negative, or neutral) in which comments cycle

2 Aims and Objectives

2.1 Aims

The principal point of the paper is to contemplate convenient and an effective technique to collect multi-language reviews data, cleaning, changing, storing on separate database, comparing, analyzing and demonstrating information to find helpful data for decision making. The motivation is to collect and extract valuable data from specific hotel, attraction and restaurant reviews. In addition, providing a simple user interface is necessary for everyone, especially necessary for those who are not good at computer operation. The possibility to collect multilingual reviews with minimal computer interaction of just entering a URL target and a single click is much needed. The motivation behind multi-languages reviews data collection is to extract valuable information and possibility to analyses the reviews depend on the languages.

2.2 Objectives

The general objective of the study is to propose a reliable, convenient and accurate reviews collection and analysis system. The study has the following specific objectives:

i) The point of the study is to optimizing the ability to aggregate multi-language reviews data obtained from TripAdvisor reviews.
ii) To develop a tool for personal user to achieve faster data collection.
iii) To collect and store multi-language reviews to separate database
iv) To Provide a simplest user interface to operate the system of just entering a URL target and single click.
v) To provide an analyze tool of both hotel and reviews comments depend on the basic languages of the reviews.
vi) To analyze the comments, ratings or anything else with innumerable options.

3 Existing Methods and Problem Statement

3.1 Using Online Survey

Researchers have utilized alternative approaches to collect consumer behavior data that mimics online behaviors. Previous studies directly captured data in a simulation lab or survey setting where research participants acted as if they were actually completing online activities [5]. Conducting large-scale research studies is possible, but many barriers make the process difficult. On the other hand, the scale of web scraped data is usually large in nature. Collecting publicly available online data is cheaper than conducting surveys, even considering the fees required for commercial services. The cost and effort required to conduct a survey increases linearly with each additional sample, whereas web scraping requires only a one-time cost. However, Hotel prices changed daily, and online reviews are posted even more frequently. In such a dynamic market, one sampling may not be enough. Conducting multiple surveys is expensive and require a lot of effort, but collecting publicly available online data allows researchers to extract as much data as they need in real time.

3.2 Manual Data Collection

A large amount of data is available on all the websites. Extraction of data from the website is difficult and manually extracting it is possible, but time-consuming and so much effort is required [6]. The manual web data extraction process has two major problems. Firstly, it can't measure costs efficiently and can escalate it very quickly. The data collection costs increase as more data is collected from each website. In order to conduct a manual extraction, businesses need to hire large number of staffs, this increases the cost of labor significantly. Secondly, each manual extraction is known to be error prone. Further, if any business process is very complex then cleaning up the data can get expensive and time consuming.

3.3 Using API

There are several alternative approaches that hospitality researchers may consider when trying to automate the collection of online data. One promising option is using an Application Programming Interface (API) provided by the firm hosting the data. However, APIs are often difficult to access, accessible only for a limited duration, or require upfront fees even when researchers are able to access an API. Furthermore, APIs generally will not expose all the required variables [7].

3.4 Third Party Scrapers Tool

Another data collection method often chosen by researchers is outsourcing the scraping task to a third-party commercial firm [8]. With the help of scraper tools, the extraction process is quick and easy, and we also monitor our data and keep it saved. Web scraping has so many purposes in today's world. This is probably the simplest option, but it has potential pitfalls. May arise due to lack of discretion in data collection process. For example, researchers often find out too late. They forgot to ask the company to scrape specific variables. As a result, they are forced to either spend additional time, effort, and money or complete their research without having access to all the desired variables or behaviors. Additionally, there are many motives to use scarper for business selling and pricing inquiries of their opponents, weather data monitoring, and research, contact scraping, and monitoring all the changes and analyzed data easily [9]. If we don't use a proper web scraping tool, we may face some challenges in completing our task. Website may crash or go down sometimes because of maintenance. These types of problems can be occurred during web scraping sessions. Renita Crystal Pereira et al., gave a summary of online scraping strategies and tools, which confront several challenges because data extraction isn't simple [10]. Because there is a great volume of data to handle and maintain, these tactics ensure that the data collected is correct [11], consistent, and has superior integrity. Although there are a few issues with functional approaches, such as the increased volume of web scraping, they can do serious harm to websites [12]. The web scraper's measurement level will differ from the original source file's measurement units, making it impossible to comprehend the data [13].

4 Proposed System

The MULARS (Multi Languages Reviews Scrapers) is based on a list of TripAdvisors website addresses (URLs) and proceeds as follows:

1. The first webpage usually a website's main page) is requested using the first address in the given URL list.
2. The first webpage which is used to store the website's extracted content, meta-data (e.g., timestamps, total number of reviews, top 3 reviews languages, number of reviews using each language, average rates, etc.).

   The main page is processed as follows:

   i) Content from the main page is extracted and stored in the collector item.

   ii) Base on the collected data from the first page, URLs which refer to next pages of the same website and website domain for different language reviews are created and stored in the collector item's URL stack.
3. The algorithm continues to request subpages of the website using URLs from the URL stack and collected reviews of the next pages continuously till last page. This process also performed for website domain for top 3 different language reviews.

   i) Content of reviews and timestamp are collected from each website page and stored in the collector item.

   ii) The next URL in the URL stack is processed.
4. The algorithm stops processing a website once all reviews page has been processed or as soon as a pre-calculated number of webpages per website have been processed.
5. The collected content of reviews is processed and exported into separate output excel files depend on the languages of the reviews.
6. The next website is processed by requesting the next URL from the URL list. The described process continues until all hotel website addresses from the list provided by the user have been processed.

The system has been developed in python by utilizing the Beautiful Soup and Request module for HTML analysis and data collection from the website. The Pandas module is used to organize and manage dataframe. The Openpyxl module is used to enter the collected data into an excel file and arrange the data to the specified format.

5 Demonstration of Experiment Result

Program executed on windows Surface notebook PC with 8GB RAM memory, Intel(R) Core(TM) i5-1035G7 CPU @ 1.20GHz, 1498 MHz, 4 quad coreand 8 logical processor. We have tried to implement the tools we have developed to download reviews of several 5 star hotels in Bali island. Figure 1 shows a shell prompt window that displays the results of the data that has been collected and the total time needed to collect the data. In addition, Fig. 2 shows the log file in csv format of the data collection process. We can get information the name of the hotel, the number of reviews, the rating of each category from excellent to terriable, the number of reviews that we have managed to collect and the number of hotel responses that have been posted and managed to collect. Figure 3 shows a log file containing the hotel name, the total number of reviews and the highest number of reviews in the top 3 languages. Figures 4 and 5 are examples of collected English and Japanese reviews data formatted in an excel files. The collected data is containing the reviewer's name, date of stay, review title of the reviews and the contents of the text reviews.

From the case of the experiment in collecting some 5 starts hotels reviews in Bali island, the MULLARS is very effective because it is able to collect mostly all of the reviews. When the effectivenes of data collection completion is defined by number of collected reviews devided by the total number of reviews should be collected. For example, in collecting english reviews of the Apurva Kempinski Bali reviews, the Mulia, the St. Regist Bali Resort, the Mulia Resort, the Amarterra Villas Bali, Samaba Bali suites, and The Ritz-Carlton Bali, the effectifeness of the data collection completition are 1,013/1,013 = 100%, 2,190/2200 = 99.5%, 1,865/1,885 = 98.9%, 3,057/3,067 = 99.7%, 1,016/1,016 = 100%, 1,090/1,090 = 100%, and 3,203/3,223 = 99.4%, respectively.

```
collected 405 reviews / 513 reviews
collected 420 reviews / 513 reviews
collected 435 reviews / 513 reviews
collected 450 reviews / 513 reviews
80.39073371887207 seconds
```

Fig. 1. Example of the system execution result

1	Name	Number of Reviews	Average Rating	Excellent	Very good	Average	Poor	Terrible	Collected Reviews	Time_dataCollection	Date	Time
2	The Apurva Kempinski Bali	1201	4.5	912	123	66	39	61	1013	202.0005012	2023-08-08	14:06:47
3	The Mulia	2843	5	2561	197	41	14	30	2190	504.8446298	2023-08-08	14:15:17
4	The St. Regis Bali Resort	2527	5	2212	199	67	21	28	1865	381.4307292	2023-08-08	14:21:40
5	Mulia Resort	4037	4.5	3185	444	220	91	97	3057	544.8570886	2023-08-08	14:30:47
6	Amarterra Villas Bali Nusa D	1359	5	1148	155	38	12	6	1016	201.2550652	2023-08-08	14:34:10
7	Samabe Bali Suites & Villas	1343	5	1155	126	33	14	15	1090	209.1278512	2023-08-08	14:37:41
8	The Ritz-Carlton, Bali	3842	4.5	3169	401	134	74	64	3203	622.5247037	2023-08-08	14:48:06
9	The Westin Resort Nusa Du	4823	4.5	3442	991	240	80	70	3985	767.3496165	2023-08-08	15:00:55
10	Sofitel Bali Nusa Dua Beach	7862	4.5	5572	1420	493	214	163	6104	1134.588831	2023-08-08	15:19:51

Fig. 2. Example of collected hotels ratings and the number of reviews

1	Name	language	Number	language(1)	Number	language(2)	Number	language(3)	Number
2	The Apurva Kempinski Bali	All languages	1201	English	1013	Indonesian	95	Japanese	17
3	The Mulia	All languages	2843	English	2200	Indonesian	232	Chinese (Sim.)	184
4	The St. Regis Bali Resort	All languages	2527	English	1885	Chinese (Sim.)	326	Chinese (Trad.)	302
5	Mulia Resort	All languages	4037	English	3067	Indonesian	447	Chinese (Sim.)	216
6	Amarterra Villas Bali Nusa [	All languages	1359	English	1016	Chinese (Sim.)	141	Chinese (Trad.)	126
7	Samabe Bali Suites & Villas	All languages	1343	English	1090	Chinese (Sim.)	157	Chinese (Trad.)	133
8	The Ritz-Carlton, Bali	All languages	3842	English	3223	Chinese (Sim.)	261	Chinese (Trad.)	246
9	The Westin Resort Nusa Du	All languages	4823	English	4025	Chinese (Sim.)	248	Indonesian	223
10	Sofitel Bali Nusa Dua Beach	All languages	7862	English	6134	Indonesian	765	French	375

Fig. 3. Example of collected data of the hotels top 3 languages of reviews and its number of reviews

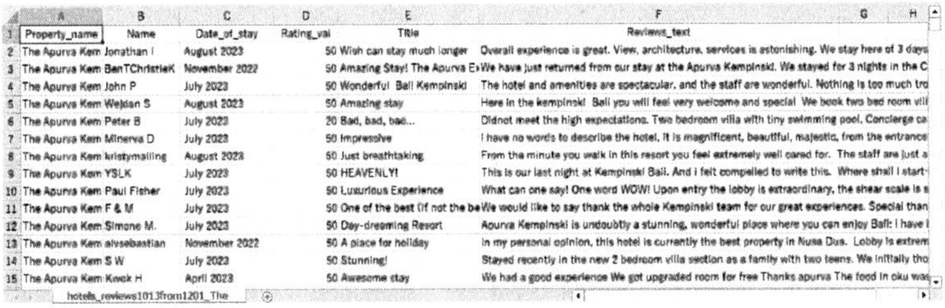

Fig. 4. Example of collected English reviews on excel formatted file.

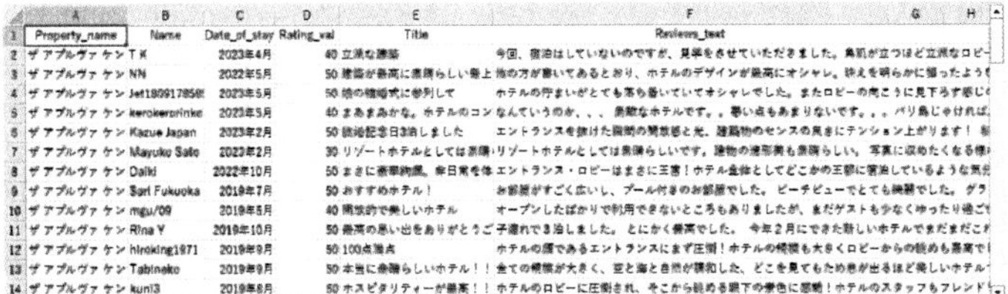

Fig. 5. Example of collected Japanese reviews on excel formatted file.

6 Advantages and Future Possibilities

Firstly, the best thing about using MULARS is effectiveness and the speed of reviews data collection with a simple computer operation. MULARS allows user to quickly collect a large number of hotel reviews without having to operate with complex menus. Just by entering the main URL and clicking a single button it will collect almost all thousands of reviews within a few minutes. From the average test that we did in the demonstration and experiment, it only takes about 0.17 s to collect each review. Secondly, the next advantage is a very simple and specific system with a very low cost. So, there is no need to invest in building complex systems or hiring additional staff for data collection tasks, because everyone can simply use the MULLARS.

Thirdly, MULARS is also able to collect ratings in real time, this allows MULARS to be used to monitor and perform sentiment analysis of thousands of customers reviews. As we know the customers publish their reviews and service experiences on online review sites every day. This vast amount of data is open to the public and can be easily extracted by the MULARS and analyses them to obtain information about the business, competitors, possible opportunities and the latest trends. In the future, if the data collection is integrated with NLP (Natural Language Processing) it can also reveal how customers react to their products and services, as well as their feedback on campaigns and products.

Fourthly, Flexibility and systematic approach is next of the advantage. The system was implemented in Python and it is highly modifiable, compatible with other python module and other scripts. The system is possible to reshape it to fit many different tasks by making only tiny changes. The system is possible to empowers users to customize the data collection and analysis process, and take full advantage of its features to fulfill all their data collection ambitions.

7 Conclusion

Our work provides a simple method of collecting online hotels reviews using the Python language and its modules. As our goal is to make hospitality researchers more comfortable collecting data online. to contemplate convenient and an effective technique to collect multi-language reviews data, cleaning, changing, storing on separate database, comparing, analyzing and demonstrating information to find helpful data for decision making. We focus on how scraping processing functions on the main travel website such as TripAdvisor.

MULARS is able to present data collection quickly and effectively with very simple interactions. It is very easy for anyone to operate even if they don't understand python programming. In addition, the effectifeness of the data collection completition is very high, because the MULARS is able to collect mostly all of the reviews data. The experient result shown us that it is able to collect more than 99% of the data. In the some case, there were able to collect 100% of the reviews data. The system has ability to collect quickly and in real time, it is allowing to analyze reviews in a more up to date manner.

Because it is implemented in Python, the system will make it easier to integrate with other modules for the future development of some required features. This system requires improvements in the future, such as creating a user interface that makes it more possible for this system to run on various device platforms. The system still uses a single threat to collect data. In the future work, it is required to use multi-thread to improve speed in downloading huge number a hotel's reviews data or collecting thousands of specific hotels reviews effectively. Finally, the future work also requires integration with the required function layers such as, smart database layer for multi-languages, the data analysis layer, visualization layer, etc.

References

1. Broucke, S., Baesens, B.: Practical web scraping for data science. Best Practices and Examples with Python. Apress, New York (2018). https://doi.org/10.1007/978-1-4842-3582-9.pdf
2. Igor, H.C., Jadranka, G., Saša, K.: ETourism: a comparison of online and offline bookings and the importance of hotel attributes, journal of information and organizational. Science **34**, 41–54 (2010)
3. Fuji, T., Fujiu, M.: Data utilization in tourism industry. Hokuriku Econ. Res. J. **468**, 10–19 Toyama (2019)
4. Satoshi, W.: Efforts toward inbound tourism using ICT by local governments and their effects. InfoCom Research Inc., Tokyo (2018)
5. Shin, H., Perdue, R.R., Pandelaere, M.: Managing customer reviews for value co-creation: An empowerment theory perspective. J. Travel Res. **59**, 792–810 (2019)
6. Landers, R., Brusso, R., Cavanaugh, K., Collmus, A.: A primer on theory-driven web scraping: Automatic extraction of big data from the internet for use in psychological research. Psychol. Methods **21**(4), 475–492 (2016)
7. Andreas, J., Tadas, R., Alberto, S., Giancarlo, S.: Towards understanding of structural attributes of web APIs using metrics based on API call responses, IFIP Advances in Information and Communication Technology, vol. 427, pp. 83-92 (2014). https://doi.org/10.1007/978-3-642-55128-4_11

8. Wu, L., Mattila, A.S., Wang, C.Y., Hanks, L.: The impact of power on service customers' willingness to post online reviews. J. Serv. Res. **19**(2), 224–238 (2015)
9. Gallagher, E.: Scraping Websites for Law Enforcement, School of Computing, Engineering & Systems, Computer Science (2018)
10. Vargiu, E., Urru, M.: Exploiting web scraping in a collaborative filtering-based approach to web advertising. Artif. Intell. Res. **2**(1), 44–54 (2013)
11. Sundas, S., Badotra, Y., Alotaibi, S.A., Khalaf, O.: Modified bat algorithm for optimal VM s in cloud computing. Comput. Mater. Continua **72**(2), 2877–2894 (2022). https://doi.org/10.32604/cmc.2022.025658
12. Zhao, B.: Web scraping. Encyclopedia of big data, pp. 1–3 (2017)
13. Sirisuriya, S., et al.: A comparative study on web scraping (2015)

Applying High-Quality Test Case Management Mechanisms to Improve Regression Testing Speed and Quality

Sen-Tarng Lai[1](✉) and Fang-Yie Leu[2]

[1] Department of Information Technology and Management, Shih Chien University, Taipei 10462, Taiwan
stlai@mail.usc.edu.tw

[2] Department of Computer Science, Tunghai University, Taichung 40704, Taiwan
leufy@thu.edu.tw

Abstract. Regression testing takes an important task in DevOps, agile software development, and software maintenance workflow. To ensure requirements quality and speed up product delivery and deployment, regression testing must overcome four challenges: identifying affected test cases correctly, reducing duplicate test cases, finding defects as early as possible, and testing with automated tools. Test case management is an important tool to assist regression testing to overcome the challenges. Applying high-quality test case management mechanisms (TCMM) is a critical way to speed regression testing tasks and assure quality. In this paper, the quality factors and characteristics of TCMM are discussed and analyzed. Quantifying the quality characteristics can identify and improve the defects and insufficiencies of TCMM. Based on quantifying characteristics, the TCMM measurement model is proposed to identify the defects and insufficiencies of TCMM, and then effectively improve the quality of TCMM. Using high-quality TCMM for regression testing makes DevOps, agile software development, and software maintenance workflow to speed up product deployment and delivery, and the required quality.

1 Introduction

Information system is an essential equipment in all walks of life to improve customer convenience and enhance market competitiveness. However, with the continuous and rapid evolution of the environment, the system in operation or development must quickly adjust and extend the existing functions according to the changing environment, to keep pace with the times and meet the requirements of customers. In the process of adjustment and extension, regression testing (RT) is a key process that affects requirements quality and product delivery speed. Several software development methodologies and processes (eg. Agile software development [1–3], DevOps [4–6], CI/CD process [6–8]) have been proposed to overcome the threat of changing environments. These development methods must once again confirm requirements can meet the specifications using RT, and they must also have the advantages of rapid delivery and deployment of new versions.

L. Barolli (Ed.): BWCCA 2023, LNDECT 186, pp. 218–227, 2024.
https://doi.org/10.1007/978-3-031-46784-4_20

The goal of RT is to ensure new version requirements quality and speed up deployment and delivery to reduce the inconvenience of change management and improve the customer's competitive advantages. RT uses the existing test cases (TC) of software repository to retest the affected software items to ensure that existing functional and non-functional requirements are not affected by change management. To confirm the requirements quality after software changes, RT must get complete and correct TC from the software repository for retesting. To speed up the delivery and deployment of changing version software, RT should reduce redundant testing, arrange testing priorities to find bugs and defects as early as possible, and adopt automated testing tools. To complete the goal, RT must overcome four challenges:

- Identify and get useful test cases from the software repository correctly to assure the new version software quality.
- Combine test case selection approaches [9, 10] to reduce redundant testing.
- Combine prioritization algorithms [11] to find the bugs and defects as early as possible.
- Evaluate and support testing tools to enhance the regression testing automation.

In the software change management process, RT is an important and necessary workflow. For the new version, customers not only pay attention to the specified requirements of the system but also pay more attention to the time of product delivery and deployment. Rapid product delivery and deployment can enhance customer trust and user competitive advantage. RT needs to make extensive use of tested TC and combine related software tools to speed up RT workflow and ensure product quality. For this, the paper plans the Test Case Management Mechanism (TCMM) for improving the speed and quality of RT. Quantified quality characteristics can identify and improve TCMM defects and deficiencies. Based on the quantified characteristics, a TCMM measurement model is proposed to identify the defects and deficiencies of TCMM, and then improve the quality of TCMM. In software maintenance and CI/CD workflows, applying high-quality TCMMs for RT can speed up product deployment and delivery with the required quality. In Sect. 2, the importance of TC and TC management for RT is described. Section 3 discusses the quality characteristics and influencing factors of TCMM. In Sect. 4, the TCMM measurement model is proposed and TCMM quality improved manners are described. Section 5 evaluates the advantages of applying high-quality TCMM for RT. In Sect. 6, the advantages of applying high-quality TCMM for RT are reiterated and the paper is concluded.

2 Regression Testing and Test Case Management

2.1 Importance of Regression Testing

RT is retesting previously working software after a change to ensure that unchanged software as it did before the change. According to IEEE, RT is the selective retesting of a system or App to verify that modifications have not caused unintended effects and that the system or App still complies with its specified requirements [12]. RT is defined as "the process of retesting the modified parts of the software and ensuring that no new errors have been introduced into previously tested code" [13].

RT needs to use numerous tested TC and test suits to ensure the original functions that not affected by requirements changes or function expansion in new versions. Information system needs to concern users' trust and users' inconvenience, RT of the change management process needs to confirm the original quality and to speed up product delivery and deployment. For this, RT Sholud has two main objectives:

(1) Find errors or defects introduced during the software change management to ensure that the new version system still complies with its specified requirements.
(2) Speed up the delivery and deployment of new versions, and improve the competitive advantage of enterprises and users.

Four challenges are derived from two main goals of RT. To complete two goals, regression testing must overtake the following challenges:

(1) Identify the affected software items correctly and completely, and acquire affected TC to assist in RT.
(2) Combine the TC selection approaches to select the most appropriate and non-repetitive test items.
(3) Combine the practical TC prioritization algorithms to find errors or defects that do not meet the software requirements as early as possible.
(4) Support testing tools with automated testing to speed up RT and ensure requirements quality.

2.2 Importance of Test Case Management for Regression Testing

The TC is an important asset for the software development and maintenance process. [1] Each TC has a specific task and objective to detect defects or bugs in the development and maintenance product. The TC must be well preserved until the software system is retired. RT needs to use many tested TC to ensure product quality. Effective use and management of TC in the repository are key to overcoming the four challenges of RT. And, TC in the repository must be timely combined with development tools, RT techniques, and automatic testing tools to ensure new version requirements quality and speed up the product delivery and deployment. Therefore, a Test case Management Mechanism (TCMM) that is based on the software repository, software development tools, RT techniques, and testing tools needs to be planned and designed to improve the RT speed and quality (shown in Fig. 1). Based on the software change management, TCMM applies software tools to identify correct and complete TC to assure new version requirements quality. Based on the software repository, TCMM should combine TC selection approaches and TC prioritization algorithms to improve RT efficiency. Using the tested test cases/suits, TCMM needs to support testing tools to enhance RT automation.

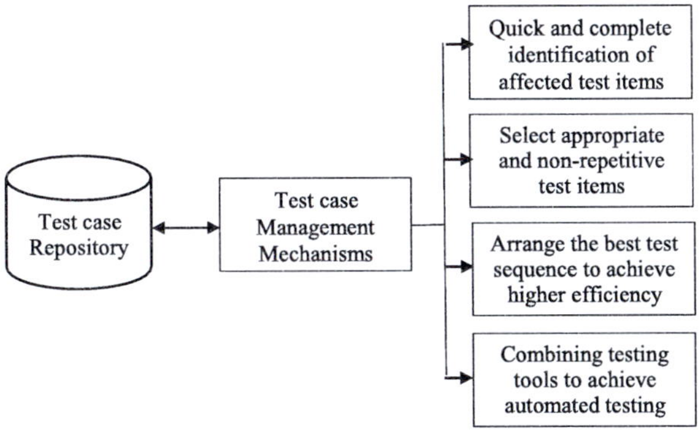

Fig. 1. Relationships among TCMM, TC repository and 4 challenges

3 Quality Characteristics and Influencing Factors of TCMM

To improve the speed and quality of RT, existing TC should have high-quality management mechanisms.

3.1 Quality Characteristics of TCMM

TC are necessary assets for software testing and quality assurance. In addition, most test cases in RT need to use existing or tested test cases of software repository. Therefore, how to effectively manage and control test cases is a critical issue in improving RT workflow speed and quality. TCMM is a TC management and control environment that needs to combine several development tools, RT techniques, and testing tools to overcome four challenges and accomplish the goal of RT [14]. Therefore, TCMM should have four high-quality quality characteristics:

(1) Based on change management, in RT workflows, TCMM needs to correctly and completely identify affected items and TC to ensure the requirements quality.
(2) Combining suitable TC selection approaches [9, 10] to avoid or reduce duplicate test cases to speed up RT and new version delivery and deployment.
(3) Combining with practical TC prioritization algorithms [11] to find errors or defects as early as possible to reduce RT operation time.
(4) TCMM has to support automatic testing tools to improve RT efficiency.

3.2 Quality Influencing Factors of TCMM

Each quality characteristic of TCMM may be influenced by several quality factors. In order to keep the quality of TCMM, quality influencing factors of TCMM characteristics are discussed as follows: (shown in Table 1).

(1) Identify affected software items and TC: In order to assure correctness and completeness of the function and non-function requirements, affected items and test cases must be identified in the change management process. Development tools such as Configuration Management (CM), Cross-Reference tables, version control system,

and software repository can help identify affected software items and TC in software change management.

(2) Combine TC selection approaches: Avoiding or reducing redundant TC can improve the RT efficiency. In order to avoid or reduce redundant TC, the suitable selection approaches for RT should be analyzed. In addition, TCMM should have the ability to evaluate and integrate selection approaches.
(3) CombineTC prioritization algorithms: In order to find the defects or errors as early as possible in RT workflow, the practical TC prioritization algorithms for RT should be introduced. TCMM should have the ability to analyze, evaluate, and integrate prioritization algorithms.
(4) Support automatic testing tools: In order to speed up RT workflow, RT needs to effectively use automatic testing tools. According to the interface of testing tools, TCMM should assist with Test Data Input Processing (TDIP), Test Results Processing (TRP), and Testing Tools Combination (TTC) to support various automatic RT tools.

The workflow of regression testing combines four quality characteristics of TCMM, and software repository shown in Fig. 2.

Table 1. Quality characteristics and quality influencing factors of TCMM

Quality Characteristics of TCMM	Influencing factors of quality characteristics
Identify affected software items and test cases	• Configuration Management (CM) • Cross-Reference Table (CRT) • Version Control System (VCS) • Software Repository (SR)
Combine TC selection approaches	• Analysis TC selection approaches • Evaluation TC selection approaches • Integration TC selection approaches
Combine TC prioritization algorithms	• Analysis TC prioritization algorithms • Evaluation TC prioritization algorithms • Integration TC prioritization algorithms
Support automatic t esting tools	• Test Data Input Processing (TIDP) factors • Test Results Processing (TRP) factors • Testing ToolsCombination (TTC) factors

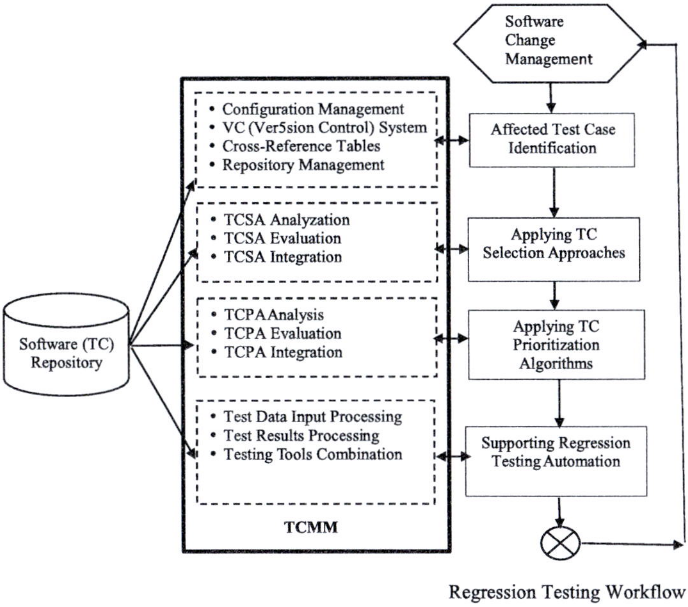

Fig. 2. Workflow of regression testing combines TCMM

4 TCMM Measurement Model and Quality Improvement

4.1 TCMM Measurement Model

In order to identify the defects of TCMM and improve its quality, the affect factors of TCMM characteristics need to be defined and quantified. Based on the linear combination model [15], this section proposes the TCMM measurement model which combines TC identification, TC selection approaches, TC prioritization algorithms, automatic testing tools four quality measurements. Inspect and review quality affect factors, software testing professionals and experienced maintainers can draw up quantitative values of basic quality items. A quantified value approaching 1 represents good quality, and a quantified value approaching 0 represents bad quality. In the model, the senior software engineers assign the weighted value between 0 and 1. A weighted value close to 1 indicates that the quality item is important for quality measurement. The TCMM measurement model is described as follows:

(1) TC Identification Measurement (TCIM): In order to identify affected items and TC from the software repository correctly and completely, Configuration Management (CM), Version Control (VC) system, Cross-Reference Tables (CRT), and Repository Management (RM)should have high quality. For this, TC Identification Measurement (TCIM) combines CM, VC, CRT, and RM four tool quality factors. The combination formula is shown as Eq. (1):

TCIM: TC Identification Measurement.

$$\begin{array}{ll} CM: CM\ Quality\ Factor & W_1: Weight\ of\ CMQF \\ VC: VC\ system\ Quality\ Factor & W_2: Weight\ of\ VCQF \\ CRT: CRT\ Quality\ Factor & W_3: Weight\ of\ CRTQF \\ RM: RM\ Quality\ Factor & W_4: Weight\ of\ RM \end{array}$$

$$TCIM = W_1 * CM + W_2 * VC + W_3 * CRT + W_4 * RM \quad W_1 + W_2 + W_3 + W_4 = 1 \tag{1}$$

(2) TC Selection Approaches Measurement (TCSAM): RT needs to use numerous TC and test suits of the repository to ensure the correctness of the original functions in new versions. Excellent TC selection approach can remove the useless or redundant TC and test suits to increase RT efficiency. TCSAM needs to consider Analysis of TC Selection Approaches (TCSAA), Evaluation of TC Selection Approaches (TCSAE), and Integration of TC Selection Approaches (TCSAI) three quality factors. Combination formula is shown as Eq. (2):

TCSAM: TC Selection Approach Measurement.

$$\begin{array}{ll} TCSAA: Analysis\ TCSA\ Quality\ Factor & W_1: Weight\ of\ TCSAA \\ TCSAE: Evaluation\ TCSA\ Quality\ Factor & W_2: Weight\ of\ TCSAE \\ TCSAI: Integration\ TCSA\ Quality\ Factor & W_3: Weight\ of\ TCSAI \end{array}$$

$$TCSAM = W_1 * TCSAA + W_2 * TCSAE + W_3 * TCSAI \quad W_1 + W_2 + W_3 = 1 \tag{2}$$

(3) TC Prioritization Algorithms Measurement (TCPAM): Excellent TC Prioritization Algorithms can arrange testing sequences to find the errors and defects early and increase RT efficiency. TCPAM needs to consider Analysis of TC Prioritization Algorithms (TCPAA), Evaluation of TC Prioritization Algorithms (TCPAE), and Integration of TC Prioritization Algorithms (TCPAI) three quality factors. Combination formula is shown as Eq. (3):

TCPAM: TC Prioritization Algorithms Measurement.

$$\begin{array}{ll} TCPAA: Analysis\ TCPA\ Quality\ Factor & W_1: Weight\ of\ TCPAA \\ TCPAE: Evaluation\ TCPA\ Quality\ Factor & W_2: Weight\ of\ TCPAE \\ TCPAI: Integration\ TCPA\ Quality\ Factor & W_3: Weight\ of\ TCPAI \end{array}$$

$$TCPAM = W_1 * TCSAA + W_2 * TCSAE + W_3 * TCSAI \quad W_1 + W_2 + W_3 = 1 \tag{3}$$

(4) Testing Tools Supporting Measurement (TTSM): In order to integrate with testing tools, TTSM needs to consider three quality factors of Test Data Input Processing (TDIP), Test Results Processing (TRP), and Testing Tools Combination (TTC). Combination formula is shown as Eq. (4):

TTSM: Testing Tools Supporting Measurement

$$\begin{array}{ll} TDIP: Test\ Data\ Input\ Processing\ Quality\ Factors & W_1: Weight\ of\ TDIP \\ TRP: Test\ Results\ Processing\ Quality\ Factors & W_2: Weight\ of\ TRP \\ TTC: Testing\ Tools\ Combination\ Quality\ Factors & W_3: Weight\ of\ TTC \end{array}$$

$$TTSM = W_1 * TDIP + W_2 * TRP + W_3 * TTCW_1 + W_2 + W_3 = 1 \tag{4}$$

(5) TCMM Quality Indicator (TCMMQI) combines four phase quality measurements which include TCIM, TCSAM, TCPAM, and TTSM. According to the influence of quality measurement, the weighted value of linear combination equations is assigned. Combination formula is shown as Eq. (5):

TCMMQI: TCMM Quality Indicator.

$$\begin{array}{ll} TCIM : TC\ Identification\ Measurement & W_1 : Weight\ of\ TCIM \\ TCSAM : TC\ Selection\ Approaches\ Measurement & W_2 : Weight\ of\ TCSAM \\ TCPAM : TC\ Prioritization\ Algorithms\ Measurement & W_3 : Weight\ of\ TCPAM \\ TTSM : Testing\ Tools\ Supporting\ Measurement & W_4 : Weight\ of\ TTSM \end{array}$$

$$TCMMQI = W_1 * TCIM + W_2 * TCSAM + W_3 * TCPAM + W_4 * TTSM \quad W_1 + W_2 + W_3 + W_4 = 1 \tag{5}$$

4.2 TCMM Quality Improvement Manners

Quality quantification can help identify the quality defects of the TCMM and take appropriate improvements. The following describes the TCMM quality improvement measures based on the TCMM measurement model:

(1) IF TCMMQI < 0.6, according to Eq. (5), the 4 quality measurements should be detected to identify defective factors, and apply the rule-based improvement mechanism:
(2) IF TCIM < 0.6* and weight value of TCIM > 0.4

THEN according to Eq. (1), CM, VC, CRT, and RM quality factors need to be inspected to identify the defective items with the quantified value < 0.6.

For the identified bad-quality items, it is necessary to find out the influencing factors one by one and plan correct improvements manners. (*threshold can be adjusted)

(3) IF TCSAM < 0.6* and weight value of TCSAM > 0.4

THEN according to Eq. (2), TCSAA, TCSAE, and TCSAC quality items need to be inspected to identify the defective items for which the quantified value < 0.6.

For the identified bad-quality items, it is necessary to find out the influencing factors one by one and plan correct improvements manners. (*threshold can be adjusted)

(4) IF TCPAM < 0.6* and weight value of TCPAM > 0.3

THEN according to Eq. (3), TCPAA, TCPAE, and TCPAC quality items need to be inspected to identify the defective items with the quantified value < 0.6.

For the identified bad-quality items, it is necessary to find out the influencing factors one by one and plan correct improvements manners. (*threshold can be adjusted)

(5) IF TTSM < 0.6* and weight value of TTSM > 0.3

THEN according to Eq. (4), TDIP, TRP, and TTC quality items need to be inspected to identify the defective items with the quantified value < 0.6.

For the defective items, it is necessary to find out the influencing factors one by one and plan correct improvement methods. (*threshold can be adjusted)

5 Evaluation of TCMM

Tested TC are an important asset to the CI/CD process and software maintenance operations. RT needs to use numerous tested TC to ensure the requirements quality of the new version. In order to improve the speed and quality of RT operations, TC need a practical set of management and control approaches. In this section, three TC management approaches are evaluated using four task characteristics. Three TC management approaches are described as follows (shown in Table 2):

(1) Simple TC Access Tool: The traditional software development and maintenance process uses a simple access tool to manage test cases. Tested test cases can be easily saved to and retrieved from the TC repository using the access tool. However, the access tool can't identify the affected TC with software change management and version control system. In addition, this tool can't support the TC selection approach, prioritization algorithm, and testing tools.
(2) TC Management Tools: Agile software development, DevOps, and maintenance operations apply CI/CD process to enhance new version delivery and deployment speed and quality. RT is an important and frequent task in CI/CD process. For this, TC management tools based on software change management and version control system can identify the affected test cases in the CI/CD process. Combining testing tools, the TC assistant tool can support automatic regression testing tools to improve regression testing (and CI/CD process) speed and quality.
(3) TCMM: TCMM has the features of TC Management Tools to support RT automation. In addition, in order to speed up the new version delivery and deployment, TCMM also combines TC selection approaches and prioritization algorithms. TCMM provides the input data of selection approaches and prioritization algorithms and uses the output results for the testing tools. Integrating CM, the VC system, selection approaches and prioritization algorithms, and testing tools for RT can speed up the new version delivery and deployment, and ensure new version requirements quality.

Table 2. Three Test Cases Management Comparison Table

Types of TC Mam Features of TC Man	Simple TC Access Tool	TC management Tools	TCMM
Identify useful test cases	Weak	Strong	Strong
Assist related algorithms	none	Weak	Strong
Support automatic tools	none	Weak	Strong
Assist CD process	none	Weak	Strong

6 Conclusion

RT plays an important role in DevOps, agile software development, and software maintenance workflow, helping to speed up the product delivery and deployment time, and ensure requirements quality. For this, RT has to overcome four challenges: identify useful test cases, apply TC selection tools, apply TC prioritization algorithms, and test with automatic tools. Applying high-quality Test Case Management Mechanisms (TCMM) is a practical way to speed up RT workflow and ensure requirements quality. In this paper, the characteristics and influencing factors of high-quality TCMM are discussed and organized. Quantifying the quality characteristics can identify and adjust the defects and insufficiencies of TCMM. Based on quantifying characteristics, a TCMM measurement model is proposed to identify the defects and insufficiencies of TCMM and make improvements. Applying high-quality TCMM in RT workflow assists DevOps, agile software development, and software maintenance workflow to speed up product deployment and delivery, and the required quality.

References

1. Schach, S.R.: Object-Oriented Software Engineering, vol. 7. McGraw-Hill, New York (2008)
2. Abrahamsson, P., Salo, O., Ronkainen, J., Warsta, J.: Agile software development methods: Review and analysis (2017). arXiv:1709.08439
3. Beck, K. et al.: Manifesto for Agile Software Development (2001). http://www.agilemanifesto.org/
4. Ebert, C., Gallardo, G., Hernantes, J., Serrano, N.: DevOps. IEEE Softw. **33**(3), 94–100 (2016)
5. Leite, L., Rocha, C., Kon, F., Milojicic, D., Meirelles, P.: A survey of DevOps concepts and challenges. ACM Comput. Surv. (CSUR) **52**(6), 1–35 (2019)
6. Shahin, M., Babar, M.A., Zhu, L.: Continuous integration, delivery and deployment: a systematic review on approaches, tools, challenges and practices. IEEE Access **5**, 3909–3943 (2017)
7. Purohit, K.: Executing devOps & CI/CD. Reduce Manual Dependency IJSDR **5**(6), 511–515 (2020)
8. Jackson, L.: The CI/CD Pipeline. In The Complete ASP. NET Core 3 API Tutorial, pp. 305–347. Apress, Berkeley, CA (2020)
9. Engström, E., Runeson, P., Skoglund, M.: A systematic review on regression test selection techniques. Inf. Softw. Technol. **52**(1), 14–30 (2010)
10. Narciso, E.N., Delamaro, M.E., Nunes, F.D.L.D.S.: Test case selection: A systematic literature review. Int. J. Softw. Eng. Knowl. Eng. **24**(04), 653–676 (2014)
11. Khatibsyarbini, M., Isa, M.A., Jawawi, D.N., Tumeng, R.: Test case prioritization approaches in regression testing: A systematic literature review. Inf. Softw. Technol. **93**, 74–93 (2018)
12. IEEE: IEEE standard for software test documentation. IEEE Std (829–1983, Revision) (1983)
13. Engström, E., Runeson, P.: A qualitative survey of regression testing practices. In: Product-Focused Software Process Improvement: 11th International Conference, PROFES 2010, Limerick, Ireland, 21–23 June 2010. Proceedings 11, pp. 3–16. Springer Berlin Heidelberg (2010). https://doi.org/10.1007/978-3-642-13792-1_3
14. Onoma, K., Tsai, W.T., Poonawala, M., Suganuma, H.: Regression testing in an industrial environment. Commun. ACM **41**(5), 81–86 (1988)
15. Fenton, N.E.: Software Metrics - A Rigorous Approach, Chapman & Hall (1991)

Case Study in Generating RFC Maps of IETF Standards

Jeng-Wei Lin[1], Yi-Ting Lin[1], and Fang-Yie Leu[2](✉)

[1] Department of Information Management, Tunghai University, Taichung 407224, Taiwan
{jwlin,g09490019}@thu.edu.tw

[2] Department of Computer Science, Tunghai University, Taichung 407224, Taiwan
leufy@thu.edu.tw

Abstract. Internet Engineering Task Force (IETF) have developed many Internet Standards, described by RFCs (Request for Comments). As time goes by, some RFCs were obsoleted, some were updated, and some were developed for new concepts, functions, or options. Today, there are more than 9000 RFCs. It is not easy to figure out how many RFCs are related to a specific Internet standard, such as Domain Name Systems (DNS) or Border Gateway Protocol (BGP), and to understand the roles these RFCs play in the standard. In this study, given an Internet Standard name, we aim to generate an IETF standard map of RFCs. An experimental system is built. RFC maps of several Internet standards are generated and discussed.

1 Introduction

Internet has significantly changed our everyday lives. As the most important standard organization for Internet standards, Internet Engineering Task Force (IETF) [1] have developed many Internet Standards, such as TCP/IP protocol suite [2–5], Domain Name Systems (DNS) [6–8], World Wide Web (WWW) [9–13], and so on.

Unlike most standardization bodies, IETF is a totally open community. There is no formal membership, and its standardization process is based on "rough consensus and running code" [14]. IETF standards are published as Request for Comments (RFC), in a form of a memorandum describing methods, behaviors, researches, or innovations applicable to Internet and Internet-connected systems. We note that not all RFCs are standard documents. Various contributors can provide input to the RFC series, such as Internet Architecture Board (IAB), IETF working groups in different areas, Internet Research Task Force (IRTF) working groups, and individual submissions [15]. RFCs are labeled with different statuses: Internet Standard, Proposed Standard, Best Current Practice, Experimental, Informational, and Historic, depending on what they cover, and their maturity level [1].

As time goes by, new technologies have emerged. Some RFCs were obsoleted (or replaced by other RFCs), some were updated, and some were developed for new concepts, functions, or options. For example, TCP was initially described in RFC 793 in 1981 [3]. Since then, several RFCs were published to enhance TCP in security, performance, and many other aspects. Over time, RFC 793 and many other RFCs have

L. Barolli (Ed.): BWCCA 2023, LNDECT 186, pp. 228–237, 2024.
https://doi.org/10.1007/978-3-031-46784-4_21

combined to serve as the core specification for TCP. As well, a large number of experimental modifications to TCP have also been published in the RFC series, along with informational notes, case studies, and other advice. A least 300 RFCs are related with TCP. A brief summary of the RFCs that define TCP is usually helpful to a newcomer, or even an old hand [16]. We also note that in 2022, RFC 9293 obsoletes RFCs 793, 879, 2873, 6093, 6429, 6528, and 6691, and updates RFCs 1011, 1122, and 5961 (see Fig. 1). It brings together all of the IETF Standards Track changes and other clarifications that have been made to the base TCP functional specification (RFC 793) and to unify them into an updated version of the specification [4].

Internet Engineering Task Force (IETF)	W. Eddy, Ed.
STD: 7	MTI Systems
Request for Comments: 9293	August 2022
Obsoletes: 793, 879, 2873, 6093, 6429, 6528, 6691	
Updates: 1011, 1122, 5961	
Category: Standards Track	
ISSN: 2070-1721	

Fig. 1. Header of RFC 9293.

Correct and efficient implementations of Internet standards are the base of the success of Internet. To implementers, experimenter, or students, a guidance is in need on the relevance and significance of standard track RFCs, informational RFCs, and best current practice RFCs that relate to a specific Internet standard.

In IETF official website, there is a RFC search page [17], on which a user can search RFCs by many different criteria, such as RFC number, title, keyword, authors, and so on. For example, when this paper was written, given TCP as a keyword, the search page returned a table of 318 RFCs and their basic information in the order of RFC numbers; given BGP as a keyword, the search page returned a table of 226 RFCs. These table were long. It was not easy to capture, organizes, and summarizes these RFCs.

In our previous case study [18] on Border Gateway Protocol (BGP) [19], we hand-painted a RFC map of BGP, as shown in Fig. 2. We listed 69 RFCs related to BGP on it, including some early obsoleted RFCs (BGP1, and BGP3), BGP4 RFCs, and RFCs in different BGP aspects in different colors. It greatly visualizes the relevance and significance of RFCs related to BGP.

In this paper, we present our preliminary study to automatically generate a RFC map to an assigned Internet standard.

2 System Architecture

Figure 3 shows the system architecture of the system. The right hand side is how we preprocess the published data related to all RFCs and store the required information in the database. The left hand side is the algorithm to generate a RFC map to an assigned Internet standard.

2.1 Preprocessing

All RFCs are free accessed on the IETF official website in several formats, as well as some metadata maintained by the RFC editors. In this beginning of this study, we first examine the RFC index [20]. It contains citations for all RFCs in numeric order, such as title, authors, published date, update and obsolete information, status, and stream. In the XML version of the RFC index, more metadata are recorded, such as abstract, keyword list, the original draft and its version number, area, errata, and so on.

We retrieved the metadata from the RFC index for all RFCs and stored them in a database.

2.2 Algorithm to Generate an RFC Map

As shown in Fig. 3, when a user gives an Internet standard name, the system will search the name in the title and keyword list in the database and retrieve an initial list of RFCs. Currently, the system adopts a simple case-insensitive exact substring match.

Later, the system will explore the update and obsolete information of all RFCs in the list. When a new RFC is found, it will be inserted into the list. As a result, the list expands. The list will expand with several iterations. The expand iteration threshold can be assigned by the user. The list expansion is in a breadth-first-search manner, and a final list of RFC is determined.

When a new RFC is found in the list expansion, it is linked with the original RFC. As a result, they together form an adjacent list, on which every link is associated with update or obsolete information.

Finally, based on the adjacent list, the system generates a RFC map where each node represents a RFC, and each link represents an update or obsolete relation. Nodes in different colors and shape are used to reflect the status of a RFC. As well, links in different colors are used to reflect the relation between nodes. All nodes are vertically aligned to their published year from left to right. When a node is pointed, title and abstract of the RFC are shown in a floating box. When the node is clicked, the RFC is retrieved and displayed in the browser.

The algorithm to expand the initial list of RFCs shown as following.

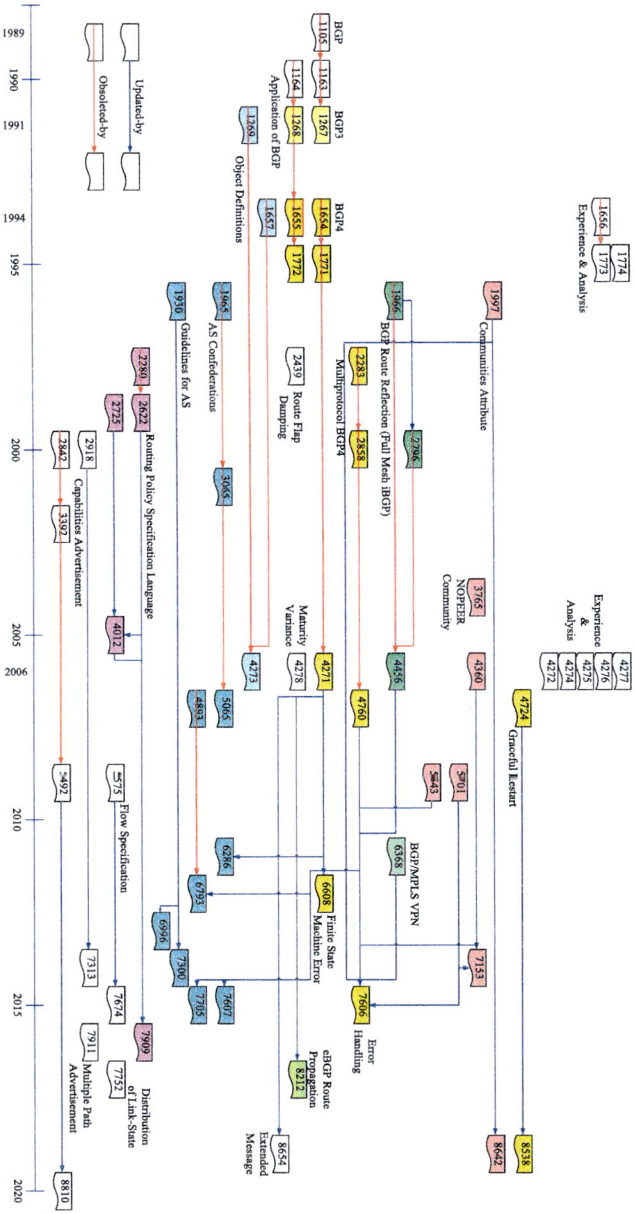

Fig. 2. A hand-painted RFC map of BGP (retrieved from [18])

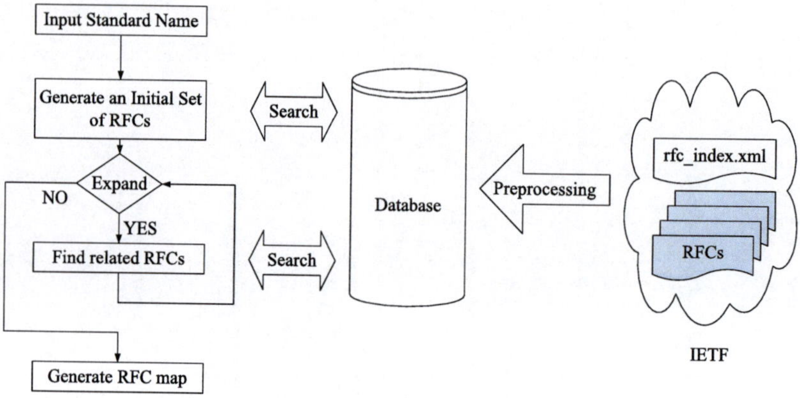

Fig. 3. System Architecture

ALGORITHM 1:

```
RFC_EXPAND(String name, Integer max)
// name is the name of the standard
// max set the maximum number of RFC expansions
begin
    list = Database_Lookup(keyword = name) // return a list of RFC numbers
    foreach (rfc_no in list)
        iteration [rfc_no] = 1      // iteration number of the RFC expansion
        ADJ [rfc_no] = {}           // ADJ is the adjacent list of the RFCs
    end of foreach
    reset(list)
    while (rfc_no = next(list))
        if (iteration [rfc_no])> max)
            break
        end of if
        foreach (rfc_related in related_rfcs[rfc_no])
            // related_rfcs[rfc_no] is the collection of RFCs
            //      updating, or obsoleting rfc_no, and
            //      updated by, obsoleted by rfc_no
            if (rfc_related is not in list)
                iteration [rfc_related ] = iteration [rfc_no] + 1
                ADJ [rfc_related] += ADJ [rfc_no]
                ADJ [rfc_no ] += ADJ [rfc_related]
                append(list, rfc_related)
            end of if
        end of foreach
    end of while
    return ADJ
end
```

3 System Implementation and Experiment Results

We use a standard Web design to build the proposed system, consisting of a generic PC running Ubuntu 20.04 SMP operation system, Apache 2 Web server, and MySQL 8 database. We adopt PHP 7.4.3 to implement the algorithm to generate RFC maps. The visualization engine is Graphviz [21]. The RFC map generated by Graphviz is in SVG (Scalable Vector Graphics) format [22], which supports user interactive in many Web browsers, such Google Chrome, and Microsoft Edge. Thus, we can embed codes in JavaScript so that when a user moves the mouse over a RFC node, its title, stream, area, workgroup acronym, and abstract is floated in an information box.

3.1 RFC Map of RPKI (Resource Public Key Infrastructure)

Figures 4 and 5 show the RFC maps of RPKI (Resource Public Key Infrastructure) [23] generated in the proposed system, with and without RFC expansion respectively. Numbers of RFCs listed in Figs. 4 and 5 are 42 and 48 respectively.

We use the string RPKI as the keyword. The system first searches it in the title and keyword list in the database, and retrieves 42 RFCs as the initial list. Without RFC expansion, the system generates RFC map of RPKI as shown in Fig. 4,

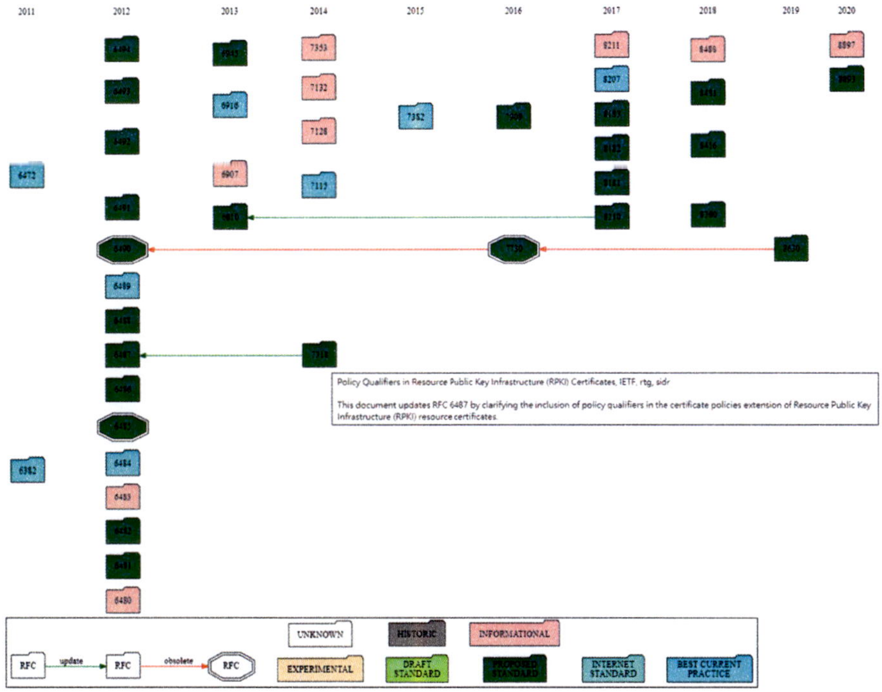

Fig. 4. RFC map of RPKI with 42 RFCs (no expansion)

In Fig. 4, it is easy to see that there were 15 RFCs published in 2012. Their RFC numbers are consecutive numbers, from 6480 to 6494. This gives us a hint they are probably the core of the RPKI. When the mouse is over any of the two RFCs (RFCs 6472 and 6382) published one year ago, we can find its title, stream, area, working group, and abstract. According to these, we can find that they both are related to AS (Autonomous System) [24]. Since the two are blue, their statuses are **Best Current Practice**. As well, we can see that some RFCs are **Proposed Standard**, and some are **Informational**. However, none of them become **Internet Standard**. When any of them is clicked, the system will retrieve and display the whole requested RFC on the screen.

We can also see that some RFCs were updated. For example, in 2014, RFC 7318 updated RFC 6487. Again, when we move the mouse over RFC 7318, we can read its abstract and know that it clarifies the inclusion of policy qualifiers in the certificate policies extension of RPKI. For further understanding, we can click RFC 7318 to read the whole document.

We can see that some RFCs were obsoleted, such as RFCs 6485, 6490, and 7730. Again, we can move the mouse over RFCs 7730 and 8630 to read their abstract for initial understanding. However, we cannot know when RFC 6485 was obsoleted and whether some RFC replaced it directly in Fig. 4.

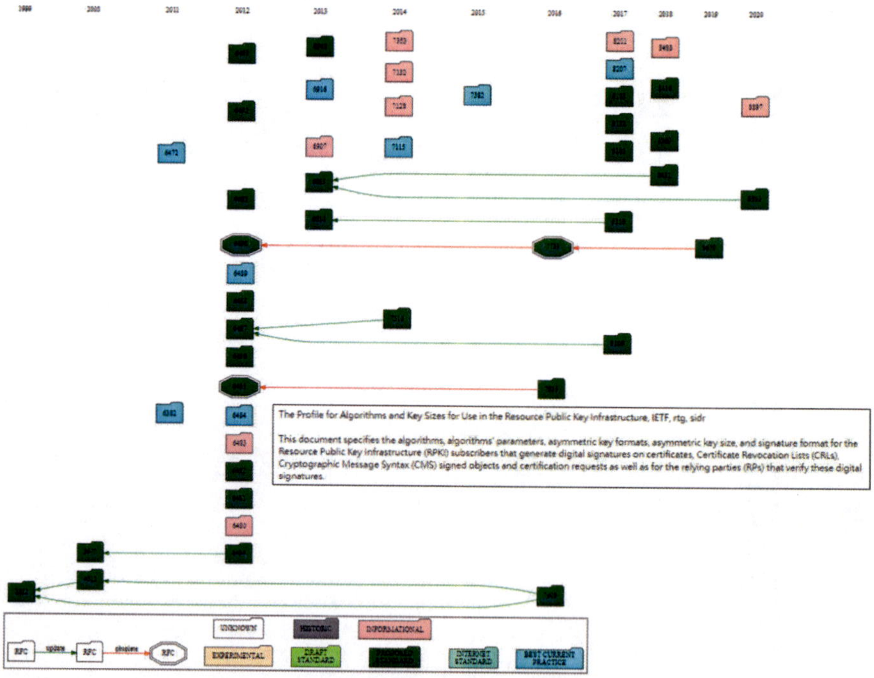

Fig. 5. RFC map of RPKI with 48 RFCs (expended once)

We then increase the number of RFC expansion to 2. The system expands once the initialize list, finds 1, 2, 1 and 1 more RFCs published in 1999, 2005, 2013, and 2017 respectively, and 6 RFCs are appended at the end of the initial list. As a result, there are 48 RFCs listed in **Fig. 5**. As a result, we can realize that RFC 7935 obsoleted RFC 6485 in 2016. We can also find that RFC 6487 is not only updated by RFC 8318, but also RFC 8209 in 2017.

3.2 Number of RFCs Listed in RFC Maps

Figure 6 shows the RFC maps of DHCP (Dynamic Host Configuration Protocol) [25–27] generated in the proposed system without RFC expansion. 79 RFCs are found in the initial list and depicted in Fig. 6. In the bottom left corner, RFCs 2131 and 2132 are found to obsolete RFCs 1531, 1541, 1533 in the 1990s. When we move the mouse over them to see their titles and abstracts, we can identify them as the base of DHCP. However, it is not so obvious to find the base of DHCPv6 (DHCP for IPv6) in Fig. 6, which are RFCs 8415, 3633 (obsoleted) and 3736 (obsoleted).

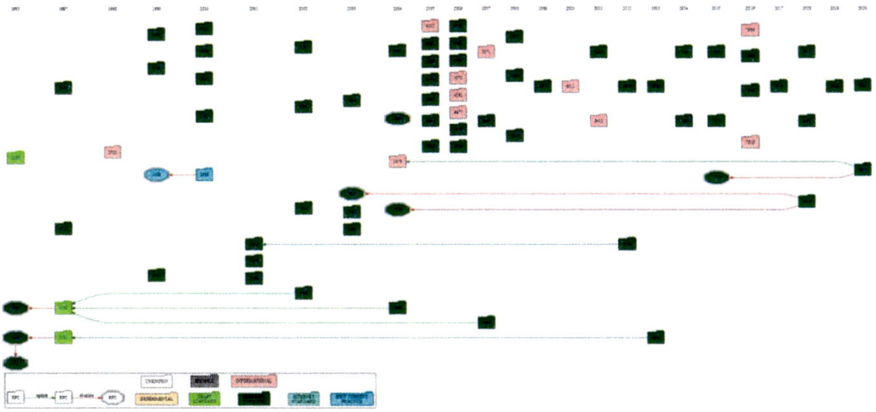

Fig. 6. RFC map of RPKI with 48 RFCs (expended once)

As we have stated above, an Internet standard is probably described by many RFCs. Some of them were updated, and some were obsoleted. Some are informational, some are experimental, and some are standards in different levels of maturity.

Table 1 shows the numbers of RFCs related to RPKI, DHCP, BGP, and DNS. They are retrieved by our system with different expansions, and from RFC search [17].

Usually, only a few RFCs are the base of an Internet standard. They may combine with some others to form the core of the standard, while some extend the standard and provide new options. Some Internet standards are independent to others, while some heavily depend on each other.

Currently, the proposed system only explores the update and obsolete information between RFCs in the map. As a result, in Figs. 4, 5, and 6, many RFCs are replaced in the upper side independently. However, in the hand-painted Fig. 2, closely related RFCs are further grouped together, and given a proper label. The RFCs in a group are placed nearby in the map with its label. Thus, it can provide more information.

Typically, key phrases may appear repeatedly in a group of deep related RFCs. We will further investigate algorithms to detect related RFCs, and assign proper labels to groups of related RFCs [28].

Table 1. Number of RFCs listed in the RFC maps of RPKI, DHCP, BGP, and DNS

Iterations of RFC expansion	Number of RFCs listed in RFC map of			
	RPKI	DHCP	BGP	DNS
1	42	79	206	280
2	48	99	229	353
3	54	127	250	540
4	55	142	285	686
5	64	156	322	810
6	68	166	366	904
Number of RFCs retrieved in RFC search [17]	47	147	226	317

4 Conclusions and Future Works

In this paper, we have built a simple system to automatically generate RFC maps of Internet standards. Given the name of an Internet standard, the proposed system utilizes the metadata in the RFC index [20] maintained by the RFC editors to generate a RFC map of the standard. RFC maps of RPKI [23] and DHCP [25–27] are shown and discussed.

Currently, the proposed system only explores the update and obsolete information between RFCs. We are currently investigating other deeper relationships between RFCs and algorithms to detect these relationships. Thus, the system can further divide the RFCs related with a given Internet standard into base, core, and many smaller but meaningful groups. Then, the proposed system can generate more informative RFC maps.

Acknowledgments

References

1. Internet Engineering Task Force. https://www.ietf.org/
2. Postel, J. (ed.): RFC 791, Internet Protocol. IETF (1981)
3. Postel, J. (ed.): RFC 793, Transmission Control Protocol. IETF (1981). Obsoleted by RFC 9293
4. Eddy, W. (ed.): RFC 9293, Transmission Control Protocol (TCP). IETF (2022)
5. Fall, K.R., Stevens, W.R.: TCP/IP Illustrated, Vol.: The Protocols, 2nd edn. Addison-Wesley, United States (2011)
6. Mockapetris, P.V.: RFC 1034, Domain Names - Concepts and Facilities. IETF (1987)
7. Mockapetris, P.V.: RFC 1035, Domain Names - Implementation and Specification. IETF (1987)
8. Liu, C., Paul Albitz, P.: DNS and BIND, 5th edn. O'Reilly Media, United States (2006)
9. Berners-Lee, T., Connolly, D.: RFC 1866, Hypertext Markup Language - 2.0. IETF. Obsoleted by RFC 2854 (1995)
10. Connolly, D., Masinter, L.: RFC 2854, The 'text/html' Media Type. IETF (2000)
11. Berners-Lee, T., Fielding, R., Frystyk, H.: RFC 1945, Hypertext Transfer Protocol -- HTTP/1.0. IETF (1996)

12. Fielding, R., Gettys, J., Mogul, J., Frystyk, H., Berners-Lee, T.: RFC 2068, Hypertext Transfer Protocol -- HTTP/1.1. IETF. Obsoleted by RFC 2616 (1997)
13. Fielding, R., et al.: RFC 2616, Hypertext Transfer Protocol -- HTTP/1.1. IETF (1999)
14. Ohta, M.: IETF and internet standards. IEEE Commun. Mag. **36**, 126–129 (1998)
15. Housley, R., Daigle, L. (eds): RFC 8729, The RFC series and RFC editor. IETF (2020)
16. Duke, M., Braden, R., Eddy, W., Blanton, E., Zimmermann, A.: RFC 7414, A Roadmap for Transmission Control Protocol (TCP) Specification Documents. IETF (2015)
17. RFC Search. https://www.rfc-editor.org/search/rfc_search.php
18. Lin J.W. (林正偉): 前瞻及完整之 IPv6 生態系推動方案, 附錄九物聯網服務與應用研析及網際網路技術標準研析. Taiwan Network Center (2019). (in Chinese)
19. Rekhter, Y., Li, T., Hares, S.: RFC 4271, A Border Gateway Protocol 4 (BGP-4). IETF (2006)
20. RFC Index. https://www.rfc-editor.org/rfc-index.html
21. Graphviz. https://graphviz.org/
22. Scalable Vector Graphics (SVG). https://www.w3schools.com/graphics/svg_intro.asp
23. Lepinski, M., Kent, S.: RFC 6480, An Infrastructure to Support Secure Internet Routing. IETF (2012)
24. Hawkinson, J., Bates, T.: RFC 1930, Guidelines for Creation, Selection, and Registration of an Autonomous System (AS). IETF (1996)
25. Droms, R.: RFC 2131, Dynamic Host Configuration Protocol. IETF (1997)
26. Droms, R., Bound, J., Volz, B., Lemon, T., Perkins, C., Carney, M.: RFC 3315, Dynamic Host Configuration Protocol for IPv6 (DHCPv6). IETF. Obsoleted by RFC 8415 (2003)
27. Mrugalski, T., et al.: RFC 8415, Dynamic Host Configuration Protocol for IPv6 (DHCPv6). IETF (2018)
28. Li, Q., et al.: A survey on text classification: from traditional to deep learning. ACM Trans. Intell. Syst. Technol. **13**, 2 (2022). Article 31

Enhancing Resource Allocation for Cloud Computation Platform with Priority Based Scheduling

Chen Lung Pin[1], Leu-Fang Yie[1(✉)], Pan Junrui[1], Chia-Chen Kuo[2], and Ming-Jen Wang[2]

[1] Department of Computer Science, Tunghai University, Taichung City, Taiwan
{lbchen,leufy}@thu.edu.tw, g10350021@go.thu.edu.tw
[2] National Center for High-Performance, Hsinchu City, Taiwan
{c00kcc00,renny}@narlabs.org.tw

Abstract. Cloud computing platforms are clusters of high-end computers in Internet-based data centers that provide elastic software and hardware products at scale. In this paper, we propose a formula for predicting the number of resources needed for executing a job in a cloud render cluster. The users of the cloud computing platforms can use this formula to better balance production costs and deadlines, thus reducing costs and improving performance. According to the result of our experiments, our approach can accurately predict the system resource consumption. Thus, it can be used for setting parameters that maintain system performance.

1 Introduction

People are becoming increasingly dependent on the Internet and multimedia due to the rapid development of science and technology. A very important application among them is 3D rendering. 3D rendering is a multimedia technology that can be applied in games, movies, animation, architecture, design, etc. 3D rendering is the process of converting a 3D scene into a 2D image. The process may take hours or even days, since it requires a lot of resources to adjust light, material, and scene parameters for the best results.

The render farm is a cluster composed of a group of high-performance computers, whose main purpose is to distribute computing tasks to speed up rendering. With the render farm, the rendering speed of the computer can be greatly accelerated, because video frames can be rendered concurrently [1].

Render farms can greatly increase rendering speed, reduce rendering time, and utilize computing resources more efficiently through reasonable scheduling. The majority of render farms charge fees. Thus, achieving the best results with limited funds is a very important topic [1, 2]. This paper proposes a method to predict the number of resources needed for executing a job per using the traditional priority-based scheduling algorithm. This paper conducts experiments to simulate and verify the method.

The organizational structure of the rest of this paper is as follows. The second section introduces the research background, and the third section introduces the formula and experimental verification. Finally, a summary is given in Sect. 4.

L. Barolli (Ed.): BWCCA 2023, LNDECT 186, pp. 238–243, 2024.
https://doi.org/10.1007/978-3-031-46784-4_22

2 Research Background

2.1 Render Farm

Render farms are clusters of computers that are capable of accelerating 3D rendering. The rendering of 3D images is a multimedia technology that can be applied to a wide range of applications, including games, movies, animation, architecture, and design. Rendering 3D scenes into 2D images is time-consuming and requires a lot of bandwidth.

AWS Deadline [4] is a commercial-grade graph farm manager that supports Windows, Linux, and macOS. As shown in Fig. 1, Deadline consists of three components: database, repository, and clients (workstations and rendering nodes). The Database component is used to store jobs, settings, and worker configuration components; the Repository component stores job-related plug-ins, scripts, logs, and auxiliary rendering material files; the client contains the Render Nodes and Workstations that perform specialized functions.

A breakdown of the Deadline render farm [4] is shown in Fig. 2. A *task* is composed of the computation task of one frame or a series of frames in the digital animation. A collection of related tasks is grouped as a *job*. During the execution of a job, the constituent tasks are assigned to different rendering nodes. Once all constituent tasks have been completed, the job is complete.

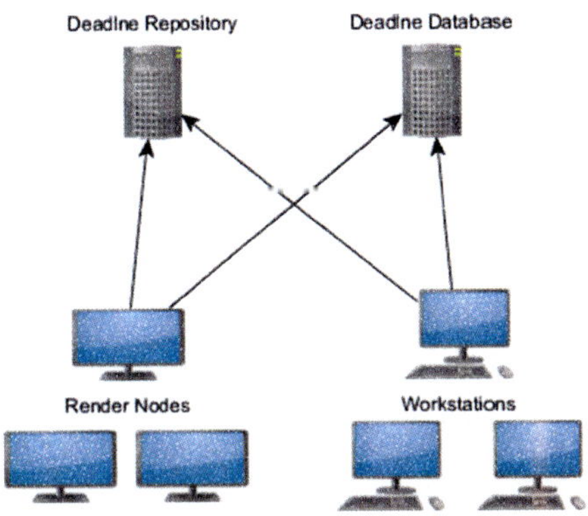

Fig. 1. Architecture diagram of AWS Deadline cloud rendering platform.

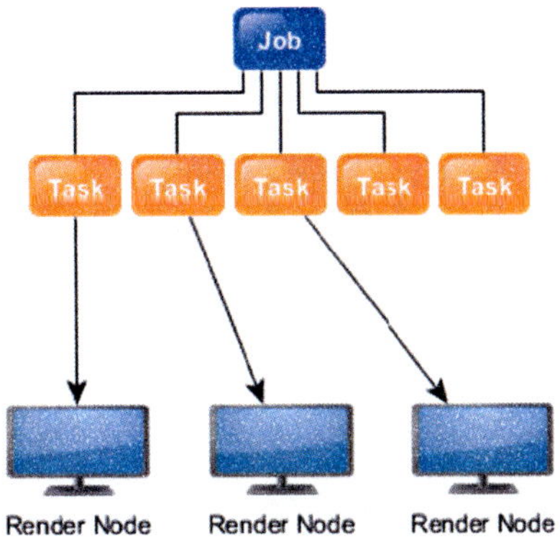

Fig. 2. Job and tasks.

2.2 Weight Priority Scheduling

For each job, its weight is calculated as follows [3]:

$$\begin{aligned} weight = & (job.Priority * PW) \\ & +(job.Errors * EW) \\ & +((NOW - job.SubmissionTimeSeconds) * SW) \\ & +((job.RenderingTasks - RB) * RW) \end{aligned} \tag{1}$$

- PW = priority weight
- EW = error weight
- SW = commit time weight
- RW = Render task weight
- RB = render task buffer length
- NOW = current system time

There are four items to determine for weight priority scheduling:

- Job priority (*job.Priority*): Depends on the user's payment amount.
- Submission time (*job.SubmissionTimeSeconds*): Job submission time, in seconds.
- Number of rendering tasks (*job.RenderingTasks*): The number of tasks of a job currently being executed.
- Errors (*job.Errors*): The number of task failures.

Jobs are sorted by weight in the job queue. Whenever a worker issues a task request, the scheduler assigns the job with the highest weight to the worker. We usually set PW and SW as positive numbers, so that users who pay more money or wait with a longer time are given priority to execution. In contracts, RW is usually set to a negative number, representing users with a large number of tasks being executed, and resources are assigned to those with fewer tasks [1].

3 Resource Prediction for a Job

The formula research in this paper is based on the popular traditional render farm [1]. In this paper, we focus on use the parameters related to job priority and ignore the other parameters such as error rate. Specifically, we set $PW = 1, RW = -1$ and $SW = 0, EW = 0, RB = 0$ for all the experiments presented in this paper.

3.1 Estimating the Number of Workers Allocated for a Job

According to the traditional priority-based scheduling [1] introduced in 2.2, when receiving a task request, the scheduler starts with the highest weight. When a job is performed more frequently, its weight gradually decreases, prompting the scheduler to select the job with the next highest weight.

In this paper, a *round* refers to the time interval between the scheduler gets a weight and the next weight. The *cumulative*(r) represents the number of scheduling events in round r. We deduce that.

$$cumulative(r+1) = cumulative(r) + \left(P^D_{r+1} - P^D_{r+2}\right) * (r+1) if r \epsilon [0, N) \cap \mathbb{N} \quad (2)$$

In formula (2), P^D_r refers to the weights that are arranged in round r. N is the total number of jobs. According to the definition of the previous formula, no worker will be assigned before the job with the highest weight is assigned, so $cumulative(0) = 0$. The scope of application of this formula $cumulative([1, N])$, that is $r \epsilon [0, N) \cap \mathbb{N}$. Since the round number is no more than N, we let $P^D_{N+1} = 0$.

Table 1. The parameters of Experiment 1

Number of Workers	30	simulation time	200	number of jobs	5
	Job 1	Job 2	Job 3	Job 4	Job 5
priority	30	20	15	10	5
Number of Tasks	200	200	200	200	200
Task length	5	5	5	5	5

3.2 Experimental Verification

This subsection develops the Python program to simulate the scheduling based on Eq. (2).

Table 1 shows the first experimental parameters. Five jobs are prioritized as follows: 19,8,3,0,0. Five jobs are divided into sufficient tasks for each, and given enough time to complete the execution, the experimental results are shown in Figs. 3 and 4.

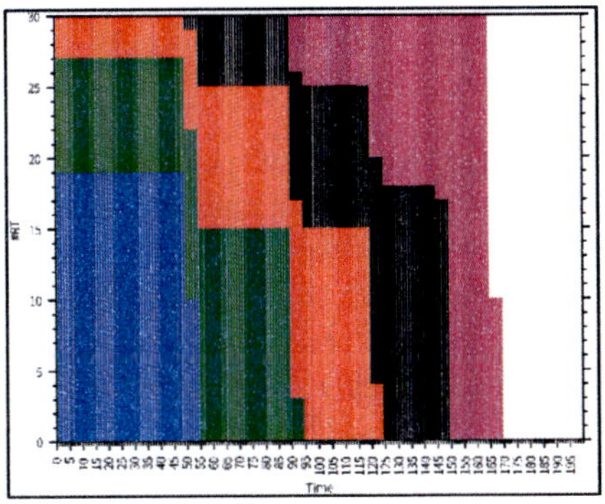

Fig. 3. The stacked Bar Chart of *RenderingTasks*.

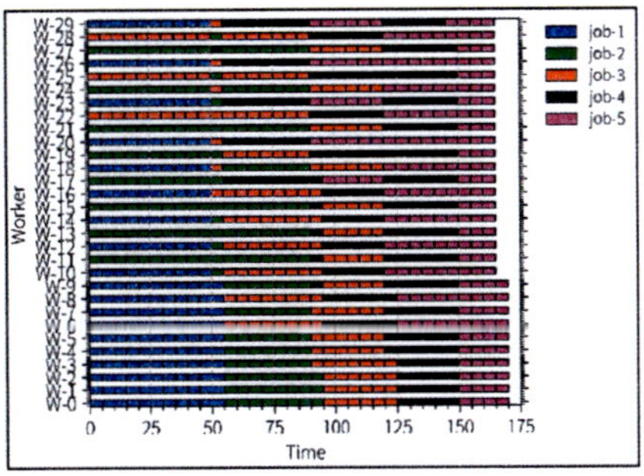

Fig. 4. The Gantt Chart of worker load distribution

Then we use the formula to calculate. The first step is to calculate the demarcation line for consuming working hosts.

$$cumulative(0) = 0$$
$$cumulative(1) = 10$$
$$cumulative(2) = 20$$
$$cumulative(3) = 35$$
$$cumulative(4) = 55$$
$$cumulative(5) = 80$$
$$cumulative(6) = \infty$$

Then according to the dividing line, it can be concluded that the jobs with the highest weight can be assigned to the workers:

$$cumulative(2) \leq 30 < cumulative(3)$$

Finally, the result consistent with the experiment can be obtained as follows:

$$worker(1) = [1 \leq 3] * \left\{ \begin{array}{c} (30 - 15) \\ +\frac{30-20}{3} \\ +[(30 - 20)mod3 \geq 1] \end{array} \right\} = 1 * (15 + 3 + 1) = 19$$

$$worker(2) = [2 \leq 3] * \left\{ \begin{array}{c} (20 - 15) \\ +\frac{30-20}{3} \\ +[(30 - 20)mod3 \geq 2] \end{array} \right\} = 1 * (5 + 3 + 0) = 8$$

$$worker(3) = [3 \leq 3] * \left\{ \begin{array}{c} (15 - 15) \\ +\frac{30-20}{3} \\ +[(30 - 20)mod3 \geq 3] \end{array} \right\} = 1 * (0 + 3 + 0) = 3$$

$$worker(4) = [4 \leq 3] * \{ \quad \cdots \quad \} = 0 * (\quad \cdots \quad) = 0$$

$$worker(5) = [5 \leq 3] * \{ \quad \cdots \quad \} = 0 * (\quad \cdots \quad) = 0$$

4 Conclusion

According to the current experimental results, when the work is sufficient, in the basic mode of weight priority scheduling, it is possible to determine the number of workers assigned to each work via the researched formula, which is equivalent to the resource occupation ratio. With this result, we can analyze the completion time of the work and evaluate whether or not it is feasible to research new work and how much to pay to complete the maximum benefit. However, the current formula is in limited use, and the objective of future research is to expand its scope and application to make it more valuable and comprehensive.

References

1. Chen, L.-P., Leu, F.-Y., Kuo, C.-C., Lin, T.-C., Wang, M.-J.: Efficient weighted and balanced resource allocation for high-performance render farms. In: Barolli, L. (ed.) Advances on Broad-Band Wireless Computing, Communication and Applications: Proceedings of the 17th International Conference on Broad-Band Wireless Computing, Communication and Applications (BWCCA-2022), pp. 292–300. Springer International Publishing, Cham (2023). https://doi.org/10.1007/978-3-031-20029-8_28
2. Chen, L.-P., Wu, I.-C., Liang, G.-Z.: Enhancing parallel game-tree searches by using idle resources of a high performance render farm. In: 2015 Conference on Technologies and Applications of Artificial Intelligence (2015)
3. AWS Thinkbox: "Job Scheduling". https://docs.thinkboxsoftware.com/products/deadline/10.1/1_User%20Manual/manual/job-scheduling.html (2020)
4. Schwarzkopf, M., Konwinski, A., Abd-El-Malek, M., Wilkes, J.: Omega: flexible, scalable schedulers for large compute clusters. In: SIGOPS European Conference on Computer Systems, pp. 351–364 (2013)
5. Savage, T. M., Vogel, K.E.: An Introduction to Digital Multimedia, p. 145, Jones & Bartlett Publishers (2014)

6. Sheharyar, A., Bouhali, O.: A framework for creating a distributed rendering environment on the compute clusters. Int. J. Adv. Comput. Sci. Appl. **4**(6), 117–123 (2013)
7. Yao, J., Pan, Z., Zhang, H.: A distributed render farm system for animation production. In: Natkin, S., Dupire, J. (eds.) ICEC 2009. LNCS, vol. 5709, pp. 264–269. Springer, Heidelberg (2009). https://doi.org/10.1007/978-3-642-04052-8_31

5G Network Base Station Timeline Scheduling

Yu-Han Chen[1], Chao-Hsiang Hsu[1], Heru Susanto[2,3], and Fang-Yie Leu[1,4](✉)

[1] Computer Science Department, Taichung, Taiwan
{S07353043,S07353030,leufy}@thu.edu.tw
[2] National Research and Innovation Agencies, Jakarta Pusat, Indonesia
heru.susanto@utb.edu.bn
[3] University of Technology Brunei, Darussalam, Brunei
[4] Emergency Response Management Center, Ming-Chun University, Taoyuan City, Taiwan

Abstract. Currently, the main applications of 5G networks are divided into three types, including: Enhanced Mobile Broadband (eMBB), Ultra-reliable and Low Latency Communications (uRLLC) and Massive Machine Type Communications, mMTC). This study modifies the base station SRA algorithm, by increasing priority and weight scheduling for packets. Among these packets, uRLLC's are the highest, and of course, are transmitted first. The priorities of the remaining two types of packets are a little lower. We divide a base station's resource blocks into multiple virtual networks according to time to practice network slicing with which to deliver packets, aiming to achieve effective network resource allocation.

1 Introduction

With the fast development of technology, people's demands for high-tech services have become more diverse. Current network systems have also advanced rapidly in the past ten years. In order to meet market demands, user's service quality and cost-effectiveness are taken into account by 5G networks. In fact wireless networks have been developed towards low-costs, low-latencies, and high-throughputs, particularly for 5G networks.

Generally, the resources of a wireless network are limited. To achieve effective resource allocation, several issues must be considered. For example, 5G networks divide their service targets into three types: Enhanced Mobile Broadband (eMBB), Ultra-reliable and Low Latency Communications (uRLLC) and Massive Machine Type Communications (mMTC) which individually use different software defined networks (SDNs). Different types of packets serve different types of users. 5G networks also adopt Network Function Virtual (NFV) to practice their functions. Edge computers are added to reduce the delay time of packet delivery and response time when events occur.

In order to efficiently utilize the limited resources, the resource blocks (RBs) of a base station need to be allocated efficiently. In this study, a base station's Scheduling and Resource Allocation (SRA) algorithm is adjusted and the packets sent by UE or switch are delivered according to the packet type and different number of available resource blocks. Especially, real-time streaming packets can be transmitted in a timely manner. For media transmission, connections that request high speed transmission are given larger bandwidth and longer transmission time in order to meet the network requirements needed

L. Barolli (Ed.): BWCCA 2023, LNDECT 186, pp. 244–252, 2024.
https://doi.org/10.1007/978-3-031-46784-4_23

by users. The SRA algorithm we developed is called 5G Timeline Scheduling (5GTS), which gives different time lengths to different types of packets from the perspective of time.

This paper is organized as follows. Section 2 presents the background and related work of this study. Section 3 introduces the system architecture. Section 4 states the experiment and verification. Section 5 concludes this study.

2 Background and Related Work

5G network [1] is the abbreviation of the 5th Generation Mobile Networks. Its technical specifications are formulated by 3GPP (3rd Generation Partnership Project). One of the goals is to achieve better cost-effectiveness and satisfying a variety of application requirements, such as low latencies, low costs, high throughputs, etc. so as to meet the needs of different users.

2.1 5G – Air – Simulator

The core functions of 5G-air-simulator [2] include: simulating multi-user, multi-cellular and heterogeneous network environments. The control and data planes are also practiced by using sophisticated and standards-compliant radio resource management methods. The models of channel propagation and data transmission (e.g. Hybrid Automatic Repeat Request (HARQ)) are also employed. Basically, the 5G-air-simulator calibrates the channel model and employs multiple-input multiple-output (MIMO) and massive MIMO transmission modes where MIMO is an important technology for improving throughputs and spectral efficiencies of wireless networks.

2.2 Packet Data Convergence Protocol (PDCP)

The Packet Data Convergence Protocol (PDCP) layer functions of 5G are similar to those of 4G. The main functions are [3]:

(1) IP header compression on the user plane (the compression algorithm is jointly determined by UE and the base station)
(2) Encryption/Decryption (Control Plane/User Plane).
(3) Control plane integrity verification (5G user plane can be selectively verified)
(4) Sorting and Duplication Detection.
(5) For the Option3X architecture under the Non-stand alones (NAS) network, the PDCP of the base station (gNodeB) performs traffic distribution and has routing functions

2.3 Scheduling and Resource Allocation (SRA) Algorithm

SRA is an algorithm used by 5G for scheduling and resource allocation. SRA users can change the scheduling methods and weights of packets. In the 5G networks, its position is placed between BS and UE.

3 Design Architecture

3.1 Architecture

This study chooses the use of a single base station to simulate the 3 types of applications, i.e., uRLLC, mMTC and eMBB, currently defined in the 5G network in an interference environment. When receiving a packet P, this system inserts P into its Data Queues and schedules P. We give priority to P, with which to determine when P will be forwarded to its destination so as to achieve the so-called resource block allocation. Figure 1 shows the mechanism of resource configuration and the transmission between a base station and UE, including the MAC layer and the physical layer. The SRA sorts three types of packets in the Data Queue of the MAC Layer for priority scheduling and processing. Usually, uRLLC packets are sent to the physical layer first.

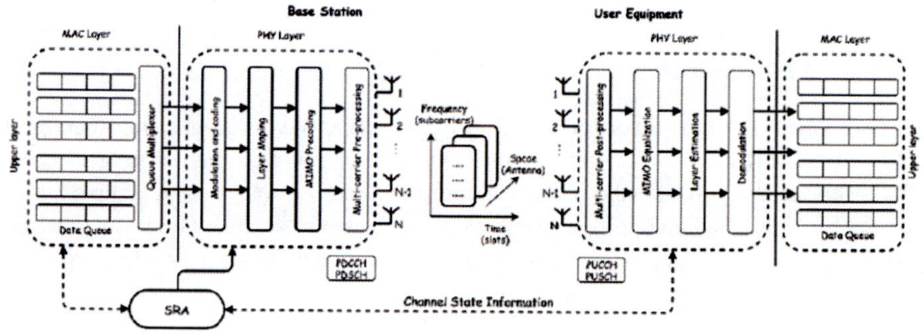

Fig. 1. Prioritization of packets in the Data Queue [4]

3.2 Implementation Method

We select and set the environment in the scenarios to the 5G air simulator, and then use the VoIP, CBR and Video packets to simulate uRLLC, mMTC and eMBB packets, respectively. Next, SRA algorithm chooses appropriate resource blocks to UE's QoS packets following this UE's QoS requirements.

```
until [ $COUNT -gt $NUMSIM ]; do
    TOTALNAME=$FILE"_"$COUNT"_"$FILENAME"_PROPOSED_"$NBUE"U"$CELS"C"".sim"
    ../5G-air-simulator SingleCellWithI $CELS $RADIUS $NBUE $NBVOIP $NBVIDEO $NBBE $NBCBR 9 $FRAME_STRUCT $SPEED $MAXDELAY $VIDEOBITRATE >$TOTALNAME
    echo FILE $TOTALNAME CREATED!
    let COUNT=COUNT+1
done
COUNT=1
```

Fig. 2. Environmental Settings.

Figure 2 shows the environmental settings. As stated above, an environment having a single base station with interference is established. A packet transmission environment and an environment for a number of users connected to this BS are also built.

```
void PROPOSEDDownlinkPacketScheduler::RBsAllocation()
{
    DEBUG_LOG_START_1(SIM_ENV_SCHEDULER_DEBUG)
    cout << " ---- PROPOSED DownlinkPacketScheduler::RBsAllocation\n";
    DEBUG_LOG_END

    FlowsToSchedule *flows = GetFlowsToSchedule();
    int nbOfRBs = (int)GetMacEntity()->GetDevice()->GetPhy()->GetBandwidthManager()->GetDlSubChannels().size();
    //create a matrix of flow metrics
    std::vector<std::vector<double>> metrics(nbOfRBs, std::vector<double>((int)flows->size()));

    double l_dAllocatedRBCounter = 0;
    bool *l_bFlowScheduled = new bool[flows->size()];
    for (int k = 0; k < (int)flows->size(); k++)
        l_bFlowScheduled[k] = false;
    int l_iScheduledFlows = 0;
    vector<double> *l_bFlowScheduledSINR = new vector<double>[flows->size()];

    // vector to track how many users are allocated on each RB
    vector<int> allocationsPerRB(nbOfRBs);
    fill(allocationsPerRB.begin(), allocationsPerRB.end(), 0);
    vector<int> HarqScheduledUEs;

    int defaultTxMode = GetMacEntity()->GetDefaultDlTxMode();
    vector<shared_ptr<arma::cx_fmat>> precodingMatrices;
    if (defaultTxMode == 11)
    {
        precodingMatrices.resize(nbOfRBs);
        for (int rb = 0; rb < nbOfRBs; rb++)
        {
            precodingMatrices.at(rb) = nullptr;
        }
    }
```

Fig. 3. Resource Block application.

Figure 3 shows the way of network slicing, i.e., to allocate Resource Blocks to packets that are required to be delivered, and to create a matrix for flow metrics.

Figure 4 shows how Resource Blocks are allocated and the arrangement of non-retransmitted data.

```
for (int rb = 0; rb < nbOfRBs; rb++)
{
    if (l_iScheduledFlows == (int)flows->size())
    {
        break;
    }

    bool RBIsAllocated = false;
    if (allocationsPerRB.at(rb) > 0)
    {
        RBIsAllocated = true;
    }
    vector<shared_ptr<FlowToSchedule>> scheduledFlows;
    vector<int> l_iScheduledFlowsIndices;

    // schedule non-retransmission flows
    double targetMetric;
    for (int i = allocationsPerRB.at(rb); i < m_maxUsersPerRB; i++)
    {
        targetMetric = 0;
        shared_ptr<FlowToSchedule> scheduledFlow;
        int l_iScheduledFlowIndex;
        for (int k = 0; k < (int)flows->size(); k++)
        {
            if (metrics[rb][k] > targetMetric && metrics[rb][k] > -INFINITY && !l_bFlowScheduled[k] &&
            flows->at(k)->IsHarqRetransmission() == false &&
            find(HarqScheduledUEs.begin(), HarqScheduledUEs.end(), flows->at(k)->GetUe()->GetIDNetworkNode()) == HarqScheduledUEs.end() &&
            find(scheduledFlows.begin(), scheduledFlows.end(), flows->at(k)) == scheduledFlows.end() && flows->at(k)->GetDataToTransmit() > 0)
            {
                HarqManager *harqManager = flows->at(k)->GetUe()->GetTargetNodeRecord()->GetHarqManager();
                if (harqManager != nullptr)
                {
                    if (harqManager->GetPidForNewTx() >= 0)
                    {
                        targetMetric = metrics[rb][k];
                        RBIsAllocated = true;
                        scheduledFlow = flows->at(k);
                        l_iScheduledFlowIrdex = k;
                    }
                }
```

Fig. 4. Resource Block application.

4 Experiment and Validation

We replace the SRA with the one we developed to allocate resource blocks to flows and measure packet loss rates, delays, and throughputs of resource block allocation. The number of UEs varies between 1 and 26.

4.1 CBR

We simulate mMTC packets with CBR packets. Figs. 5, 6 and 7 show the throughputs, packet loss rates and delays, respectively. When the number of users of each algorithm increases, the throughputs are also higher. The packet loss rates are not very different. But the delays also increase. Compared with other algorithms, the throughputs of the 5GTS are moderate, and the packet loss rates and delay rates are high. The reason is that it is performed purely from a time perspective, meaning that it is necessary to add a frequency perspective to improve it.

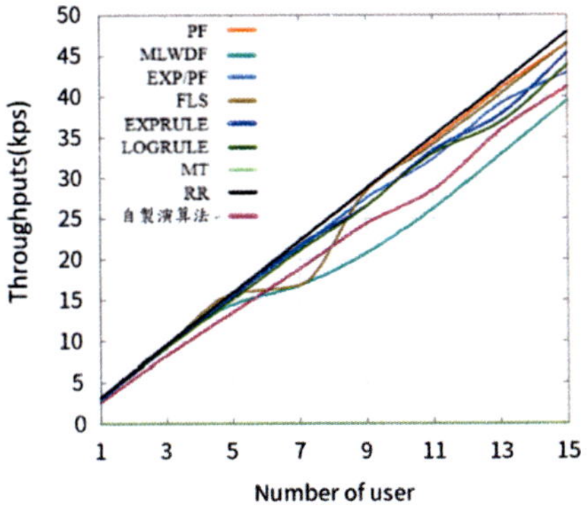

Fig. 5. Throughputs of CBR packets.

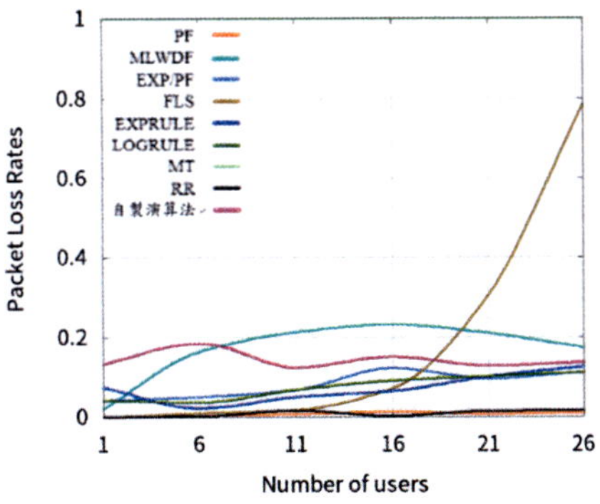

Fig. 6. Packet loss rates of CBR packets.

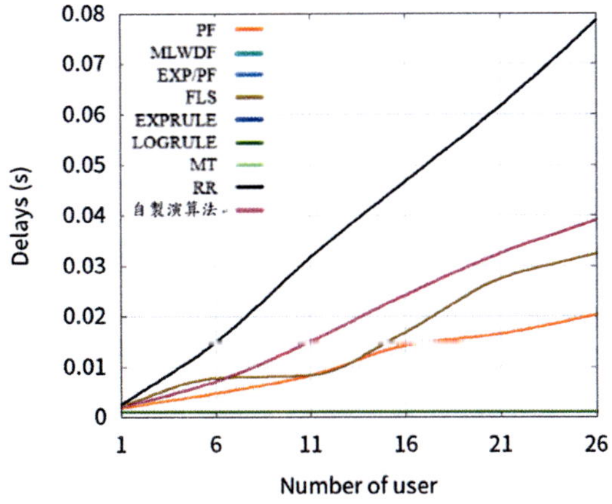

Fig. 7. Delays of CBR packets.

4.2 VoIP

VoIP packets are utilized to simulate eMBB packets. Figures 8, 9 and 10 show throughputs, packet loss, and delays, respectively. Generally speaking, when the number of users of each algorithm increases, the throughputs, packet loss rates and delays will correspondingly higher. In Fig. 10, due to the Y-axis scale, delays look approaching zero. Actually, the performance of 5GTS is not better than that of others.

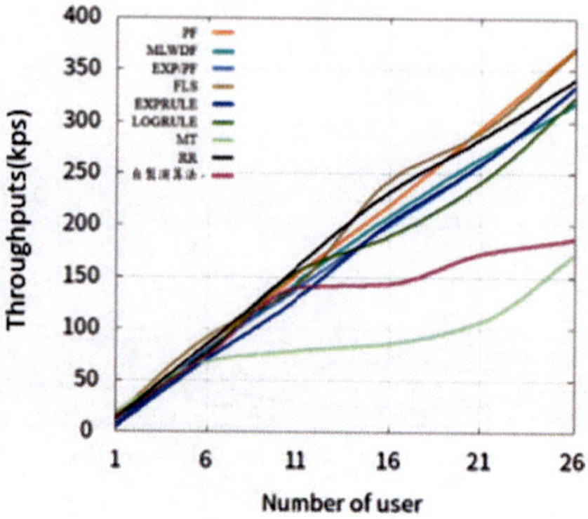

Fig. 8. Throughputs of VoIP packets.

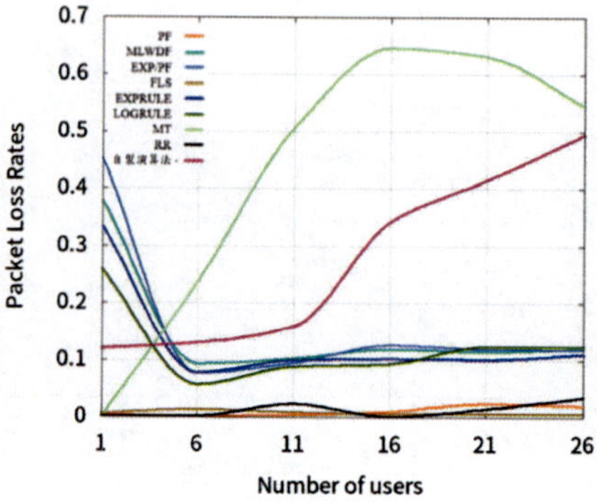

Fig. 9. Loss rates of VoIP packets.

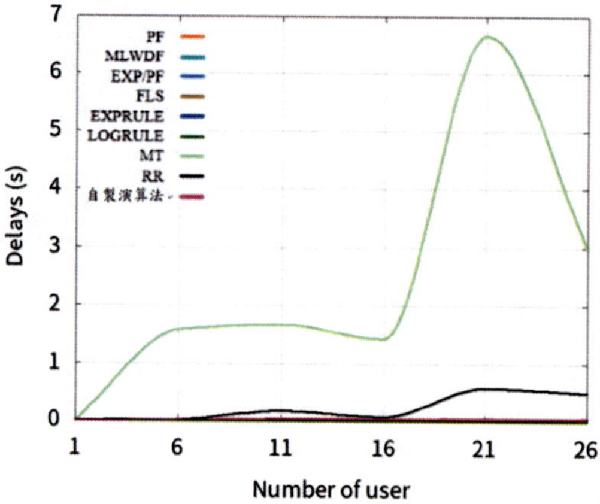

Fig. 10. Delays of VoIP packets.

4.3 Video

We simulate uRLLC by using Video packets. Figures 11, 12 and 13 show the throughputs, packet loss rates and delays of this experiment, respectively. As shown, the 5GTS's throughputs, packet loss rates and delays are not better than those of other algorithms.

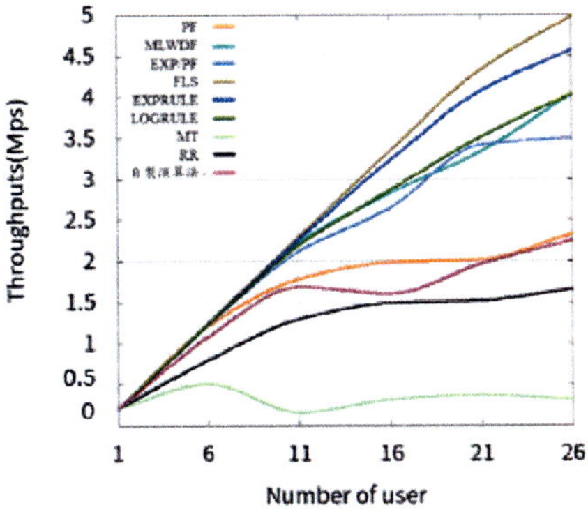

Fig. 11. Throughputs of Video packets.

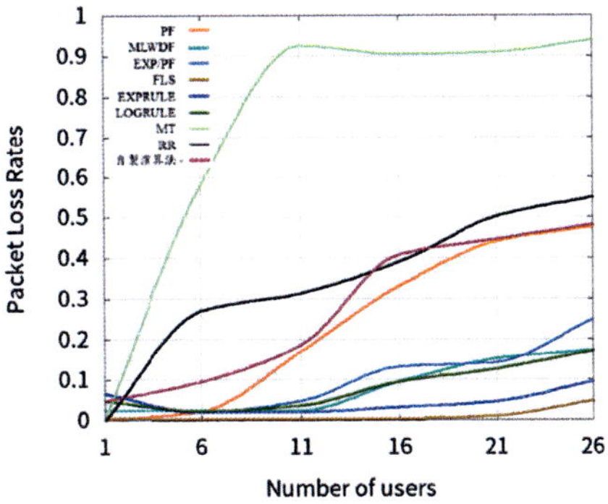

Fig. 12. Packet loss rates of Video packets.

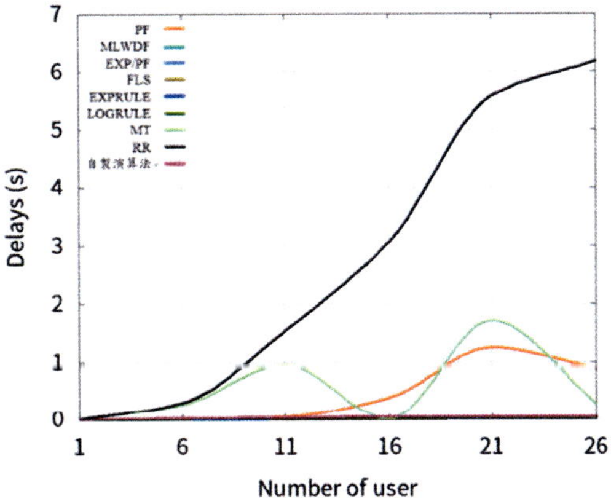

Fig. 13. Delays of Video packets.

5 Conclusion and Future Studies

With the rapid development of science and technology, for supporting diversified applications and services, 5G networks in the field of wireless networks are trying to satisfy this. The 5G network uses the concept of network slicing to yield multiple virtual networks on the physical network logically. The employed equipment, data transmission, network access and core-network operations in each network slice operate independently from others. When any virtual network, i.e., a slice, fails, it will not affect the operation of other virtual networks, thus accelerating our entry into the Internet of Things era where everything is connected.

In this study, 5GTS is used to realize the network slicing based on the time line. In 5G-air-simulator, a single base station acted as is the main base is employed, and other surrounding base stations do not serve the UE. The 5GTS simulates uRLLC, eMBB and mMTC with different types of packets. Experimental results indicate that the SRA we designed is not as efficient as other SRAs in transferring data, because other algorithms, besides time domain, also deal with frequency domain. We have proved that 5GTS is feasible to design SRAs with the concept of time as the main parameter, and performance does not improve. In other words, SRA's consideration on frequency is necessary.

In the future, we will add more functions to 5GTS. For example, the allowable delays of packets in different uRLLC QoS flows are different, and the bandwidths required by different AV/VR sessions vary. We must fine adjust the allocation policies of sliced resources so that 5GTS can work appropriately adapted to user's everyday lives. In addition, to realize the slicing of base stations in 5G network, it is also necessary to face various coefficient calculations, the fairness of spectral efficiency and performance. In terms of fairness, it is also necessary to adapt to various types of packets in our real world, such as Router Information Protocol (RIP) packets and icmp packets. Both are also transmitted through base stations. Therefore, a more sophisticated SRA algorithm is needed to create a network slice that can actually be used on base stations.

Acknowledgments. This study is financial support in part by Ministry of Science and Technology, Taiwan under the grants MOST 108-2221-E-029-009 and MOST 109-2221-E-029-017-MY2.

References

1. 5G network: https://en.wikipedia.org/wiki/5G
2. Martiradonna, M.S., Grassi, A., Piro, G., Boggia, G.: Understanding the 5G-air-simulator: a tutorial on design criteria, technical components, and reference use cases. Comput. Netw. **177**, 1–4 (2020)
3. Hu, S., Chen, Q., Ma, H.: PDCP/RLC/MAC layer is responsible for LTE Layer 2 protocol to check QoS. New Communication Components Magazine. https://www.2cm.com.tw/2cm/zh-tw/magazine/-Technology/A64431A9FDDC48F4AFFFC99688A8743D
4. Hong, Z.: A Q-learning based downlink scheduling in 5G system, Master Thesis of Tunghai University, Taiwan (2021)

Assessment of RIWM and FC-RDVM for Small and Middle Scale WMNs Considering Stadium Distribution and UNDX-M Crossover Method

Leonard Barolli(✉)

Department of Information and Communication Engineering, Fukuoka Institute of Technology, 3-30-1 Wajiro-Higashi, Higashi-Ku, Fukuoka 811-0295, Japan
barolli@fit.ac.jp

Abstract. In this paper, we present WMN-PSOHCDGA simulation system and implement two router replacement methods: RIWM and FC-RDVM for Wireless Mesh Networks (WMNs). We consider Stadium distribution of mesh clients and UNDX-m crossover method, and carry out a comparison study for small scale and middle scale WMNs. The simulation results indicate that for small scale WMNs, the SGC for RIWM and FC-RDVM is 100%, thus all mesh routers are connected. Considering NCMC, for RIWM two mesh clients are not covered, while for FC-RDVM are not covered four mesh clients. The load balancing of RIWM is better than FC-RDWM. Also for middle scale WMNs for both methods all mesh routers are connected. So, the SGC is maximized. The FC-RDVM can cover all mesh clients. However, for RIWM five clients are not covered. The load balancing for both methods is almost the same, but RIWM has slightly better load balancing than FC-RDVM.

1 Introduction

The Wireless Mesh Networks (WMNs) have many applications. They are cost-effective and have good scalability. However, they also have some drawbacks. Especially, the optimization of the locations of mesh routers is a challenging problem because there are needed many parameters which make the problem NP-Hard. Therefore, intelligent and heuristic approaches are good solutions to deal with this problem.

For mesh router node placement problem, the objective is to have a good allocation of mesh routers in the considered area in order to achieve optimal network connectivity, client coverage and load balancing. In order to measure these metrics, we consider three parameters: SGC (Size of Giant Component), NCMC (Number of Covered Mesh Clients), and NCMCpR (Number of Covered Mesh Clients per Router).

Many research works deal with the mesh node placement in WMNs [2,4–6,12] and some intelligent algorithms are already discussed and investigated [1,3,7,8].

In [9,10], some intelligent simulation systems are implemented for solving the node placement problem in WMNs. However, the authors consider only single intelligent algorithms such as PSO (Particle Swarm Optimization), HC (Hill Climbing), GA

L. Barolli (Ed.): BWCCA 2023, LNDECT 186, pp. 253–262, 2024.
https://doi.org/10.1007/978-3-031-46784-4_24

(Genetic Algorithm), and SA (Simulated Annealing). In some other works, the authors implemented a hybrid simulation system based on PSO and Distributed GA (DGA).

In this paper, by combining PSO, HC and DGA, we present a hybrid intelligent system for WMNs called WMN-PSOHCDGA. We implemented two mesh router replacement methods: Random Inertia Weight Method (RIWM) and Fast Convergence Rational Decrement of Vmax Method (FC-RDVM).

We compare the results of these methods considering Stadium distribution of mesh clients, UNDX-m crossover method and small and middle scales WMNs. The simulation results indicate that for small scale WMNs, the SGC for RIWM and FC-RDVM is 100%. While for RIWM two mesh clients are not covered, for FC-RDVM there are four mesh clients that are not covered. The load balancing of RIWM is better than FC-RDWM. Also for middle scale WMNs for both methods all mesh routers are conneted. So, the SGC is maximized. The FC-RDVM can cover all mesh clients. However, for RIWM five clients are not covered. The load balacing for both methods is almost the same, but RIWM has slightly better laod balancing than FC-RDVM.

We organize the paper as follows. After introduction section, the intelligent algorithms are described in Sect. 2. Then, WMN-PSOHCDGA simulation system is Sect. 3. The simulation results are discussed in Sect. 4. Finally, conclusions and future work are given in Sect. 5.

2 PSO, HC and DGA Intelligent Algorithms

In this section, we give a short description of PSO, HC and DGA algorithms, which are integrated to design and implement the hybrid intelligent system.

2.1 Particle Swarm Optimization

The PSO algorithm considers a group of particles to explore the search space and find optimal solutions. Each member of the group can benefit from the experience of others by the interaction between the particles, thus a better solution can be found.

By considering two locations: the best-known location and the best location achieved by other flock members each particle moves toward the optimal solution. This process is repeated iteratively untill all particles converge to the global optimum or a good local optimum that satisfies the application requirement.

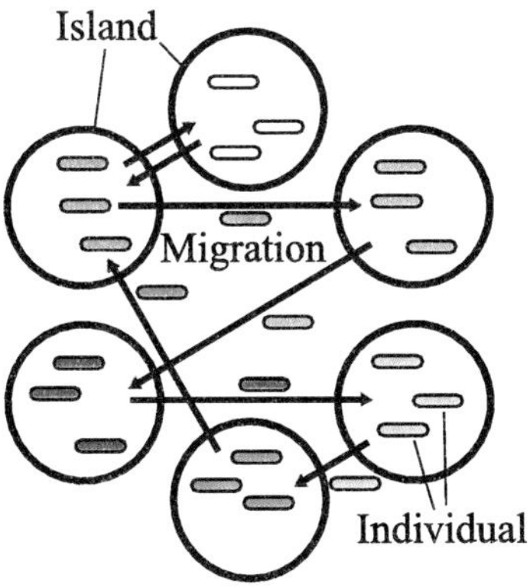

Fig. 1. Model of Migration in DGA.

2.2 Hill Climbing Algorithm

The HC algorithm is a heuristic approach that searches a sufficiently good solution to the problem, but this may not be the global optimal solution. It moves in the direction to find the mountain peak, which is the best solution for a given problem.

The HC algorithm can be used for different optimization mathematical problems. It is also called greedy local search because considers only good immediate neighbor state. Furthermore, it does not need to handle the search tree or graph because the algorithm only keeps a single current state.

2.3 Distributed Genetic Algorithm

Genetic Algorithms (GA) is a powerful search technique, which has the ability to avoid the local optimum. While, Distributed GA (DGA) is a better approach and can escape from local optimum by considering multiple islands. As shown in Fig. 1, each island operates independently, but the individuals are migrated between islands.

The main operators of GA are Selection, Crossover and Mutations. Also, there are different methods for implementing these operators. In this paper, we consider UNDX-m and SPX crossover methods. The UNDX crossover method is shown in Fig. 2. We implement in our simulation system UNDX-m, which is the multi-parental version of the UNDX. This crossover method can improve the search performance of the algorithm.

In UNDX-m are selected m + 2 parents and the first m + 1 parents are used to span the m-dimensional sub-space. The distribution of children is done by normal distribution. By the third parent is given perturbation to the children in order to remaining in

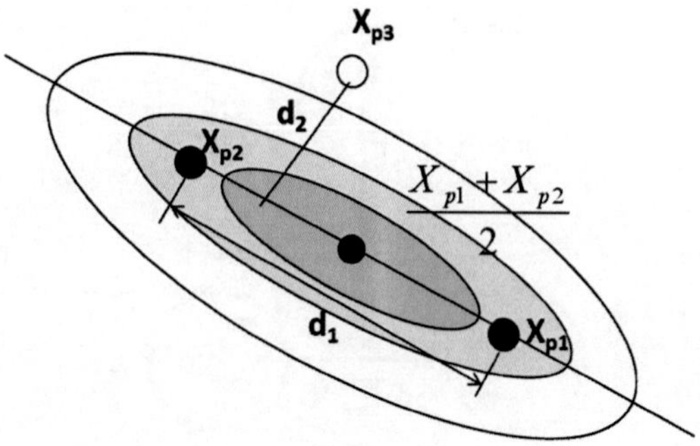

Fig. 2. UNDX crossover method.

the sub-space. When *m* is one (UNDX-1), the method is the same with original UNDX. By using UNDX-m, the mean vector and the covariance matrix of the population can be preserved. Therefore, another model should be used to search through the different individuals in the population.

3 WMN-PSOHCDGA Simulation System Design

We show the flowchart of WMN-PSOHCDGA simulation system in Fig. 3. The proposed system generates randomly the initial solution by *ad hoc* methods [13]. Then, it optimizes each island by using DGA and PSO algorithms. Finally, the HC algorithm is used to improve the convergence of PSO algorithm. The most crucial aspect of HC is defining the neighbour solution, which directly impacts the performance of the HC algorithm. In our WMN-PSOHCDGA system, we use the next locations of particle-pattern as the neighbour solutions.

In WMN-PSOHCDGA system, the initial solution and velocity of particles are generated randomly and a particle is considered as a mesh router. We consider the position of mesh routers and mesh clients to calculate the fitness value of a particle-pattern. A WMN is represented by a gene and each individual is a combination of mesh routers.

The fitness function of WMN-PSOHCDGA is shown by Eq. (1), where the weight coefficients are α, β, and γ ($\alpha+\beta+\gamma=1$), while SGC, NCNC and NCMCpR are metrics for network connectivity, client coverage and load balancing, respectively.

$$Fitness = \alpha \times SGC(\boldsymbol{x}_{ij},\boldsymbol{y}_{ij}) + \beta \times NCMC(\boldsymbol{x}_{ij},\boldsymbol{y}_{ij}) + \gamma \times NCMCpR(\boldsymbol{x}_{ij},\boldsymbol{y}_{ij}). \quad (1)$$

In this work, we distribute the mesh clients by Stadium distribution as shown in Fig. 4.

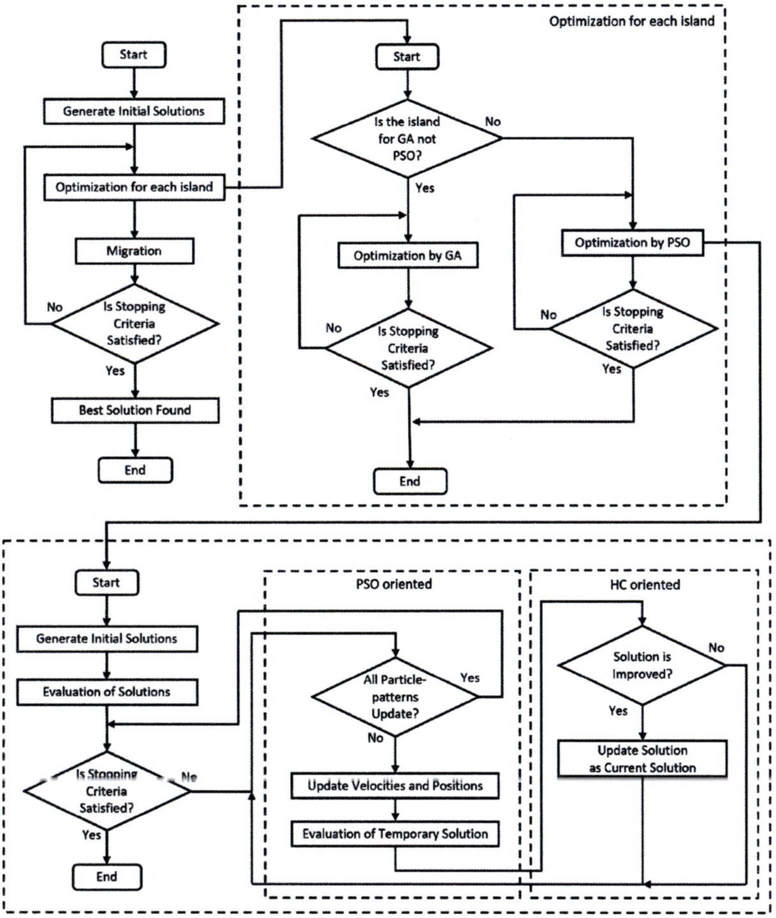

Fig. 3. Flowchart of WMN-PSOHCDGA system.

In this study, we consider RIWM and FC-RDVM router replacement methods. In RIWM, the ω is changed randomly from 0.5 to 1.0 and is estimated by the weak stable region. The average value of ω is 0.75. While, the C_1 and C_2 are kept 2.0 as constant values. In FC-RDVM [11], V_{max} is maximum velocity of particles and is calculated by Eq. (2).

$$V_{max}(k) = \sqrt{W^2 + H^2} \times \frac{T-k}{T+\delta k} \tag{2}$$

In Eq. (2), W and H are width and height of the considered area. While, T is the total number of iterations and k is the current number of iterations changing from 1 to T. Finally, δ is the curvature parameter.

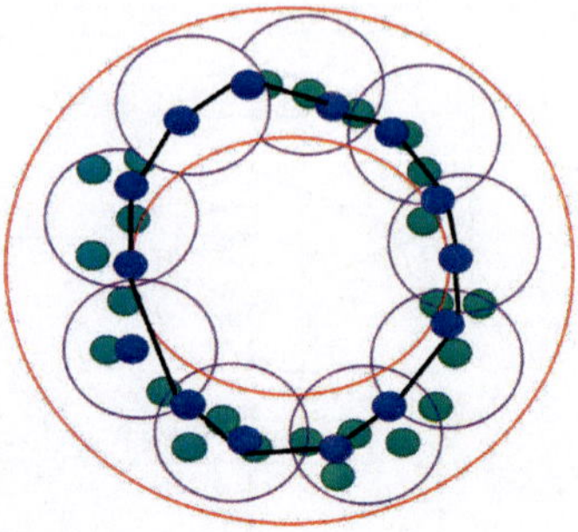

Fig. 4. A case of Stadium distribution of mesh clients.

4 Simulation Results

We evaluate RIWM and FC-RDVM router replacement methods by computer simulations. The coefficients of the fitness function are decided by many simulations: $\alpha = 0.8$, $\beta = 0.1$, and $\gamma = 0.1$. The other simulation parameters are shown in Table 1.

Table 1. Parameters used for simulations.

Parameters	Values	
	Small Scale WMN	Middle Scale WMN
$\alpha : \beta : \gamma$	8 : 1 : 1	
Number of GA Islands	16	
Number of Evolution Steps	10	
Number of Migrations	200	
Number of Mesh Routers	16	32
Number of Mesh Clients	48	96
Mesh Client Distribution	Stadium Distribution	
Selection Method	Rulette Selection Method	
Corssover Method	UNDX-m	
Mutation Method	Uniform Mutation	

We show the visualization results for small scale WMNs in Fig. 5. In Fig. 5(a) and Fig. 5(b), we show the SGC for RIWM and FC-RDVM, respectively. We can see that for both methods the SGC is 100%, which indicates that all mesh routers are connected.

We show in Fig. 6 the NCMC for small scale WMN for each mesh router. Considering Fig. 5 and Fig. 6, for RIWM two mesh clients are not covered, while for FC-RDVM there are not covered four mesh clients.

In Fig. 7, we show r parameter for small scale WMNs, which is the correlation coefficient used for load balancing. The load balancing of RIWM is better than FC-RDWM because the standard deviation is decreased with the increase of the number of updates.

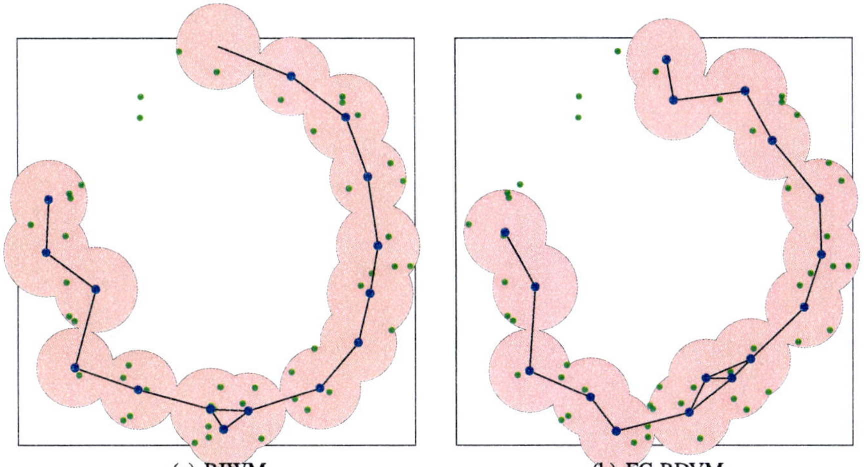

(a) RIWM (b) FC-RDVM

Fig. 5. Visualization results after optimization for small scale WMNs.

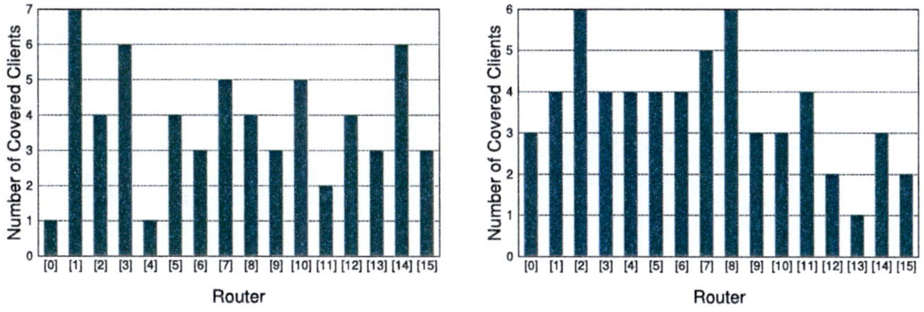

(a) RIWM (b) FC-RDVM

Fig. 6. Number of covered mesh clients for small scale WMNs.

In Fig. 8 are shown the visualization results for middle scale WMNs. Also, for middle scale WMNs, for both methods all mesh routers are conneted. So, the SGC is maximized.

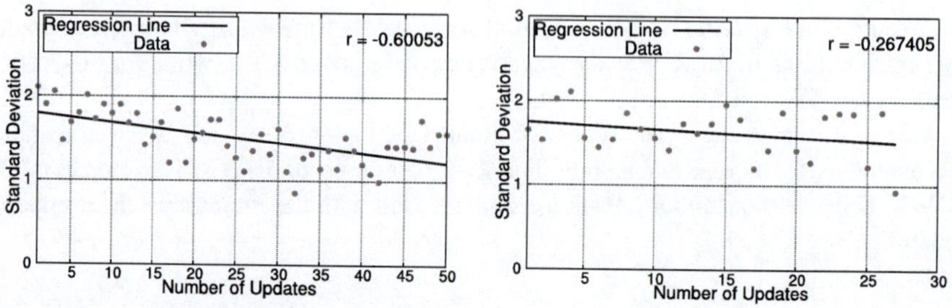

(a) RIWM (b) FC-RDVM

Fig. 7. Standard deviation for small scale WMNs.

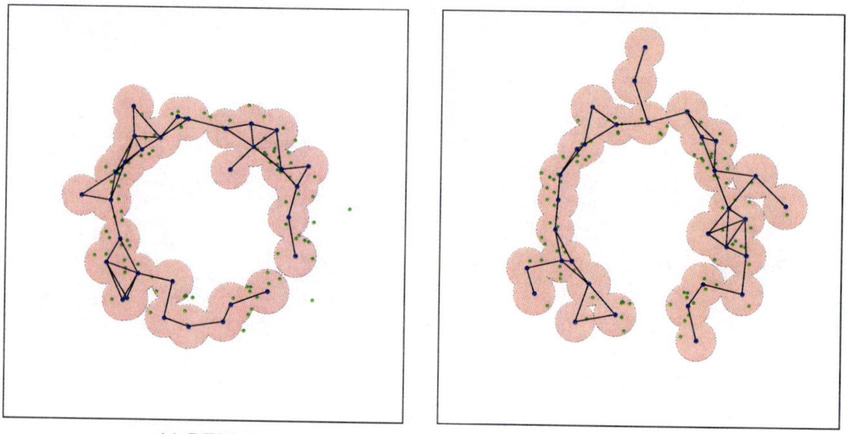

(a) RIWM (b) FC-RDVM

Fig. 8. Visualization results after optimization for small scale WMNs.

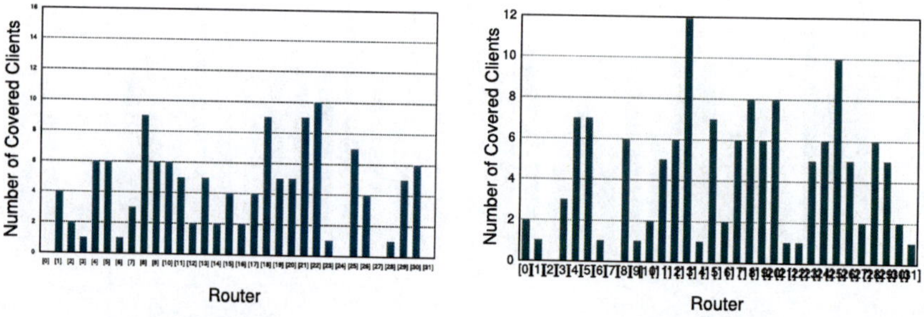

(a) RIWM (b) FC-RDVM

Fig. 9. Number of covered mesh clients for middle scale WMNs.

We show in Fig. 9 the NCMC for middle scale WMNs. Considering Fig. 8 and Fig. 9, we can see that FC-RDVM can cover all mesh clients. However, for RIWM five clients are not covered.

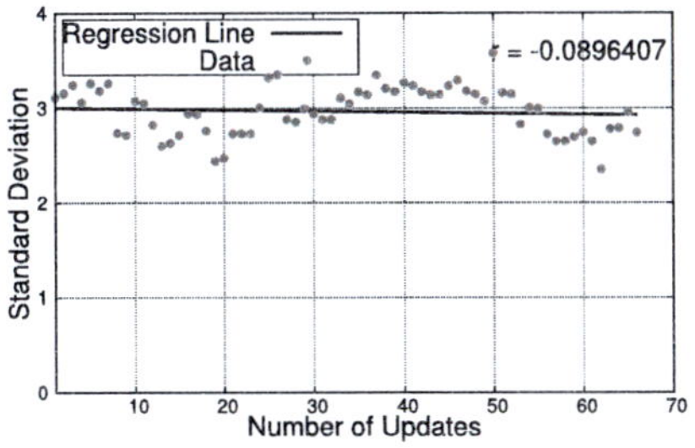

(a) RIWM

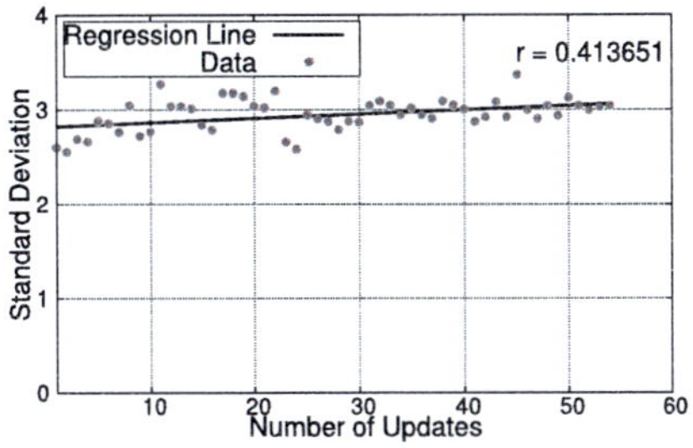

(b) FC-RDVM

Fig. 10. Standard deviation for middle scale WMNs.

We show the indicator of load balancing for middle scale networks in Fig. 10. We can see that the load balacing for both methods is almost the same, but RIWM has slightly better laod balancing than FC-RDVM.

5 Conclusions

In this study, we considered Stadium distribution of mesh clients and carried out a comparison study of RIWM and FC-RDVM for small and middle scale WMNs. From the simulation results, we conclude as follows.

- For small scale WMNs, the SGC for RIWM and FC-RDVM is 100%. While for RIWM two mesh clients are not covered, for FC-RDVM there are four mesh clients that are not covered. The load balancing of RIWM is better than FC-RDWM.
- Also for middle scale WMNs for both methods all mesh routers are conneted. So, the SGC is maximized. The FC-RDVM can cover all mesh clients. However, for RIWM five clients are not covered. The load balacing for both methods is almost the same, but RIWM has slightly better laod balancing than FC-RDVM.

In future work, we will implement in WMN-PSOHCDGA simulation system other crossover, mutation and router replacement methods.

References

1. Barolli, A., Sakamoto, S., Ozera, K., Barolli, L., Kulla, E., Takizawa, M.: Design and implementation of a hybrid intelligent system based on particle swarm optimization and distributed genetic algorithm. In: Barolli, L., Xhafa, F., Javaid, N., Spaho, E., Kolici, V. (eds.) EIDWT 2018. LNDECT, vol. 17, pp. 79–93. Springer, Cham (2018). https://doi.org/10.1007/978-3-319-75928-9_7
2. Franklin, A.A., Murthy, C.S.R.: Node placement algorithm for deployment of two-tier wireless mesh networks. In: Proceedings of Global Telecommunications Conference, pp. 4823–4827 (2007)

3. Girgis, M.R., Mahmoud, T.M., Abdullatif, B.A., Rabie, A.M.: Solving the wireless mesh network design problem using genetic algorithm and simulated annealing optimization methods. Int. J. Comput. Appl. **96**(11), 1–10 (2014)
4. Lim, A., Rodrigues, B., Wang, F., Xu, Z.: k-center problems with minimum coverage. Theoret. Comput. Sci. **332**(1–3), 1–17 (2005)
5. Maolin, T., et al.: Gateways placement in backbone wireless mesh networks. Int. J. Commun. Netw. Syst. Sci. **2**(1), 44–50 (2009)
6. Muthaiah, S.N., Rosenberg, C.P.: Single gateway placement in wireless mesh networks. In: Proceedings of 8th International IEEE Symposium on Computer Networks, pp. 4754–4759 (2008)
7. Naka, S., Genji, T., Yura, T., Fukuyama, Y.: A hybrid particle swarm optimization for distribution state estimation. IEEE Trans. Power Syst. **18**(1), 60–68 (2003)
8. Sakamoto, S., Kulla, E., Oda, T., Ikeda, M., Barolli, L., Xhafa, F.: A comparison study of simulated annealing and genetic algorithm for node placement problem in wireless mesh networks. J. Mob. Multimed. **9**(1–2), 101–110 (2013)
9. Sakamoto, S., Kulla, E., Oda, T., Ikeda, M., Barolli, L., Xhafa, F.: A comparison study of hill climbing, simulated annealing and genetic algorithm for node placement problem in WMNs. J. High Speed Netw. **20**(1), 55–66 (2014)
10. Sakamoto, S., Oda, T., Ikeda, M., Barolli, L., Xhafa, F.: Implementation and evaluation of a simulation system based on particle swarm optimisation for node placement problem in wireless mesh networks. Int. J. Commun. Netw. Distrib. Syst. **17**(1), 1–13 (2016)
11. Sakamoto, S., Barolli, A., Liu, Y., Kulla, E., Barolli, L., Takizawa, M.: A fast convergence RDVM for router placement in WMNs: performance comparison of FC-RDVM with RDVM by WMN-PSOHC hybrid intelligent system. In: Barolli, L. (ed.) CISIS 2022. LNNS, vol. 497, pp. 17–25. Springer, Cham (2022). https://doi.org/10.1007/978-3-031-08812-4_3
12. Wang, J., Xie, B., Cai, K., Agrawal, D.P.: Efficient mesh router placement in wireless mesh networks. In: Proceedings of IEEE International Conference on Mobile Adhoc and Sensor Systems (MASS-2007) pp 1–9 (2007)
13. Xhafa, F., Sanchez, C., Barolli, L.: Ad hoc and neighborhood search methods for placement of mesh routers in wireless mesh networks. In: Proceedings of 29th IEEE International Conference on Distributed Computing Systems Workshops (ICDCS-2009), pp. 400–405 (2009)

A New DTN Routing Strategy Considering Age of Information

Fuka Isayama and Tetsuya Shigeyasu(✉)

Prefectual University of Hiroshima, 1-1-71 Ujinahigashi, Minamiku, Hiroshimashi, Hiroshima-ken 734-8558, Japan
o012012vp@ed.pu-hiroshima.ac.jp, sigeyasu@pu-hiroshima.ac.jp

Abstract. Recently, DTN (Delay/Disruption Tolerant Networking) is getting much attentions as a communication architecture at a disaster affected area. Immediately after a disaster, countermeasure office must collect fresh disaster information for reducing disaster damages. This paper discusses how to improve freshness of disaster information. Then, we propose new method considering AoI (Age of Information). Mainly, our method employs two mechanisms: relay buffer management based on message originator, and changing order of message transmissions according to the message priority. By the results of simulation, we confirm that our proposal effectively shorten the AoI by comparison with the traditional DTN relay methods. In addition, this paper also discusses the situations that the network contains multiple senders having same priority. In order to cope with the situations, this paper also proposes to decide the transmission order according to a message creation time.

1 Introduction

It is obviously that acquiring fresh information relating disaster affected area is most important for rapid disaster relief activities minimizing disaster damages. For collecting such the emergency information, DTN (Delay/Disruption Tolerant Networking) [1–3] attracts much attentions from network researchers as a communication infrastructure under the unstable state of existing public communication infrastructures. The DTN realizes robust communications by its mechanisms: Store, Carry and Forward (SCF).

Needless to say, after the occurence of a disaster, disaster damages may extent in accordance with the time progress. Hence, it is highly required to grasp continuously the conditions of disaster information, especially for the monitoring information (e.g. health condition of evacuee and situation of fire site, etc.) among disaster countermeasure office and disaster affected areas.

In general, network quality have been evaluated in terms of packet (message) based evaluation items without considering a continuously of the information. For examples, network throughput, packet delivery rate and delay are the some of those parameters.

For well grasping a disaster monitoring information changing continuously, using AoI (Age of Information) [4,5] measuring freshness of information, is adequate. The conventional DTN routing algorithms, however, have been developed without considering the AoI.

L. Barolli (Ed.): BWCCA 2023, LNDECT 186, pp. 263–272, 2024.
https://doi.org/10.1007/978-3-031-46784-4_25

In this paper, we discuss the disaster information system based on DTN for collecting monitoring information in terms of AoI. For improving the AoI of DTN, we propose two methods: 1) allocating receiving buffer according to a message originator, and 2) transmission order of messages according to priority of message contents. By the results of evaluations, this paper clarifies that AoI of messages having higher priority are well improved by the above two methods.

In addition to the above, this paper also discusses the situation that the network contains multiple message originators having same message priority, and, reports that this situation degrades the AoI by consumption of the relay buffers. Then, to deal with the issue, this paper further proposes to use parameter, message creation time, for deciding transmission order.

2 Measuring Freshness of Information by AoI

AoI indicates the freshness of information changing continuously in accordance with the time progress. AoI is adequate to use as an index for evaluating quality of monitoring information. For example, a room temperature is one of good examples for monitoring information. It changes its value continuously although the ID of the information (e.g. temperature of room 3) does not change. AoI is derived using both of a message creation time and a current time. AoI A_t at the time t is calculated by the following:

$$A_t = t - \eta_t \tag{1}$$

Here, η_t is a creation time of the latest message received at the client. Naturally, if the client does not receive the other subsequent latest message, A_t increases at the time passes. Figure 1 shows the image of a behavior of AoI. AoI behaves like a *saw wave*. As shown in the figure, a value of AoI rapidly decreases when the latest message is received at the monitor client. Even though the instantaneous value of AoI is calculated by the

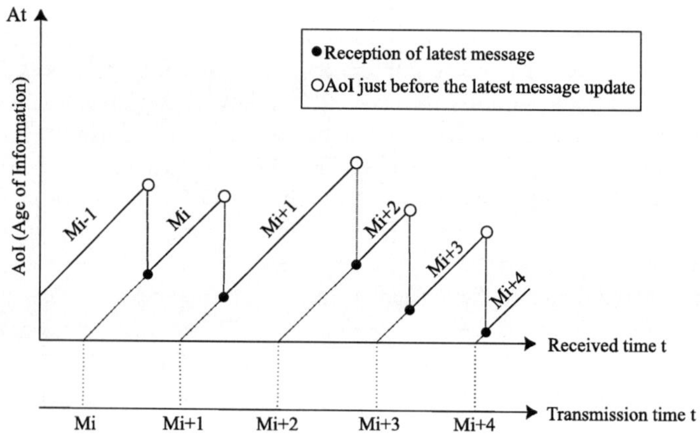

Fig. 1. An Image of AoI.

Eq. (1), whole AoI characteristics of the focused system is evaluated by the following equation:

$$m_A := \lim_{T \to \infty} \frac{1}{T} \int_0^T A_t dt \quad (2)$$

In the following sections, we use the values m_A derived from the above equation as a communication quality of the each comparative methods.

3 Disaster Information System Considering AoI

3.1 Network Model of Disaster Affected Area

Figure 2 shows the network model of disaster information system of disaster affected area. This model constructed on a filed of a square with sides of 3 km. A node put on the center of the field is a receiver node that is assumed as a countermeasure office or a shelter. 3 sender nodes, s_1, s_2 and s_3 transmit their message as a updated disaster information to the center placed node. Nodes, named $w_i (1 \leq i \leq 100)$ are the relay nodes. Those nodes relay the received message from senders while moving.

3.2 Message Relay Methods for Disaster Information

This paper assumed that all methods including comparatives relay messages based on the Epidemic Routing [6].

When the nodes contact each other by movement, nodes exchange messages each other. On a message exchange, node transmits a messages in order defined by method. In this paper, we use 5 comparative methods: RANDOM, FORWARD, OLDEST, NEWEST.

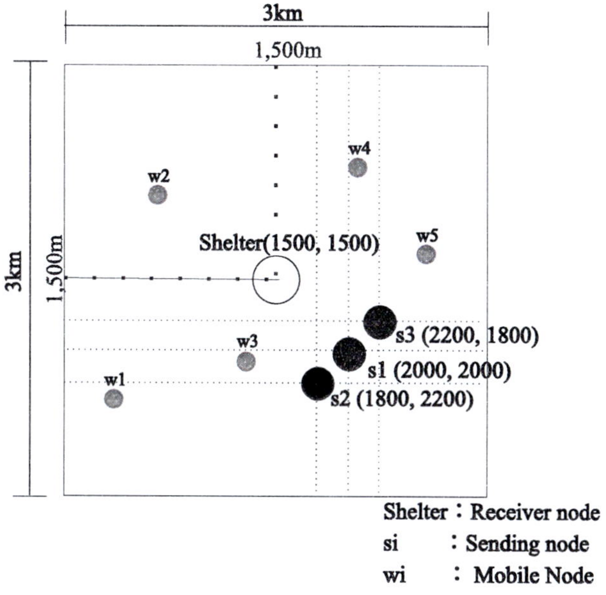

Fig. 2. Evaluation network to pology.

- RANDOM transmits messages in order of randomly selected.
- FORWARD transmits messages in order of smaller number of forwarded.
- OLDEST transmits messages in order of longer elapsed time since it created.
- NEWEST transmits messages in order of shortest elapsed time since it created. NEWEST is a reverse order of the OLDEST.

Meanwhile, the buffer management methods for these comparative methods employ method intimately related to the transmission order method.

3.3 A New DTN Routing Algorithm Considering AoI

In the followings, this paper proposes a new DTN routing algorithm considering AoI. The proposal consists of 2 parts: buffer management for received message, and, transmission order management based on the message priority.

3.3.1 Buffer Management for Received Message

As the Sect. 2 described that a instantaneous value of AoI is decreased when the reception of latest message which is created newer than the receiver has ever received. On the contrary, the AoI is not improved by the reception of a massage older than the messages the client has already received. So, the message relays of older messages than already received, is worthless.

Hence, in this paper, our proposal discards messages having the same ID except the latest message if the relay node has multiple messages having same ID by a reception, for the benefit of improving AoI while reducing network traffic.

3.3.2 Transmission Order Management Based on the Message Priority

Needless to say, information relating to the disaster, covers a wide range of topics. For examples, there are information both of injuries and diseases those caused by disaster. These types of information must be deal with sensitively if the condition is changed. In contrast, there are also information like number of relief materials, number of completed evacuees, weather information. These types of information has relatively plenty of time to deal.

Hence, these information requires different freshness. To deal with the issue, this paper proposes to change transmission order based on the message priority [7,8].

In this proposal, we grade *RANK* on each message according to its type of information.

On the transmission, relay node transmits the message in a order of higher *RANK* of messages.

On the reception of messages, if the sum of number of newly received messages and messages stored in buffer exceeds the total buffer size, receiver node discards messages in the order of lower *RANK* and keep the buffered messages less than maximum buffer size.

For simplicity of discussions, hereafter, this paper assumes that *RANK* is assigned to each message originator.

4 Performance Evaluations

This paper reports results of performance evaluations. For the evaluations, we use the network simulator One (The Opportunistic Network Environment Simulator) [9].

Table 1. Simulation parameters.

item	value
Field size	3×3 [km^2]
Number of sending nodes (fixed)	3
Number of receiving nodes (fixed)	1
Number of mobile nodes	100
Simulation time	43,200 [s]
Communication distance	100 [m]
Communication speed	2 [Mbps]
Message Size	1 [Mbyte]
Buffer size	10 [Mbyte]
Message generation interval	100–150 [s]
Movement model	RandomWaypoint [8]
Movement speed of mobile node	3.0–4.0 [m/s]

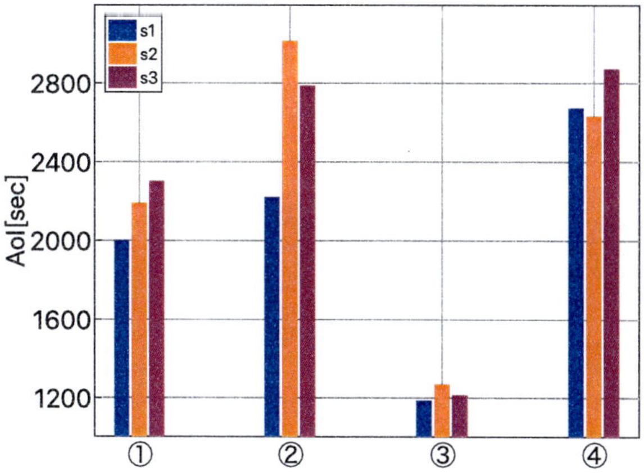

Fig. 3. Characteristics of AoIs on conventional relay methods.

4.1 Conditions of Evaluation

Table 1 shows the parameters used for the evaluations. As shown in the table, we placed the 3 message originators. The message originators, named, s_1, s_2, s_3 generate the messages with the *RANK* 1, 2 and 3, respectively. On the evaluations, we assume that the smaller value of the *RANK* has more higher priority. We use the value of evaluation

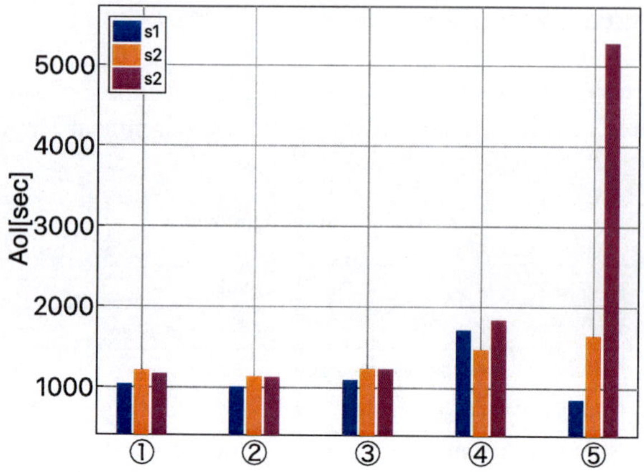

Fig. 4. Characteristics of AoIs by methods employing proposed buffer management.

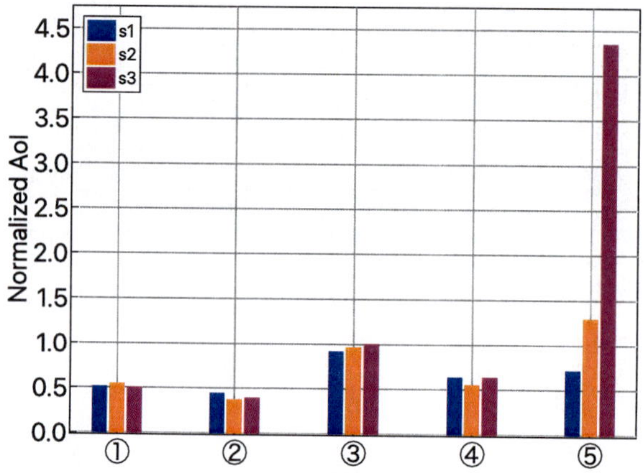

Fig. 5. Characteristics of normalized AoIs by methods employing proposed buffer management.

with the average of 10 times trials. In the followings, we indicate the method name with two strings that are way of transmission order and way of buffer management, respectively. For example, RANDOM/FIFO means this method selects the transmission order in random, and manage the receiving buffer by FIFO (First-In-First-Out). In the following figures, RANDOM/FIFO, RANDOM/RANDOM, NEWEST/OLDEST, FORWORD/FORWORD, RANK/RANK are indicated by the IDs, ①, ②, ③, ④, ⑤, respectively.

4.2 Basic Evaluation of Traditional Relay Method

Figure 3 shows the results of basic performance evaluations of traditional relay methods no consideration of AoI. As shown in this figure, method ③, NEWEST/OLDEST outperforms the other methods in AoI. As described before, NEST/OLDEST selects both of transmission order and a message to drop, based on the creation time of the messages. According to the results, we can confirm that for improving the AoI, it is key to prioritize the message creation time prior than the other parameters (e.g. reception time at relay node, number of forwarded.). However, prioritization of message creation time at the message relay is not enough as a decision parameter for disaster information system. The reason of that is message priorities (*RANK*) are not reflected to AoI. AoI of s_1 is not smaller than s_2 even though the *RANK* of s_1 is higher than s_2. If we can not prioritize a highly emergency information, it may not reduce the damages of disaster, effectively.

4.3 Performance Evaluation of Effects of Buffer Management and Transmission Order Control Considering AoI

This section evaluates the effects of proposed buffer management and transmission order control mentioned in Sects. 3.3.1 and 3.3.2. This methods manage the received buffer and transmission order on each message originator and *RANK*. Especially for the buffer management, a relay node stores only one latest message on each message originator. This method reduces both total volume of traffic and number of transmissions. The results of the evaluations are shown in Fig. 4. As shown in the figure, entire AoIs except the s_3 on method ⑤ are improved by the proposed buffer management. It is thought that these improvements are the benefit of buffer management based on a message originator. However, among the methods ①–④, we can not conclude that the AoI of each originator is not well controlled. In other words, AoIs are simply shorten regardless the consideration of the priority $s_1 \leq s_2 \leq s_3$. On the contrarily, in the case of method ⑤, values of AoI is well controlled while keeping priority of message originator though the AoI of s_3 are enlarged. The value of s_1 on method indicates that the method employing both of buffer management and changing transmission order based on *RANK* of each message originator. Hence, it can be conclude that our proposed buffer management contributes to shorten the AoI, and our proposed transmission order control by based on *RANK* contributes to keep the priority of AoI on each message originator. Figure 5 shows the normalized AoI of each method employing proposed buffer management. Each value of AoI is normalized by the value of AoI without our proposal. For the method ⑤, valued of ③ (which is the method achieved most shortest AoIs) on Fig. 3 is used. As shown in the figure, almost normalized AoIs are smaller than 1.0. Then, the AoIs are well reduced by our proposal. In case of the ⑤, normalized *AoI* of s_1 is also smaller than 1.0 (approximately 0.7). So, it can be conclude that we can control AoI of important message based on its priority by introducing our proposal.

4.4 Advanced Buffer Management for Large Number of Message Originators

Figure 6 shows the network topology which consists of large number of message originators. In this topology, there are 10 message originators: 5 *RANK* 3 nodes (s_1, s_2, s_3, s_4, s_5), 3 *RANK* 2 nodes (s_6, s_7, s_8), and, 2 *RANK* 1 nodes (s_9, s_{10}).

At the evaluation, buffer size of the relay nodes are set to 5 [Mbyte], so, the nodes only messages from 5 message originators can store at most. For improving AoI of messages according to the priority of messages, these 5 relay buffers must be allocated according to the *RANK* of message originators.

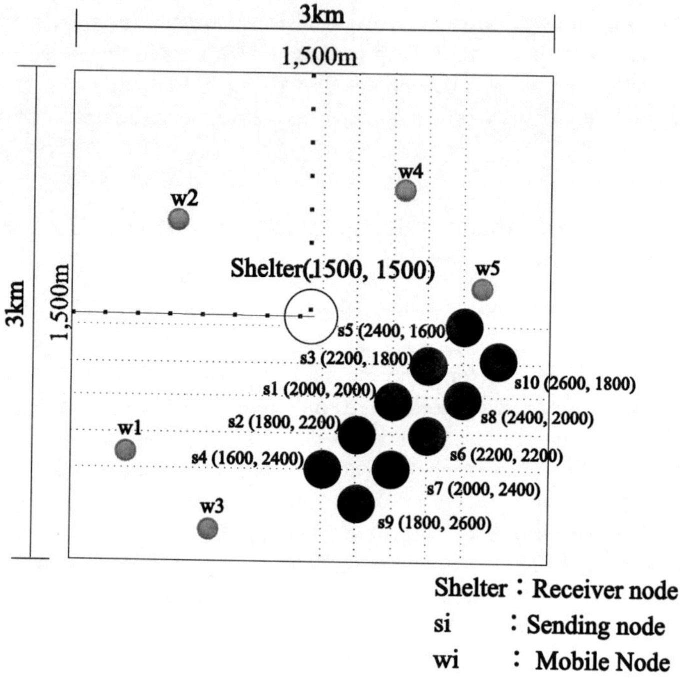

Fig. 6. The network topology which consists of large number of message originators

Hence, in the case of the situation described the above, buffers of relay nodes should be firstly allocated to s_9 and s_{10}. The number of remained buffers is 3, then, it equals to the number of message originators having next higher *RANK* 2. Then, our proposal secondly allocates the remained buffers to 3 nodes: s_6, s_7, s_8.

Figure 7 shows the results of AoIs with considering both of *RANK* and creation time. As shown in the figure, by considering both of *RANK* and creation time, AoI of nodes having higher *RANK* are correctly measured while rejecting the messages from nodes having lowest *RANK*.

In order for further investigate the effects of the above proposal, Fig. 8 shows the results of AoIs where all message originators are allocated same *RANK*. Although all created messages are relayed by considering both of *RANK* and the creation time, these AoI are varied widely. Unfortunately, these results confirmed that fairness among the nodes are not achieved even though all nodes having same *RANK*. The reason for the above is thought that the AoIs are great affected by the location of message originators. For more detail, s_6 to s_{10} are in the back row, then those nodes locate relatively distant than s_1 to s_5 and it takes longer time to reach to the same distant relay node.

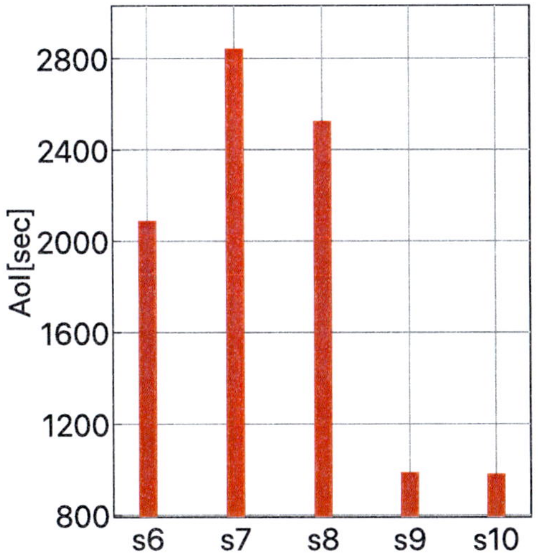

Fig. 7. The results of AoIs with considering both of *RANK* and creation time

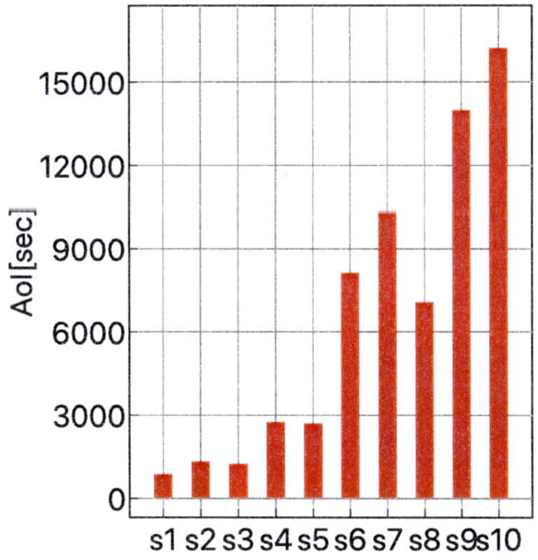

Fig. 8. The results of AoIs where all message originators are allocated same *RANK*

Hence, it is needed to consider an another condition to judge the priority of the messages in addition to both of the *RANK* and message creation time, for achieving fairness of AoI among message originators when the number of originators having same *RANK* is larger than the number of relay butter.

5 Conclusion

Under the disaster, it is highly required to deliver fresh information for effective disaster relief activities. A type of information updated continuously, is not adequate to be evaluated by traditional network performance indicators, such as throughput and delay, jitter, etc. Hence, we discussed the how the freshness of such the disaster information is influenced using AoI. From the results of evaluations of existing DTN routing algorithm, AoI can be improved by considering information priority (the is to say, *RANK*) and message creation time. However, if the number of message originators having same *RANK* is greater than number of relay buffer size, it is also clarified that the additional parameter for selecting a message to store, is needed. In the future, we will deal with the issue and will examine the proposal under the various conditions.

References

1. Farrell, S., Cahill, V., Geraghty, D., Humphreys, I., McDonald, P.: When TCP breaks: delay- and disruption-tolerant networking. IEEE Internet Comput. **10**(4), 72–78 (2006)
2. Farrell, S., Cahill, V.: Delay and Disruption Tolerant Networking. Artech House (2006)
3. Fall, K.: A delay-tolerant network architecture for challenged inter-nets. In: ACM SIGCOMM 2003 Conference on Applications, Technologies, Architectures, and Protocols for Computer Communication, pp. 27–34 (2003)
4. Talak, R., Karaman, S., Modiano, E.: Improving age of information in wireless networks with perfect channel state information. IEEE/ACM Trans. Netw. **28**, 1765–1778 (2020)
5. Inoue, Y., Takine, T.: Age of informatioin: basic concepts and its theoretical analysis. IEICE Fundam. Rev. **13**(3), 197–208 (2020)
6. Vahdat, A., Becker, D.: Epidemic routing for partially connected ad hoc networks. Technical report, CS-2000-06. Duke University, Durham (2000)
7. Tajima, H., Yusuke, N., Mizuno, O.: Design of information system in disaster-robust zero energy vehicle. In: 2019 International Conference on Information and Communication Technologies for Disaster Management (ICT-DM), pp. 1–8 (2019)
8. Broch, J., Maltz, D.A., Johnson, D.B., Hu, Y.-C., Jetcheva, J.: A performance comparison of multihop wireless ad hoc network routing protocols. In: ACM/IEEE International Conference on Mobile Computing and Networking, pp. 85–97 (1998)
9. Keranen, A.: The ONE –The Opportunistic Network Environment simulator. https://akeranen.github.io/the-one/

DTN Routing Method Based on Records of Data Transmission Paths

Kazunori Ueda[1(✉)] and Kouki Yano[2]

[1] Koeki University (Tohoku University of Community Service and Science), 3-5-1 Iimoriyama, Sakata, Yamagata 998-8580, Japan
ueda-kazunori@koeki-u.ac.jp

[2] Kochi University of Technology, 185 Miyanokuchi, Tosayamada, Kami City, Kochi 782-8502, Japan

Abstract. The authors focus on the message ferrying method, which is one of the route selection methods on DTN (Delay/Disruption-Tolerant Networking). In the message ferrying method, mobile nodes called ferry nodes transfer data among clusters in which there are many nodes including sink nodes. In the previous research of the message ferrying method, messages therefore are aggregated in nodes close to routes of ferry nodes. However, in the research, the method for data transferring to nodes in the clusters was not discussed. In this paper, the authors proposed a method that used records of the data transferred paths when data are aggregated to decide a route.

1 Introduction

In the event of a large-scale disaster, networks based on conventional technologies may be disrupted due to collapsed or disconnected telecommunications infrastructure. In such a situation, an alternative network that is not affected by network failures is necessary to quickly provide or get information such as the location of evacuation centers and safety confirmation.

In recent years, Delay Tolerant Networking (DTN), which can construct networks even in poor networking environments by being tolerant of delays and disconnections, has been proposed and studied. The message ferry method, which is one of the routing methods in DTN, is expected to ensure stable forwarding routes. In this method, a node that is called a ferry node goes around clusters in which there are many mobile nodes. The ferry nodes receive data from source nodes and transfer the data to destination nodes that do not exist in the communication range of the source nodes. However, the message ferrying method has the problem that it is hard to transfer from a node that is not near the ferry node because routes to ferry nodes are not established. For this reason, one of the authors proposed the intensive message ferry routing method, in which data are collected to nodes that are likely to be able to communicate with the ferry node in advance. By this method, data can be transferred from nodes outside the ferry node's communication range. However, it is not effective in transferring data from the ferry node to each node.

L. Barolli (Ed.): BWCCA 2023, LNDECT 186, pp. 273–281, 2024.
https://doi.org/10.1007/978-3-031-46784-4_26

In this research, the proposed routing method is modified to reduce the number of hops and to improve the data arrival rate in two-way communication for long-term operation for network environments. The proposed method is implemented on a simulator and evaluated by comparing the proposed method with a conventional method, that is the intensive message ferry routing method.

2 Delay Tolerant Networking

This section shows an overview of DTN and the message ferry method, which is one of the routing methods for DTN. In addition, the aggregated message ferry method, which is based on the message ferry method and was proposed by the authors, is explained.

2.1 Background of DTN

DTN technology was proposed for realizing communications in poor environments such as satellite communication networks and environmental monitoring, where extreme delays and intermittent connections. However, as networks are used for various usage, DTN technology is now attracting attention as a temporary network in the event of damage to communications infrastructure due to disasters or as a solution to the digital divide caused by geographical conditions.

2.2 Data Transferring in DTN

DTN adopts the method called store-and-forward transfer, in which each node stores the transferred data and starts transferring it if the neighboring node is available for communication. In this communication method, each node can transfer data to other nodes by moving, therefore, it is possible to transfer data in an environment where there is no stable communication path. First, data are transferred from the source node to the mobile nodes in the communication range. After that, mobile nodes receive the data and repeat the procedures of transferring the data to a mobile node that has entered the communication range until the transfer to the destination node is completed. In this way, it is possible to transfer data to the destination node via other nodes even in situations where direct communication to the destination node is not possible.

2.3 Routing Methods for DTN

There are several schemes for selecting the forwarding path to the destination node in DTNs, each of which has its advantages and disadvantages. An example is Epidemic Routing [1], which forwards data to all neighboring nodes and spreads them to other clusters to ensure that the data reaches the destination node, but this method consumes a lot of resources such as batteries and buffers [2]. Therefore, a method to limit the number of hops was proposed, but the data may not reach the destination node or may take a long time to reach the destination node. Spray and Wait [3] was proposed as a method to solve such a problem. This method predicts the optimal number of data replicas based on the network size and number of nodes and has achieved stable data arrival rates while preventing the generation of large amounts of data. However, it is difficult to set parameters based on the network size and the number of nodes because the network size and the number of nodes increase or decrease in places where many people are coming and going.

2.4 Message Ferry Method

The message ferry method [4] is one of the routing methods of DTNs and relays data via ferry nodes that systematically go around among clusters including mobile nodes. Since this method adopts ferry nodes that modes on a fixed route, the ferry nodes enable to establishment of a stable communication route, unlike other methods where the movement of the radio nodes is unpredictable. In the event of a disaster, it is expected that a DTN network with a high data arrival rate to the disaster response headquarters located outside will be constructed by ferry nodes going around among evacuation centers.

The ferry node moves among clusters including mobile nodes and receives data from the mobile nodes in the cluster when the ferry node reaches the cluster. If the destination of the data in its possession is in the cluster, it forwards the data. The message ferry method repeats these steps to transfer data. Since ferry nodes move around the network area based on clusters including mobile nodes, efficient data forwarding can be achieved by forming clusters along the moving routes of ferry nodes or by optimizing the patrol routes of ferry nodes.

However, since in this method data are relayed by only the ferry node, if accesses from multiple nodes may be concentrated, it causes congestion. Also, there is a problem that data cannot be obtained from nodes outside the communication range of the ferry node.

2.5 Aggregated Message Ferry Method

The aggregated message ferry method [5] is a technique that aggregates data to nodes that have a high likelihood of communicating with the ferry node to obtain data held by nodes that are outside the communication range of the ferry node. Each node forwards data to the node with the highest ferry score, which indicates the likelihood of communicating with the ferry node, and the ferry score is increased when the node communicates with the ferry node. When the ferry node comes, the node that has aggregated data forwards the data to the ferry node. Therefore, data can be transferred to the destination node via the ferry node even when the source node exists outside the communication range of the ferry node, as shown in Fig. 1.

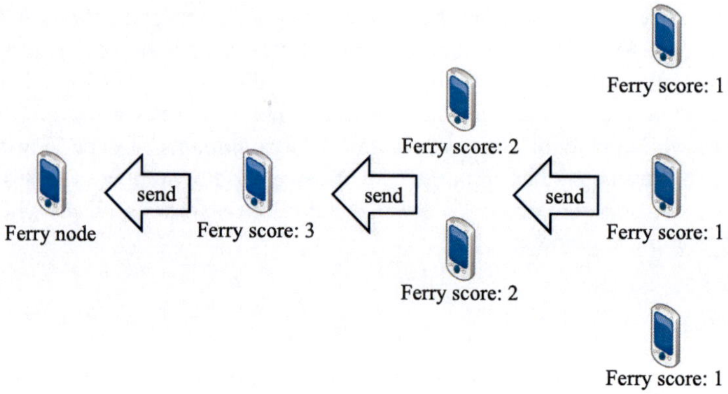

Fig. 1. Aggregate message ferry method and ferry score

3 Routing Method Based on Records of Data Transmission Paths

In the aggregated message ferry method, to increase the probability of forwarding data to a ferry node, a ferry score is set according to the number of times communicated, and data are forwarded to the node with the higher value. However, the aggregated message ferry method adds one ferry score regardless of the communication partners, which may lead to an increase in the number of hops when receiving data from a ferry node, since the data may be forwarded in a different direction from the destination. In the proposed method, each node records the nodes that are communicated with the node and sets a ferry score for each recognized node to forward data in the same direction as the destination.

3.1 Assumption of the Proposed Method

Based on the assumed environment of the aggregated message ferry method, the proposed method assumes a situation where the communication infrastructure is disrupted after a major disaster. The details are as follows:

- Difficulty in accessing telephone and network communications due to damage to the telecommunications infrastructure.
- More than one person has a wireless node capable of WiFi communication, e.g. a smartphone.
- There is a concentration of wireless nodes in several locations, e.g. in evacuation centers.
- There are ferry nodes that can move systematically within the area of communication.
- There is a travel route connecting the evacuation center with the outside of the disaster area and the ferry node can travel on the route.
- The addresses of the radio nodes distinguish between the radio nodes in the evacuation centers and other radio nodes.

The message contains the addresses of the nodes on a route, and the node that receives the message registers the recorded address of the message in a list of ferry scores. When selecting a route, the forwarding destination is decided by referring to the ferry score from the list.

3.2 Adding Ferry Score

The objective of the proposed method is to forward the data received from the ferry node in the same direction as the destination node. Therefore, the nodes that were on the path when the data was forwarded to the ferry node are assumed to be likely to be in the same direction as the ferry node, and the ferry scores are added.

Each node has a list of $n \times n$ arrays consisting of the n nodes it has recorded, and 1 is added to all combinations of two of the n nodes on routes when each node receives a message. In the Figure, the message passed nodes a, b, and c to reach the ferry node. The newly recognized node a is registered in the list and 1 is added to each of $(a,b)(a,c)(b,a)(b,c)(b,c)(c,a)(c,b)$ in the list (Fig. 2).

3.3 Methods for Determining the Communication Destination

When sending data to a ferry node (upstream), the data are transferred to the node with the highest total ferry score in the list. If the number of nodes passed is m, then each time a node is passed $(1,4,9,16,\ldots,m^2)$, a difference in the total ferry score is generated by adding and adding, so that each ferry score is different. The difference in total ferry scores makes the ferry score of the node closer to the ferry node high.

When receiving data from a ferry node (downstream), the node refers to the destination row (horizontal axis) from the list and forwards the data to the node in the column (vertical axis) with the highest value that can be communicated. This makes it possible to forward to the node that is most likely to communicate with the destination node (Fig. 3).

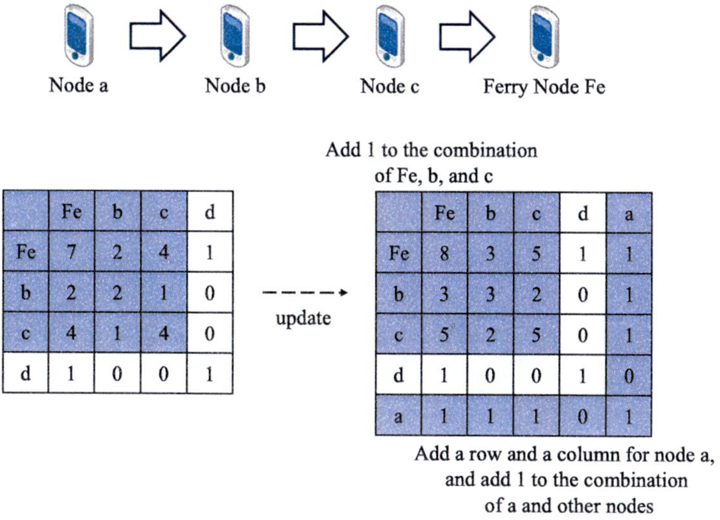

Fig. 2. Ferry score updating by the proposed method

4 Evaluations

In this section, we evaluate the proposed method by comparing the simulation results in terms of the number of hops and the data arrival rate. In addition, the simulations were divided into two cases: one in which the wireless node transfers data to the ferry node (uplink communication) and the other in which the wireless node and the ferry node transfer data together (two-way communication).

In the simulation, we verify the number of hops and the data arrival rate of the forwarding from the ferry node to the wireless node. In addition, to verify the effectiveness of the aggregated message ferrying method and the proposed method.

5 Simulation Conditions

The proposed method is implemented and evaluated on the network simulator The ONE (The Opportunistic Network Environment simulator) [6]. To confirm that the proposed method is effective regardless of the size of the area, simulations are carried out in two scenarios with different sizes of the area. Scenario 1 is 200 m square and Scenario 2 is 300 m square.

	a	b	c	n
a	7	2	4	1
b	2	2	1	0
c	4	1	4	0
n	1	0	0	1

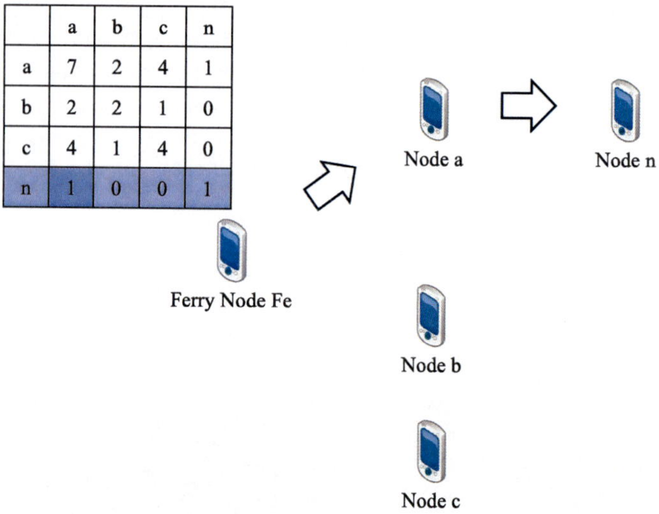

Fig. 3. Determining of next hop

The parameters used in the simulations are as follows. As evacuee nodes, 30 nodes are placed that move in random directions and 30 nodes are placed that stop in place. The data to be transferred are generated by random nodes other than the ferry node every 60 s for the up-link communication and by the ferry node every 60 s for the down-link communication.

Ferry nodes appear in the area after 1500 s and communicate with nodes for 600 s. After 600 s, they disappear from the simulation area again for a cycle time of 1500 s. In addition, messages generated between 10500 s, when the ferry node disappears from the simulation area for the last time, and 10800 s, the end time, are not included in the reachability rate because they do not reach their destinations. The communication speed and radius of the ferry node are assumed to be the same as those of the evacuee nodes, and each node is assumed to have a sufficient buffer.

5.1 Simulation Results

5.1.1 Transmission to a Ferry Node The results of the data transfer from the mobile node to the ferry node are shown in Table 1.

Table 1. Simulation results (one way)

	200 (m) square		300 (m) square	
	Previous	Proposed	Previous	Proposed
Messages	175			
Arrival rate (%)	100	99.7	100	100
Hops (average)	2.58	2.47	2.72	2.46
Hops (total)	5470	7023	5129	6353

From the simulation results, both scenarios transfer data with fewer hops than in the previous method.

5.1.2 Transmission of both Directions The results of the ferry node and mobile node bidirectional communication are shown in Table 2.

Table 2. Simulation results (two way)

	200 (m) square		300 (m) square	
	Previous	Proposed	Previous	Proposed
Messages	350			
Arrival rate (%)	99.7	94.0	99.7	94.9
Hops (average)	2.45	2.18	2.37	2.23
Hops (total)	15663	9828	15255	10290

From the simulation results, it is confirmed that the total number of hops decreases in both scenarios. Therefore, the system is expected to be used when sending a large number of messages when resources are limited, or when operating for a long time.

5.2 Discussions

The simulation results show that the total number of hops was reduced compared to the previous method during two-way communication.

In the proposed method, the communication paths are recorded when data are aggregated near the ferry node to improve the probability of data reaching the destination node and to reduce the number of hops during two-way communication. The proposed method records the surrounding nodes and in addition assigns a ferry score to each of them, which may consume a large number of buffers. In addition, the ferry score concentrates the communication to some nodes and causes a difference in battery consumption, which may hinder the long-term operation of the system. Therefore, it is necessary in the future to consider the buffers and batteries of the mobile nodes and to carry out an evaluation that is closer to the real environment.

6 Summary

In this paper, we focus on the DTN, which has attracted attention as a temporary network when an existing network fails, and the message ferry method, which is one of the route selection methods.

In the message ferry method, a mobile node that is called a ferry node carries data. The nodes outside the communication range of the ferry node could not transfer data. Therefore, in the aggregate message ferry method, which is the previous proposed method, data are aggregated to nodes that are most likely to communicate with the ferry node. However, the aggregate message ferry method cannot be effective when data is transferred to nodes outside the ferry node's communication range.

Therefore, in the proposed method, data are forwarded in the direction where the destination node exists by recording the communication path when data were aggregated, the number of hops was reduced and the data arrival rate was also improved. As a result, the proposed method can reduce the number of hops in bi-directional communication and is expected to operate for a long time during disasters.

As the future works, evaluations of the buffer and battery are necessary to show the usefulness of the proposed method in a case closer to the real environment in which a disaster occurs.

References

1. Vahdat, A., Becker, D.: Epidemic routing for partially-connected ad hoc networks. Technical report, CS-200006 (2000)
2. Grossglauser, M., Tse, D.: Mobility increases the capacity of ad-hoc wireless networks. In: Proceedings IEEE INFOCOM 2001, pp. 477–486 (2001)
3. Psounis, K., Spyropoulos, T., Raghaven, C.S.: Spray and wait: an efficient routing scheme for intermittently connected mobile networks. In: Proceedings of WDN 2005, pp. 252–259 (2005)

4. Ammar, M., Zhao, W., Zegura, E.: A message ferrying approach for data delivery in sparse mobile ad hoc networks. In: Proceedings of MobiHoc 2004, pp. 187–189 (2004)
5. Ikenoue, K., Ueda, K.: Routing method based on data transfer path in DTN environments. In: Barolli, L., Hellinckx, P., Enokido, T. (eds.) BWCCA 2019. LNNS, vol. 97, pp. 544–552. Springer, Cham (2020). https://doi.org/10.1007/978-3-030-33506-9_49
6. Information | The ONE. https://akeranen.github.io/the-one/

Effect of Polarization Loss on Channel Capacity

Trasma Yunita[1,2](✉), Chairunnisa[1], A. Adya Pramudita[2], and Achmad Munir[1]

[1] School of Electrical Engineering and Informatics, ITB, Bandung, West Java, Indonesia
33220304@std.stei.itb.ac.id, chairunnisa@itb.ac.id, munir@ieee.org

[2] School of Electrical Engineering, Telkom University, Bandung, West Java, Indonesia
pramuditaadya@telkomuniversity.ac.id

Abstract. This paper will discuss polarization mismatch, the wave's polarization transmitted by the transmitting antenna changes within the radio channel. When the radio waves reach the receiving antenna, the polarization does not match the antenna's polarization. It will reduce the received power level and the system's channel capacity. This study examined the effect of polarization loss on SISO, SIMO, and MISO channels. This aims to determine how much influence it has on system performance, especially receiving power and channel capacity. Based on the results of measurements and simulations, it was found that the mismatch between the polarization of the incident wave and the polarization of the receiving antenna causes a decrease in the system's receiving power and channel capacity. When a polarization loss occurs, the received power decreases by approximately 25.14 dBm for linear polarization and 12.69 dBm for circular polarization, and it also reduces channel capacity.

1 Introduction

In the wireless channel, depolarization occurs, namely a change in polarization waves due to reflection, refraction, and diffraction. As a result of depolarization in the channel, the polarization of the wave that reaches the receiver becomes incompatible with the polarization of the receiving antenna polarization loss factor (PLF) and decreases the quality of the received signal. As a result of multipath, fluctuations in receiving power occur on the receiving side [1–3]. The multiple input multiple output (MIMO) system is the sum of several single input single output (SISO) channels. If, on the SISO channel, there is a decrease in the receiving power and channel capacity, then this will have an impact on the MIMO system as a whole [4, 5]. So, it is necessary to investigate the SISO channel. Then, it is enriched with the influence of PLF on single input multiple output (SIMO) and multiple input single output (MISO) channels. Here, we analyze the impact of the propagation channel from different perspectives based on the effect of PLF.

2 Simulation and Measurement

The influence of electromagnetic wave polarization in the wireless channel will affect the received power level at the receiver side. The received power at the receiver is the sum of the transmitter transmit power affected by the channel and the interference. This study

L. Barolli (Ed.): BWCCA 2023, LNDECT 186, pp. 282–291, 2024.
https://doi.org/10.1007/978-3-031-46784-4_27

tried to determine whether depolarization's effect significantly impacts the system's receiving power and channel capacity. There are several scenarios investigated: (1) the condition of the SISO channel configuration, (2) the SIMO channel configuration, and (3) the MISO channel configuration. In [6], for a time-invariant AWGN channel with bandwidth (B) and received signal-to-noise ratio (SNR) (ρ), since the maximum input distribution is Gaussian, the channel capacity (C) is

$$C = B\{log_2[(1 + \rho)]\}. \tag{1}$$

Then, using the polarization loss influence approach, C in Eq. (1) can be following Eq. (2),

$$C = B\{log_2[(1 + \rho.PLF)]\}, \tag{2}$$

where PLF is a polarization loss factor. PLF has a range of values between 0–1, with 0 (∞ dB) meaning that the channel has a significant polarization loss and 1 (0 dB) meaning that the channel has no PLF.

Equation (3) and Eq. (4) are used for the SIMO channel. Then, the Eq. (5) and Eq. (6) are used for MISO channels. Equation (3) and Eq. (5) are channel capacity formulas that do not consider the occurring PLF values. However, Eq. (4) and Eq. (6) already account for the PLF on the receive antenna [7]. Channel capacity of SIMO configuration are

$$C = \{log_2[1 + \rho.N_R)]\}, \tag{3}$$

$$C = \{log_2[1 + \rho.PLF.N_R)]\}, \tag{4}$$

where N_R is the number of receiver antenna, and PLF is the receiver antenna's polarization loss factor. Then, the capacity of the MISO channel configuration is calculated using Eq. (5) and Eq. (6), with N_T is the number of transmitter antenna [7],

$$C = \{log_2[1 + \rho.N_T]\}, \tag{5}$$

$$C = \{log_2[1 + \rho.PLF.N_T)]\}. \tag{6}$$

First, using an antenna with linear and circular polarization, an experiment has been carried out on the effect of polarization loss factor on the receiving power level at the receiver, with measurement illustration as in Fig. 1. Receive power level measurements are only carried out using the SISO configuration. The measurement result is shown in Fig. 2 for linear polarization configuration and Fig. 3 for circular polarization configuration. The received power is obtained when the polarization of the transmitting and receiving antennas corresponds to the maximum value, both for antennas with linear and circular polarization. The electromagnetic waves sent by the transmitting antenna will interact with the environmental conditions of the channel before reaching the receiving antenna. The state of the propagation channel will significantly influence the propagation of electromagnetic waves [8], including polarization.

Figure 1 demonstrated simple measurements using SISO scenarios in an anechoic chamber, using antennas with several types of polarization (linear and circular). The two antennas are spaced 1.5 m apart to meet the antenna's far-field requirements. The transmitting antenna is connected to the signal generator, and the receiving antenna is connected to the spectrum analyzer. The measurement is used to determine the effect of polarization compatibility between the transmitting and receiving antennas. The received power level is highest if the polarization between the transmitting and receiving antennas is the same in the line of sight (LOS) and non-line of sight (NLOS) channels. This happens for both polarization of antenna scenarios, linear and circular. When a polarization loss occurs, the received power decreases by approximately 25.14 dBm for linear polarization and 12.69 dBm for circular polarization. It is shown in Fig. 2 and Fig. 3, respectively.

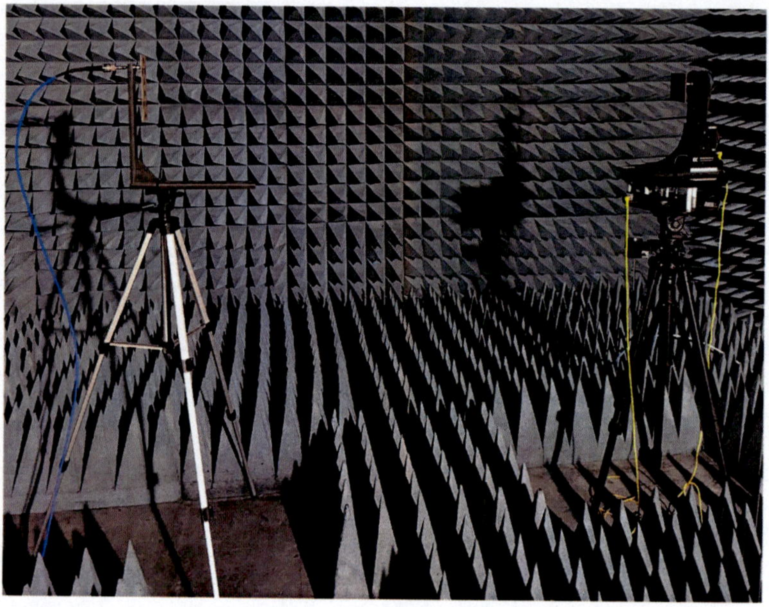

Fig. 1. Power Receive Measurement in the SISO Channel Configuration.

With the presence of PLF, the receiving power on the LOS channel conditions is no better than the NLOS channel. Polarization loss factor is believed to occur when the incoming wave has a different polarization than the receiving antenna. Figure 2 shows the VH and HV polarization scenarios for linear polarization. The circular polarizations shown in Fig. 3 are RL and LR. On the other hand, if the polarization of both waves is the same, it is thought that PLF will not occur. Linear polarization scenarios are denoted by HH (both horizontal) and VV (both vertical), while circular polarization scenarios are denoted by RR (both right) and LL (both left). Both measurements were carried out at 3.4 and 3.6 GHz, with the center frequency at 3.5 GHz.

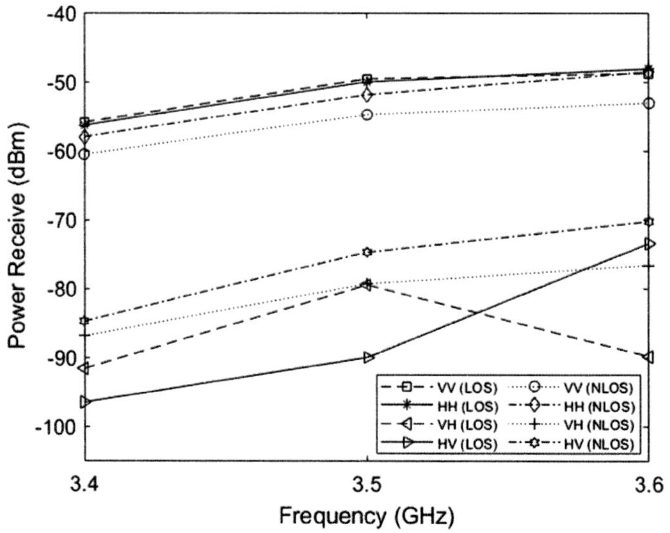

Fig. 2. Power Measurement using a Linear Polarization Antenna in the SISO Channel.

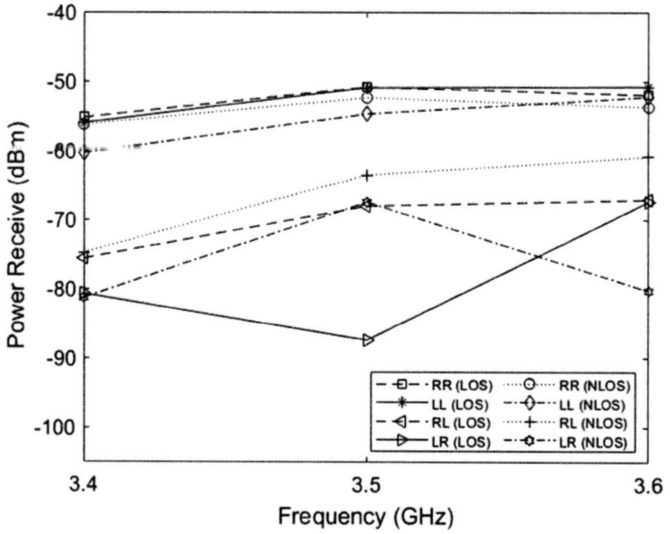

Fig. 3. Power Measurement using a Circular Polarization Antenna in the SISO Channel.

3 Result and Discussion

After measuring the received power based on the polarization loss factor, a channel capacity calculation simulation was carried out for the SISO, SIMO, and MISO channel scenarios. The simulation uses Rayleigh channel conditions and SNR in the 0–20 dB range. To represent the PLF value used is 0.1, 0.3, 0.5, 0.8, and 1. The proposed study determines the effect of polarization mismatch on system performance from the perspective of channel capacity. The simulation result of the SISO channel capacity affected by PLF is shown in Fig. 4. The channel capacity obtained at an SNR value of 10 dB without PLF equals 3.45943 bps/Hz and 1 bps/Hz with 0.1 PLF decreases in proportion to the PLF value.

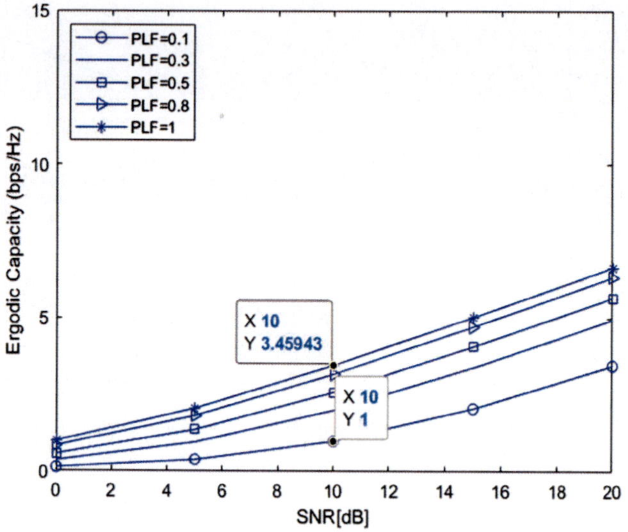

Fig. 4. Channel Capacity of SISO Channel based on PLF.

In comparison, the simulation results of the SIMO channel capacity affected by PLF are shown in Fig. 5, Fig. 6, Fig. 7, and Fig. 8. According to simulation results, the channel capacity of the 1 × 2 SIMO channel without PLF is 4,39232 bps/Hz, and with 0.1 PLF is 1.58496 bps/Hz. The channel capacity of the 1 × 4 SIMO channel without PLF is 5.35755 bps/Hz and with 0.1 PLF of 2.32193 bps/Hz. The capacity of the 1 × 8 SIMO channel without PLF is 6.33985 bps/Hz, with 0.1 PLF of 3.16993 bps/Hz. The capacity of the 1 × 16 SIMO channel without PLF is 7.33092 bps/Hz, with 0.1 PLF of 4.08746 bps/Hz. Increasing the number of receiving elements in the SIMO scenario can improve the system channel capacity, but this decreases due to the influence of PLF.

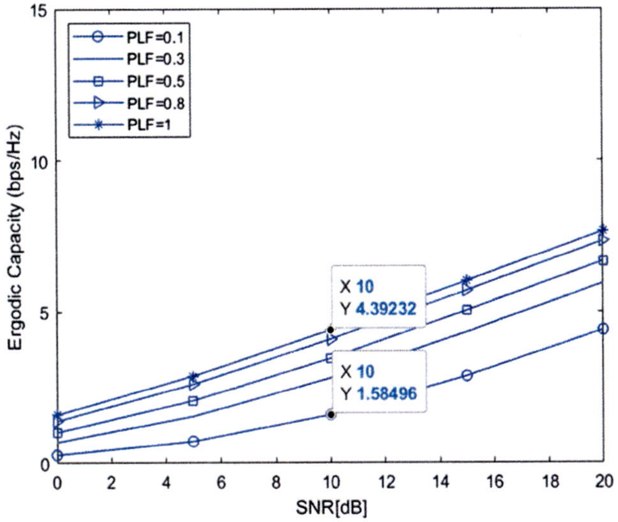

Fig. 5. Channel Capacity of 1 × 2 SIMO Channel based on PLF.

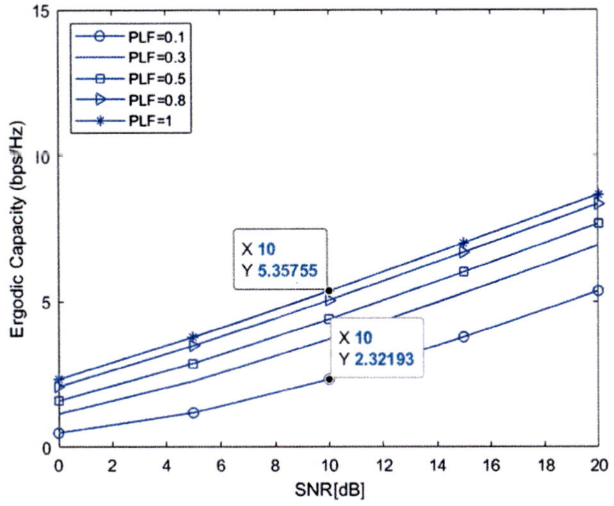

Fig. 6. Channel Capacity of 1 × 4 SIMO Channel based on PLF.

The decrease in channel capacity for each SIMO scenario is approximately 2.8–3.24 bps/Hz with PLF. This will undoubtedly affect the overall performance of the channel. The initial aim of adding elements is to increase channel capacity. It would be better if system performance could be maintained to optimal value.

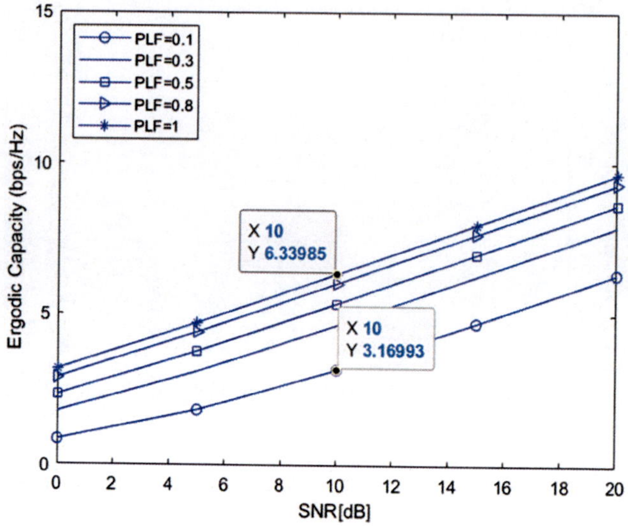

Fig. 7. Channel Capacity of 1 × 8 SIMO Channel based on PLF.

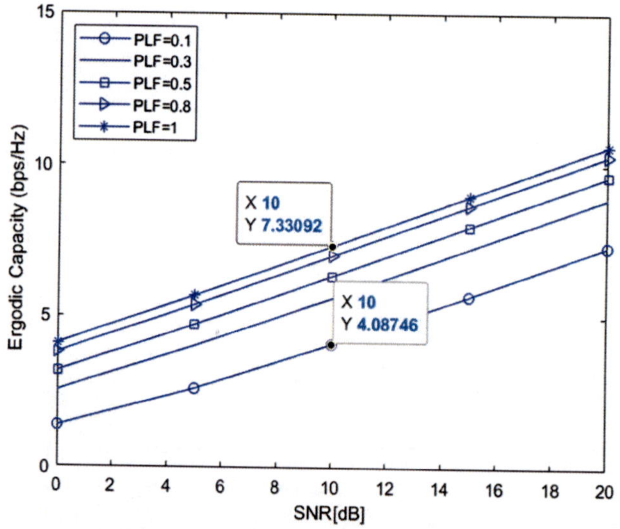

Fig. 8. Channel Capacity of 1 × 16 SIMO Channel based on PLF.

Then, the simulation results of MISO channel capacity, which PLF influences are shown in Fig. 9, Fig. 10, Fig. 11, and Fig. 12. Based on simulation results, the channel capacity of the 2 × 1 MISO channel without PLF is 3.45943 bps/Hz, with 0.5 PLF is 2.58496 bps/Hz, and with 0.1 PLF is 1 bps/Hz. The capacity of the 4 × 1 MISO channel without PLF is 5.35755 bps/Hz, and with 0.1 PLF is 2.32193 bps/Hz. The capacity of the 8 × 1 MISO channel without PLF is 6.33895 bps/Hz, and with 0.1 PLF is 3.16993 bps/Hz. The capacity of the 16 × 1 MISO channel without PLF is 7.33092 bps/Hz, and with 0.1 PLF is 4.08746 bps/Hz.

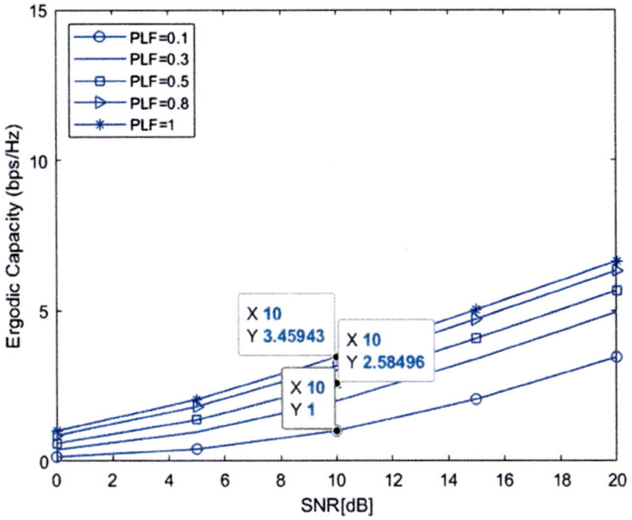

Fig. 9. Channel Capacity of 2 × 1 MISO Channel based on PLF.

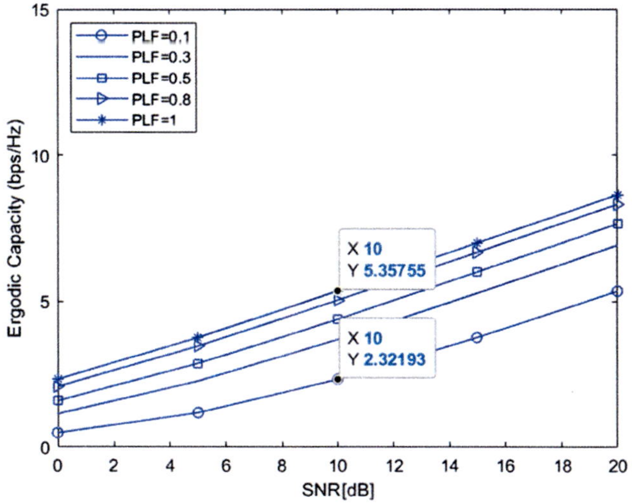

Fig. 10. Channel Capacity of 4 × 1 MISO Channel based on PLF.

Based on Eq. (3) and Eq. (5), SIMO and MISO channel capacity calculation is influenced by the SNR value and number of antenna elements; the capacity value is the same for the same number of elements. Increasing the antenna elements in the transmitter or receiver section significantly improves the system channel capacity. However, PLF still influences the acquisition of SIMO and MISO channel capacity calculations.

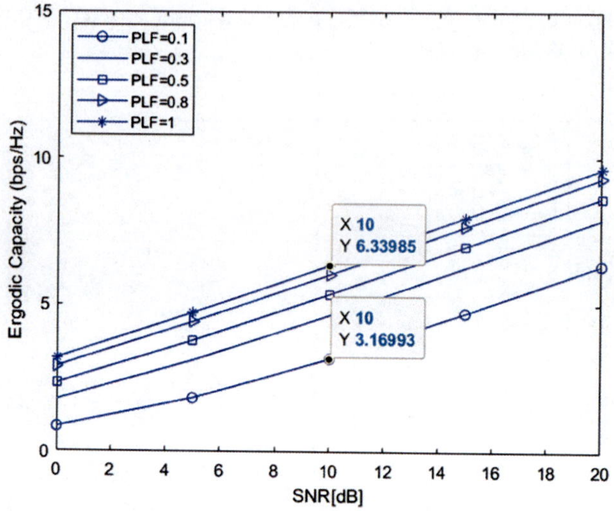

Fig. 11. Channel Capacity of 8 × 1 MISO Channel based on PLF.

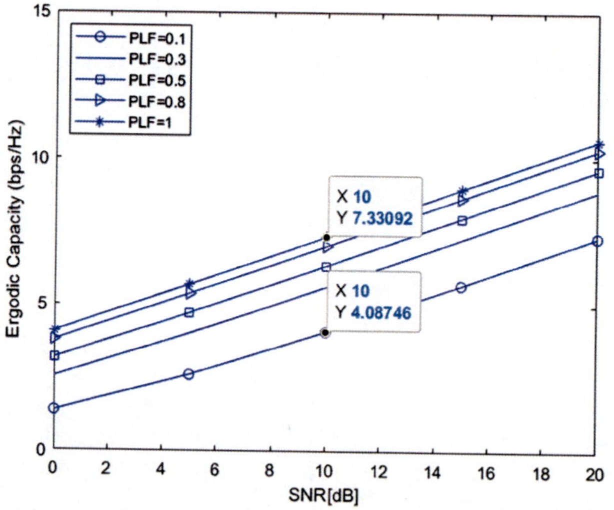

Fig. 12. Channel Capacity of 16 × 1 MISO Channel based on PLF.

4 Conclusion

Due to the polarization mismatch, the obtained SISO, SIMO, and MISO channel capacities, based on the effect of PLF on the antenna reception, are significantly reduced, along with the receiving power level at the receiver. The risk of elements experiencing polarization mismatch is even greater if multiple antenna elements exist. Therefore, when the polarization adaptation of each antenna element can be optimally tuned, it improves the channel capacity efficiency in the system. Power reduction becomes even more critical with the increasing number of antenna elements used in MIMO systems. In future work, the maximum receive power and optimal channel capacity will be achieved by adjusting the polarization of the receiving antenna with the polarization of the incident wave in the MIMO system. Efforts are underway to change the polarization direction to maintain the effect of increasing the number of antennas in increasing channel capacity, focusing on solving polarization problems. It will have a good impact if implemented in MIMO and large MIMO antenna systems. If the performance of a single parallel channel can be maintained at an optimal level, the capacity of the MIMO channel will also be held.

References

1. Samimi, M.K., Sun, S., Rappaport, T.S.: MIMO channel modeling and capacity analysis for 5G millimeter-wave wireless systems. In: 2016 10th European Conference on Antennas and Propagation, EuCAP (2016)
2. Mani, F., Quitin, F., Oestges, C.: Accuracy of depolarization and delay spread predictions using advanced ray-based modeling in indoor scenarios. EURASIP J. Wirel. Commun. Netw. **2011**(1), 1–9 (2011)
3. Turbic, K., Correia, L.M., Beko, M.: A channel model for polarized off-body communications with dynamic users. IEEE Trans. Antennas Propag. **67**(11), 7001–7013 (2019)
4. De Flaviis, F., Jofre, L., Romeu, J., Grau, A.: Multiantenna Systems for MIMO Communications. Morgan Claypool Publisher, San Rafael (2008)
5. Tsoulos, G.: The Electrical Engineering and Applied Signal Processing Series MIMO System Technology for Wireless Communications (2006)
6. Biglieri, E., Calderbank, R., Constantinides, A., Goldsmith, A., Paulraj, A., Poor, H.V.: MIMO Wireless Communication. Cambridge University Press, Cambridge (2007)
7. Cho, Y.S., Kim, J., Yang, W.Y., Kang, C.G.: MIMO-OFDM Wireless Communications with MATLAB Ò (2010)
8. De Doncker, P.: Channel modeling for polarized MIMO systems (2011)

FSALT: A Fuzzy-Based System for Assessment of Logical Trust and Its Performance Evaluation

Shunya Higashi[1(✉)], Phudit Ampririt[1], Ermioni Qafzezi[2], Makoto Ikeda[2], Keita Matsuo[2], and Leonard Barolli[2]

[1] Graduate School of Engineering, Fukuoka Institute of Technology, 3-30-1 Wajiro-Higashi, Higashi-Ku, Fukuoka 811-0295, Japan
{mgm23108,bd21201}@bene.fit.ac.jp

[2] Department of Information and Communication Engineering, Fukuoka Institute of Technology, 3-30-1 Wajiro-Higashi, Higashi-Ku, Fukuoka 811-0295, Japan
qafzezi@bene.fit.ac.jp, makoto.ikd@acm.org, {kt-matsuo,barolli}@fit.ac.jp

Abstract. Recently, the human-to-human and human-to-things connections are becoming significantly more complicated and less trustworthy in decision-making for diverse scenarios. As a result, the trust computing is getting interest from many study domains. The Logical Trust is one of the trust concepts. In this paper, we consider three parameters (Belief (Be), Experience (Ep), and Rationality (Ra)) for the implementation of a fuzzy-based system to evaluate the LT. We evaluate by simulation the proposed system. According to the simulation findings, the LT parameter increases when Be, Ep, and Ra are increasing. All LT values are greater than 0.5 when Ep values range from 0.5 to 0.9, Be is 0.9 and for any value of Ra. In this case, the persons or things are trustworthy.

1 Introduction

The social and digital worlds need to be interwoven into everyday life for humans. The exchange of knowledge in the digital world also has a significant influence on the social domain. In addition, as digital technologies advance, the relationships and connections between people and things are getting increasingly complex and problematic. Therefore, in order to safeguard digital transactions in the digital world, security algorithms or trust computing are required [1,2].

In many circumstances, trust is the cornerstone of decision-making and can be utilized in an ethical way. Therefore, by considering these qualities of trust in the digital world, digital operators can identify people and ascertain their relationships. In order to ensure a connection, it should be taken into account the acceptable risk of building the trust. For instance, when two persons communicate between them, they should use a medium that can be trusted in order to secure the information and privacy (human-to-thing trust). Additionally, they will have mutual trust before starting the interaction.

L. Barolli (Ed.): BWCCA 2023, LNDECT 186, pp. 292–301, 2024.
https://doi.org/10.1007/978-3-031-46784-4_28

Trust should be taken into account in digital surroundings. The trust concept, for instance, is used in 5G and beyond networks to ensure security and quantify the degree to which a system can be relied in order to maintain a particular level of performance. Additionally, trustworthiness in cloud, fog, and edge computing is crucial for lowering transaction risk and reducing data and privacy leaks [3–6].

Various networks and systems have employed the concept of trust. P2P networks, which connect users directly can allow the file transfer. These networks measure users' skill levels or quality in addition to posing a security risk [7,8].

Social theories may not be immediately applicable to modeling trust in the digital world, but social observations are still crucial for network-like digital societies. By considering these social ideas, a more accurate model of trust may be developed. This is challenging, because the trust is only adequately articulated by humans in their qualitative, hazy, and complex forms. Many parameters are required to make a trust judgment, which makes the task NP-Hard. For this reason, heuristic and intelligent algorithms like Fuzzy Logic (FL), Genetic Algorithms (GA), Particle Swarm Optimization (PSO), and others are required [9,10].

In this paper, we present a fuzzy-based system for assessment of Logical Trust (LT) that takes into account three parameters: Belief (Be), Experience (Ep) and Rationality (Ra). By computer simulations we assess the performance of the proposed system. The simulation results show how the Be, Ep, and Ra effect on the LT, which is the output parameter.

The paper is structured in 6 sections. After the introduction, in Sect. 2 is given a short description of trust computing. Then, in Sect. 3, we present FL overview. In Sect. 4, we describe the proposed Fuzzy-based system and its components. In Sect. 5, we discuss simulation results. Finally, in Sect. 6, we conclude the paper and give the future work.

2 Trust Computing Overview

Digital devices and data are expanding daily, enabling communication between sensors, smartphones, and other devices. The integration of the physical and social worlds with the digital world occurs when various technologies link to one another. The sharing of knowledge through the digital world has a significant effect on society. Therefore, the digital exchanges must be more reliable and secure.

The combination of the physical, social, and digital worlds creates new and complicated vulnerabilities. As a result, the authenticated access control and security mechanisms might not work. In digital networks, hostile actors can also lead to a wide range of concerns and problems. As a result, trust computing is a good strategy [11].

The trust computing can be used for interactions between people and things. The digital entities may check others and decide how to engage with them, much as in the real world, while taking into account the social qualities of trust in the digital realm. The perceptual mechanisms of the real and social world cannot be applied in digital contexts, therefore trust computing is very difficult to quantify. As a result, trust modeling may be used to real-world challenges for evaluation and decision-making [12,13].

The trust computing can assess the trust by taking into account a variety of factors, including the reasons why the trustor and trustee want to maintain a long-term relationship. From Fig. 1, the fundamental idea of trust is to identify trust as a relationship between a Truster and a Trustee. The person that needs a service is known as the trustor, and he or she puts their trust in the trustee, an alternate business, in order for it to be provided. Depending on the circumstances, the relationship may not always be one person to one person but may instead be one person to the group of persons or the group of persons to the group of persons [14].

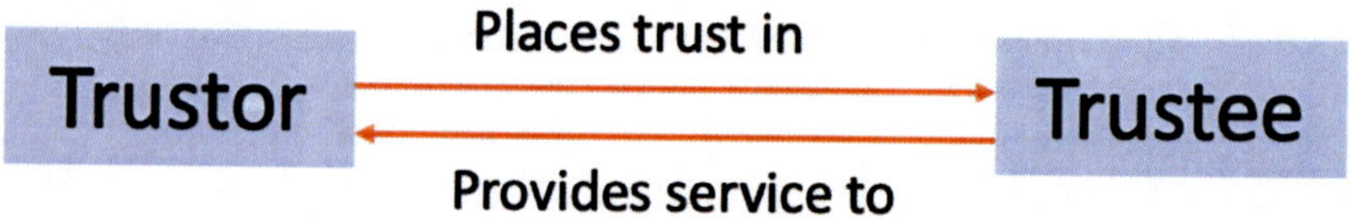

Fig. 1. Trustor and trustee relationship.

The trust assessment can be divided into Individual and Relational Trust as shown in Fig. 2 [15].

- **Individual Trust Assessment:** A person's personal attributes as a trustee are taken into account when evaluating a trust's worth. The ability to trust someone can be gauged by their emotional sense, including sentiments and inclinations.
- **Relation Trust Assessment:** It assesses the value of a trust by taking into account the trustor and trustee interactions.

3 Fuzzy Logic Overview

An Fuzzy Logic (FL) system executes nonlinear mapping of the input data vector into the scalar output and can handle both linguistic and numerical data separately. Additionally, it may deal with requests or views that may be truthful, false, or have an intermediate degree of truth value [16,17]. Additionally,

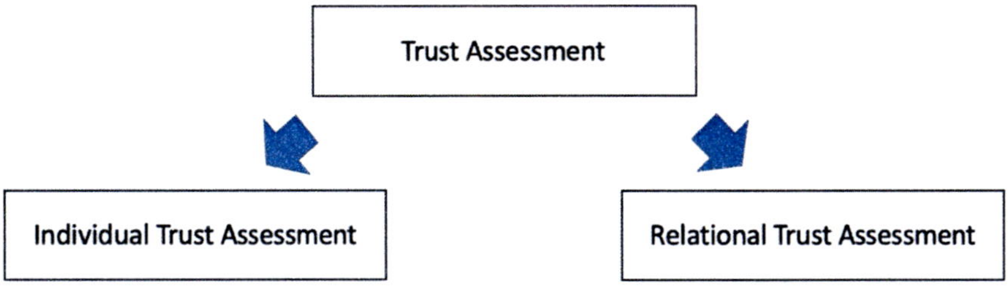

Fig. 2. Trust assessment process.

FL is far more similar to human thinking and natural language than standard logical systems and can describe nonlinear functions of any complexity based on natural language [18].

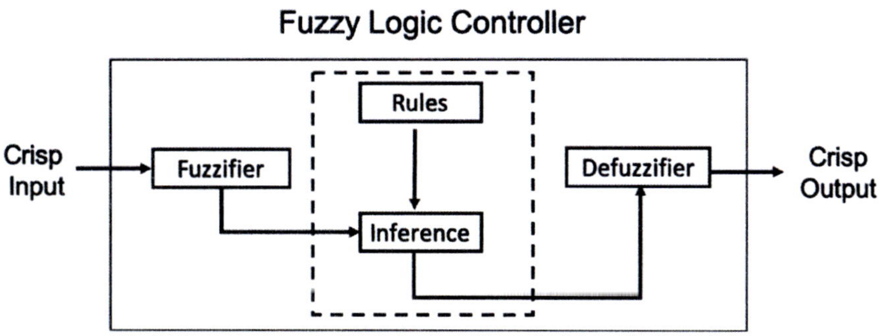

Fig. 3. FLC structure.

As illustrated in Fig. 3, a Fuzzy Logic Controller (FLC) structure has four parts: fuzzifier, inference engine, Fuzzy Rule Base (FRB), and defuzzifier. To create fuzzy inputs, the Fuzzifier uses linguistic variables and links the fuzzy sets with crisp values. The Inference Engine then infers fuzzy output using fuzzy rules and fuzzified input values. Sugeno, Mamdani, and Tsukamoto fuzzy inference are three types of fuzzy inference techniques that are often used. The fuzzy rules could come from a specialist or be developed from numerical data. Finally, the defuzzification transforms a fuzzy output into a crisp control output [18,19].

4 Proposed Fuzzy-Based System

In this paper, we consider FL to implement the Fuzzy-based System for Assessment of Logical Trust (FSALT). The FSALT structure is depicted in Fig. 4. We consider three input parameters for the implementation of our system: Belief (Be), Experience (Ep), and Rationality (Ra). The output parameter is Logical Trust (LT).

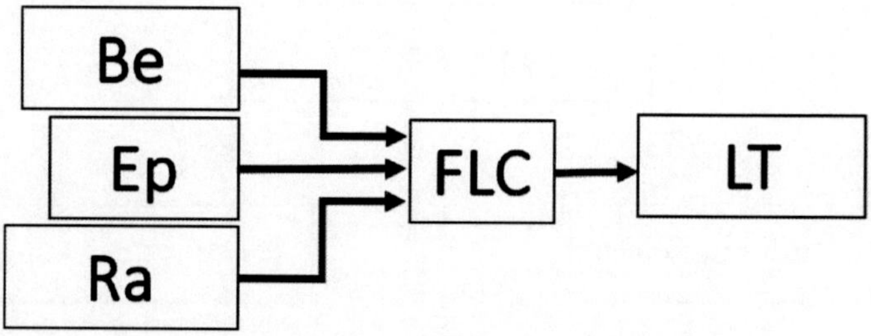

Fig. 4. Proposed system structure.

Table 1. Parameter and their term sets.

Parameters	Term set
Belief (Be)	Low (Lo), Medium (Me), High (Hi)
Experience (Ep)	Inexperienced (In), Normal (No), Good (Gd)
Rationality (Ra)	Low (Lw), Medium (Md), High (Hg)
Logical Trust (LT)	LT1, LT2, LT3, LT4, LT5, LT6, LT7

Table 2. FRB.

Rule	Be	Ep	Ra	LT
1	Lo	In	Lw	LT1
2	Lo	In	Md	LT1
3	Lo	In	Hg	LT2
4	Lo	No	Lw	LT1
5	Lo	No	Md	LT2
6	Lo	No	Hg	LT3
7	Lo	Gd	Lw	LT2
8	Lo	Gd	Md	LT3
9	Lo	Gd	Hg	LT4
10	Me	In	Lw	LT2
11	Me	In	Md	LT3
12	Me	In	Hg	LT4
13	Me	No	Lw	LT3
14	Me	No	Md	LT4
15	Me	No	Hg	LT5
16	Me	Gd	Lw	LT4
17	Me	Gd	Md	LT5
18	Me	Gd	Hg	LT6
19	Hi	In	Lw	LT4
20	Hi	In	Md	LT5
21	Hi	In	Hg	LT6
22	Hi	No	Lw	LT5
23	Hi	No	Md	LT6
24	Hi	No	Hg	LT7
25	Hi	Gd	Lw	LT6
26	Hi	Gd	Md	LT7
27	Hi	Gd	Hg	LT7

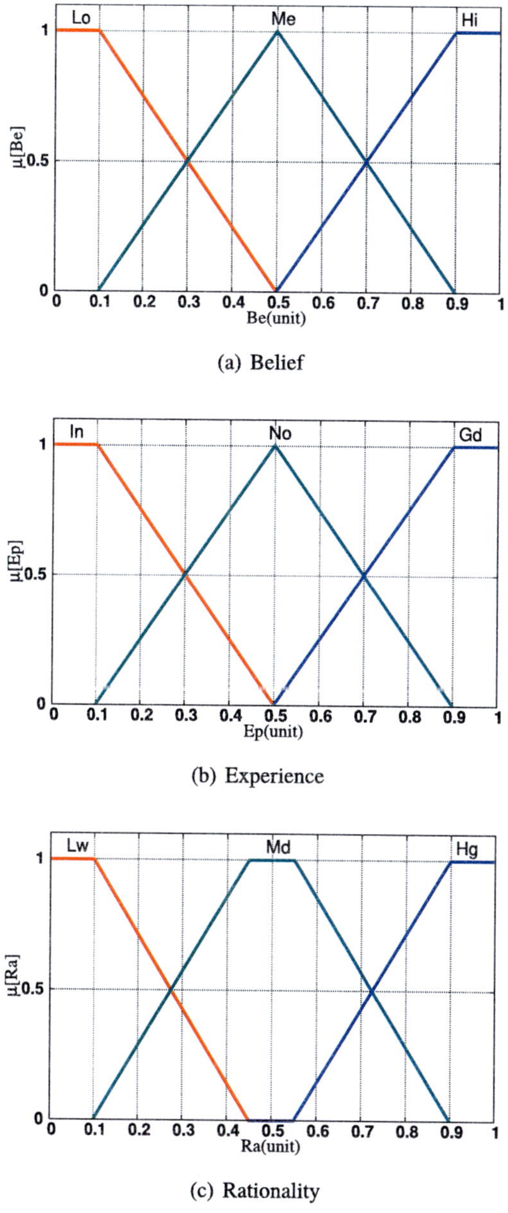

Fig. 5. Membership functions.

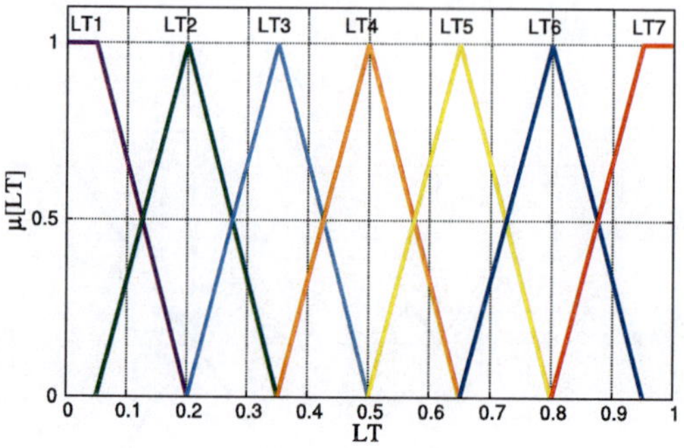

(d) Logical Trust

Fig. 5. (*continued*)

The membership functions are shown in Fig. 5. In Table 1 are shown the parameters are their term sets. Table 2 shows the FRB, which comprises of 27 rules. The control rules syntax is "IF Condition THEN Control Action". For instance, the Rule 27 meaning is "IF Belief is High, Experience is Good, and Rationality is High, THEN LT is LT7".

5 Simulation Results

In this section, we present the simulation result of our proposed FSALT system. The simulation results are shown in Fig. 6, Fig. 7 and Fig. 8. They depict the relationship between LT and Ra for various Ep values while considering Be as a fixed parameter.

We consider the Be value 0.1 in Fig. 6. For Ep 0.5, the LT increases by 13% and 15%, respectively, when Ra is increased from 0.1 to 0.5 and 0.5 to 0.9. This indicates that the trustee will have more logical trust in the trustor when the trustor Ra value is increased. For all Ep values, the LT values are below 0.5, which shows that the trustee will not trust the trustor.

To see how Be has impacted LT, we compare Fig. 6 with Fig. 7. The LT is increasing by 30% (when Ep value is 0.5 and Ra is 0.5) by increasing the Be value from 0.1 to 0.5. When the trustor has high belief to the trustee, the LT possibility will be higher.

We raise the value of Be to 0.9 in Fig. 8. We can see that the ET values have grown significantly higher compared to the results of Fig. 6 and Fig. 7.

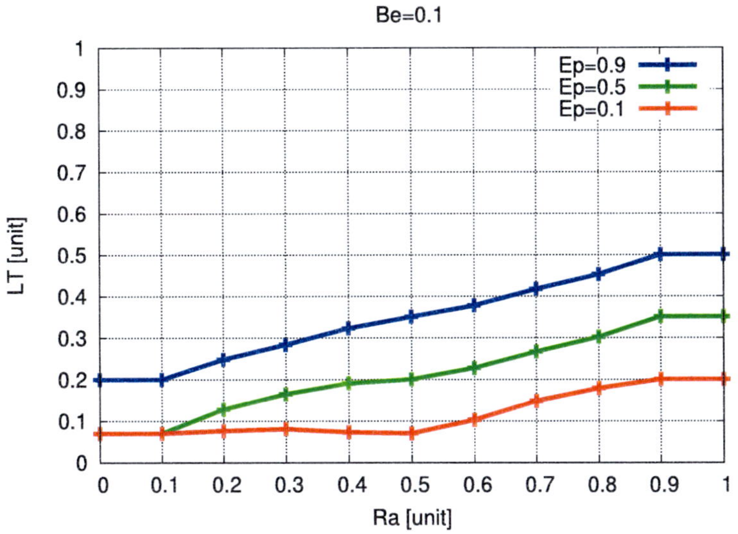

Fig. 6. Simulation results for Be = 0.1.

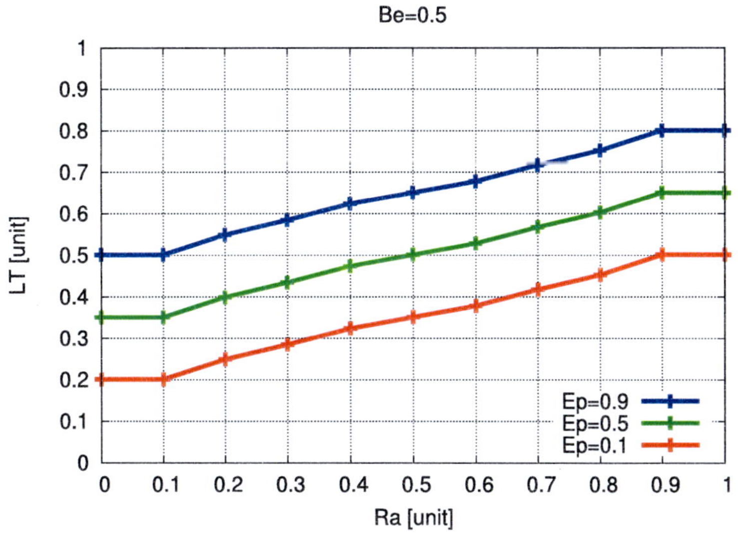

Fig. 7. Simulation results for Be = 0.5.

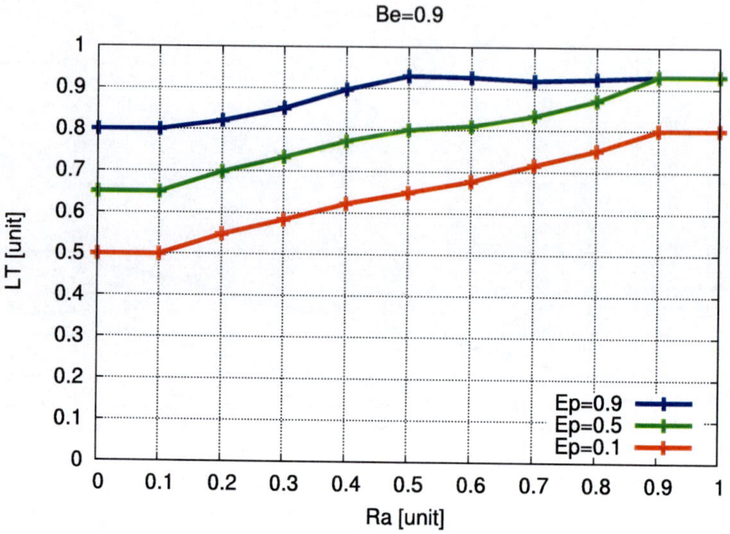

Fig. 8. Simulation results for Be = 0.9.

6 Conclusions and Future Work

In this paper, first we provided a brief overview of FL and trust computing. Then, we presented the proposed FSALT. We conclude as following from the simulation results.

- By increasing Ep values, the LT value is increased because trustor and trustee have more experience with each other.
- The belief value positively influences the LT. When Be is increased, the LT value is increased.
- When Ep values are between 0.5 and 0.9 and Be is 0.9, for any Ra value, all LT values are higher than 0.5. This indicates that the things or persons are trustworthy.

In the future work, we will take into account additional variables and perform extended simulations.

References

1. Ting, H.L.J., Kang, X., Li, T., Wang, H., Chu, C.-K.: On the trust and trust modeling for the future fully-connected digital world: a comprehensive study. IEEE Access **9**, 106 743–106 783 (2021). https://doi.org/10.1109/ACCESS.2021.3100767
2. Wang, D., Muller, T., Liu, Y., Zhang, J.: Towards robust and effective trust management for security: a survey. In: 2014 IEEE 13th International Conference on Trust, Security and Privacy in Computing and Communications, pp. 511–518 (2014)

3. Benzaïd, C., Taleb, T., Farooqi, M.Z.: Trust in 5G and beyond networks. IEEE Netw. **35**(3), 212–222 (2021)
4. Rahman, F.H., Au, T.-W., Newaz, S.S., Suhaili, W.S., Lee, G.M.: Find my trustworthy fogs: a fuzzy-based trust evaluation framework. Futur. Gener. Comput. Syst. **109**, 562–572 (2020)
5. Uslu, S., Kaur, D., Durresi, M., Durresi, A.: Trustability for resilient internet of things services on 5G multiple access edge cloud computing. Sensors **22**(24), 9905 (2022)
6. Cai, H., Li, Z., Tian, J.: A new trust evaluation model based on cloud theory in e-commerce environment. In: 2011 2nd International Symposium on Intelligence Information Processing and Trusted Computing, pp. 139–142 (2011)
7. Wang, Y., Vassileva, J.: Bayesian network-based trust model. In: Proceedings of the IEEE/WIC International Conference on Web Intelligence (WI 2003), pp. 372–378 (2003)
8. Zhou, P., Gu, X., Zhang, J., Fei, M.: A priori trust inference with context-aware stereotypical deep learning. Knowl.-Based Syst. **88**, 97–106 (2015). https://www.sciencedirect.com/science/article/pii/S095070511500307X
9. Zhang, D., Yu, F.R., Yang, R.: A machine learning approach for software-defined vehicular ad hoc networks with trust management. In: 2018 IEEE Global Communications Conference (GLOBECOM), pp. 1–6 (2018)
10. Jayasinghe, U., Lee, G.M., Um, T.-W., Shi, Q.: Machine learning based trust computational model for IoT services. IEEE Trans. Sustain. Comput. **4**(1), 39–52 (2019)
11. Hu, W.-L., Akash, K., Reid, T., Jain, N.: Computational modeling of the dynamics of human trust during human-machine interactions. IEEE Trans. Hum.-Mach. Syst. **49**(6), 485–497 (2019)
12. Zolfaghar, K., Aghaie, A.: Evolution of trust networks in social web applications using supervised learning. Procedia CS **3**, 833–839 (2011)
13. Kumar, S., Shah, N.: False information on web and social media: a survey (2018)
14. Braga, D.D.S., Niemann, M., Hellingrath, B., Neto, F.B.D.L.: Survey on computational trust and reputation models. ACM Comput. Surv. **51**(5), 1–40 (2018). https://doi.org/10.1145/3236008
15. Cho, J.-H., Chan, K., Adali, S.: A survey on trust modeling. ACM Compu. Surv. (CSUR) **48**(2), 1–40 (2015)
16. Jantzen, J.: Tutorial on fuzzy logic. Technical University of Denmark, Depertment of Automation, Technical report (1998)
17. Zadeh, L.A.: Fuzzy logic. Computer **21**(4), 83–93 (1988)
18. Lee, C.-C.: Fuzzy logic in control systems: fuzzy logic controller. I. IEEE Trans. Syst. Man Cybern. **20**(2), 404–418 (1990)
19. Mendel, J.M.: Fuzzy logic systems for engineering: a tutorial. Proc. IEEE **83**(3), 345–377 (1995)

Time Series Mean Normalization for Enhanced Feature Extraction in In-Vehicle Network Intrusion Detection System

Yusupov Kamronbek[1], Islam Md Rezanur[1], Insu Oh[2], and Kangbin Yim[2](✉)

[1] Department of Software Convergence, Soonchunhyang University, Asan, Korea
{yuskamron,arupreza}@sch.ac.kr

[2] Department of Information Security Engineering, Soonchunhyang University, Asan, Korea
{catalyst32,yim}@sch.ac.kr

Abstract. The growth of electronic control units (ECUs) has led to modern automobiles being more convenient and technologically advanced. In order to communicate amongst these control units, the widely used Controller Area Network (CAN) communication protocol is used. CAN buses do not, however, come with built-in security features while being reliable and cost-effective. Because of this weakness, they are vulnerable to several attacks from prospective enemies. The installation of an Intrusion Detection System (IDS) is, nonetheless, a practical answer to this issue. A CAN bus system's security can be considerably improved by the use of an IDS. We suggest a Long Short-Term Memory (LSTM) based intrusion detection system that is capable of effectively addressing the detection and security needs of CAN bus systems. We think that using Long Short-Term Memory (LSTM) models has a number of benefits, including highly accurate classification and effective anomaly detection. Our research findings have shown that our Long Short-Term Memory (LSTM) based IDS successfully identifies assaults on CAN systems with an accuracy rate of 99% and little loss. By providing this solution, we want to aid in the creation of reliable intrusion detection systems that can successfully protect CAN bus systems.

1 Introduction

A variety of tools and technology have helped the contemporary automobile sector advance significantly over time. ECUs, or electronic control units, are a possible addition to this list. When compared to today, modern cars can include anywhere between 50 and 100 ECUs [1]. Previously, cars had significantly less of this type of electrical technology. They facilitate driving, aid with control, and enhance comfort.

The CAN (Controller Area Network) bus is used by these integrated into the body of the vehicle components to transmit and receive crucial data. Due to its affordable, effective, and dependable design, the CAN bus is a standard communication protocol that is widely used in the automobile sector [2]. These innovations are making automobiles smarter and more interconnected. But in addition to the comfort and connection, the CAN bus has developed into a target for hackers. It is susceptible to possible incursions

L. Barolli (Ed.): BWCCA 2023, LNDECT 186, pp. 302–311, 2024.
https://doi.org/10.1007/978-3-031-46784-4_29

because of its flexible structure and simplistic design. Modern vehicles are a target for attackers because as they become increasingly interconnected with the outside world, new potential for hacks opens. The CAN bus contains built-in weaknesses such as lax access control, insufficient encryption, and inadequate authentication. By taking advantage of these weaknesses, attackers may take control of ECUs and use them to send malicious messages across the internal network of the car. The driver's and passengers' safety might be seriously jeopardized as a result of severe repercussions such as unlawful control of the vehicle, abrupt braking, or engine shutdown. It is crucial to install an intrusion detection system (IDS) in automobiles to reduce these hazards and avoid such accidents. An IDS enables the detection and prevention of several assaults, both locally and remotely from the vehicle. It monitors CAN bus data integrity, analyzes network traffic, spots suspicious behavior, and reacts to threats immediately. Vehicles may be protected and possible dangers from auto break-ins can be reduced by adopting an IDS.

This article focuses on an intrusion detection system that uses LSTM (Long Short-Term Memory), which is an anomaly detection technique relevant to automotive systems. A recurrent neural network with long-term memory that can handle sequential data is called an LSTM. The LSTM model has benefits including the ability to handle sequential data, retain knowledge over time, draw conclusions from past data, and adjust to changing environmental conditions. These features enable the real-time identification of assaults or anomalous activity on the CAN bus as well as the detection of novel attack types without the need for tight rules. The article also looks at other intrusion detection techniques for the CAN network, which could produce better outcomes but have drawbacks including a larger chance of false positives. These techniques may be quite sensitive, which might result in false positives and add to the stress on the system.

In this work, we provide a powerful LSTM-based intrusion detection technique for automobile CAN bus systems, focusing on attacks like DoS, Fuzz, Malfunction, and Replay. The paper is divided into many sections, including related work (Sect. 2), a description of the CAN protocol and its structure, as well as attack dataset (Sect. 3), LSTM, feature extraction and data pre-processing (Sect. 4), Evaluation performance and results (Sect. 5), and a commentary (Sect. 6) that wraps up the subject.

2 Related Work

The protection of the CAN system against intrusions and attacks is a top priority when it comes to system security in the field of mechanical engineering. Hackers who take over the CAN system might modify data, putting the driver's and passengers' safety in peril and having serious repercussions. In order to maintain the safety and security of CAN systems, it is essential to deploy an intrusion detection system (IDS). Modern attack classification and intrusion detection methods have been suggested by researchers.

In one research, Md. Delwar et al. investigated the effectiveness of an IDS model based on convolutional neural networks (CNN) to detect network assaults [3]. They made use of data from real Toyota, Subaru, and Suzuki cars that have been the targets of DoS, Fuzz, RPM, and Gear Spoofing assaults. Their model performed exceptionally well in classifying assaults across various vehicle kinds, obtaining a 99.99% detection accuracy. Markus et al. presented CANet, a noteworthy additional model [4]. This unsupervised model performed exceptionally well at picking up signal alterations that are difficult to pick up using more traditional methods. Using both actual and synthetic data, it effectively recognized several unidentified assaults, achieving a high negative rate (0.99). In order to identify attacks, Abdul et al. developed CANintelliIDS [5], a technique based on deep recurrent network units (GRUs) and convolutional neural networks (CNNs). Their research revealed an F1 score for class 0 (normal) of 98.21% and class 1 (attack) of 97.38%. Hyun et al. carried out three different sorts of assaults using CAN communications from several known automobiles [6]. They discovered that the interval between CAN broadcasts was a key factor in the detection of attacks. Across all experiments, their suggested IDS had a 100% detection accuracy rate with no false positives. Min et al. suggested a successful deep neural network (DNN)-based intrusion detection method to protect the CAN system [7]. Their system allowed the IDS to recognize any hacker assault on a vehicle by classifying packets into categories that were benign and malicious. The method has a 98% detection success rate. Additionally, Md. Delwar et al. created an IDS model based on LSTM [8] to recognize DoS, Fuzz, and spoofing threats. Their model correctly identified and classified anomalies with 99% accuracy and few false positives. They experimented with several LSTM setups to obtain excellent detection speed and accuracy. For clustering assaults and routine data in a CAN system, Harini et al. developed an IDS model based on unsupervised learning. They trained patterns for both CAN classes using a two-phase methodology that included an autoencoder and a mixed Gaussian mixture model (GMM) [9]. Using this method, it was possible to recognize CAN system assaults. Thien et al. offered the ResNet-based CANPerFL architecture and provided an IDS for binary encoding CAN ID categorization [10]. The temporal and geographical organization of the CAN ID sequence was represented by a matrix created by this method. The accuracy attained with this method was around 99%. Heatmaps were utilized as input for a number of models, such as VGG-16, AlexNet, and ResNet-50, in the universal intrusion detection system developed by Md Rezanur et al. The CAN ID, time interval, and Hamming distance were used to create these heatmaps [11]. By attaining a 99% accuracy rate, the ResNet-50 model surpassed the competition.

3 CAN Bus and Attack Datasets

3.1 Controller Area Network (CAN)

Controller Area Network is a well-known communication protocol [12] that is widely used in the mechanical engineering field. Using this communication protocol, communication is established between Electronic Control Units (ECU) and other vehicle systems that are located inside the vehicle, this communication method is considered reliable and efficient. Convenient architecture allows you to interact with different devices through one channel, which makes it convenient to use. Basically, CAN is used to establish communication between ECUs. The CAN frame format, as shown in Fig. 1, includes several parts. Such as Start of Frame (SOF) which demonstrates the start of a message. After SOF demonstrated the CAN ID is displayed which is the next part of the CAN frame. This part helps to identify the sender, the ECU which is responsible for a certain element. The standard CAN arbitration field is 11 bits in size, and the extended form is 29 bits. Remote Transmission Request (RTR) is part of a Data Frame that is used to transmit data to the network by another node. By using RTR, you can avoid aggregate network traffic. Data Length Code (DLC) shows the number of bytes of data contained in a message. After it there is a Data Field which has useful information depending on what information is being transmitted. Cyclic Redundancy Check (CRC) is responsible for error detection. The last part is End of Frame (EOF).

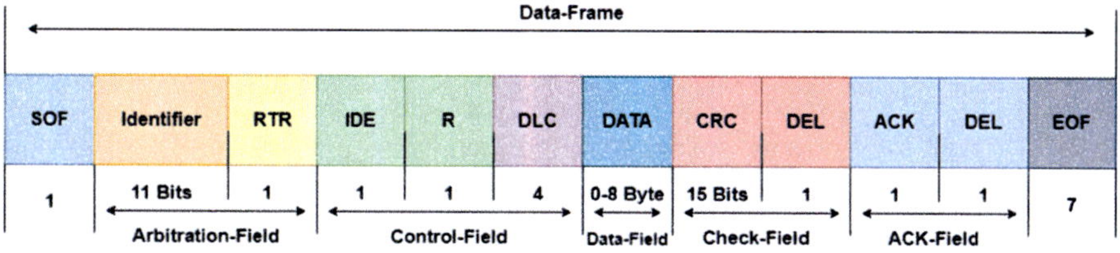

Fig. 1. Standard Data-Frame CAN [12]

3.2 Denial of Service (DoS)

Denial of Service (DoS) [13] attacks are nefarious attempts by cybercriminals to obstruct a system's regular operation, making it difficult or impossible for users to access or utilize the system. A DoS attack on the CAN bus seeks to stop the system from processing or transmitting legitimate instructions and messages, disrupting its normal operation. If a DoS attack is successful against the CAN bus, there may be serious repercussions, including engine or brake system failures that might result in accidents. In our situation, the DoS attack was executed by periodically injecting messages into the system with the CAN ID "0000" at intervals of 0.3 and 0.5 ms. The most typical identifier employed in this kind of attack is represented by the selected CAN ID "0000". The attack's goal is to flood the CAN bus with many messages bearing the targeted ID, overloading the system and resulting in a denial of service.

3.3 Fuzzing

A fuzz attack, sometimes referred to as fuzzing, is used to test a system's robustness by flooding it with a lot of false, inconsistent, or unexpected input [13]. Hackers can create fuzzing tools for the CAN bus that produce false and random messages while adhering to the CAN protocol. We used a fuzzing attack on our dataset, delivering messages with CAN ID and DATA values that were fully random at intervals of 0.3 and 0.5 ms. A fuzzing attack aims to cause ambiguity and confusion on the CAN bus. The attack seeks to prevent the system from operating normally by delivering messages with arbitrary identities and contents. This attack may result in unpredictable behavior since automotive systems are often not built to manage such random and unexpected input. This unpredictability may reveal weaknesses or result in system failures in the automobile, leaving them open to unanticipated problems.

3.4 Replay

Replay attacks are a particular kind of attack that hackers employ to target a system. A signal supplied to the CAN [14] communication system is intercepted and retransmitted in this attack by the attacker. To attack the CAN bus, perform the following steps: In order to intercept and record messages, the hacker first employs specialized equipment. The hacker then picks the messages that can be relayed back to the CAN system after sorting through the captured conversations. Following that, the target messages are picked once the chosen messages have been processed. Finally, these chosen messages are sent again, sometimes with their original sender names and contents, to deceive the system into considering them to be legitimate data and acting accordingly. The hacker inserted new traffic onto the CAN bus during the replay attack by removing normal traffic that normally occurs at a certain moment. CAN communication was seen during the assault via an actual cars OBD-II connector. In order to build the datasets used in the attack, messages were added. Each incident of message injection lasted between three and five seconds, and the captured data included information on 300 such occasions. Each dataset has between 30- and 40 min worth of CAN traffic in total.

3.5 Malfunction

Malfunction is one type of cyber-attack that is used to cause crashes or unexpected system behavior. In our case, the Malfunction attack can cause attacks on the CAN system with the aim of interfering with the normal operation of the communication. This attack is made by introducing incorrect or dangerous signals associated with the CAN [15]. Using flaws in the CAN protocol, attackers can control the ECU or monitor sensors and can change the messages that are used in vehicles. For our data set, the attack was carried out in the following order: a specific CAN ID was chosen, which is used in KIA SOUL CAN ID 0x153. The main goal was to change the values in the 8-byte data field. In order to carry out the attack, the number 00 and random numbers were discovered. Another injection was done by randomly choosing a CAN ID and combining the result with data fields. As a result of the attack, the car showed non-standard reactions after the injection, which eventually led to failures. The injections were given at a time interval of 5 s every 20 s.

4 Mean Normalization and Time Series for LSTM

4.1 Long Short-Term Memory (LSTM)

Long Short-Term Memory (LSTM) is a type of Recurrent Neural Network (RNN) that was developed to eliminate the vanishing gradient problem. In simple words, LSTM is an extended version of RNN that copes well with input data using sequential dependencies. The strengths of this model are it collects and can use different time stages. With the help of memory cells, LSTM models can recall or forget data during processing. In order to control information later, there are cell gates; the forgetting gate decides which information should be skipped further and which should be forgotten. The update gate decides which data needs to be changed inside the cell and the input gate decides which information needs to be entered into the cell. With this method, all layers affect the result of the network. Thanks to this structure, LSTM can effectively cope with long-term dependencies. LSTM models are good at prediction and classification tasks (Fig. 2).

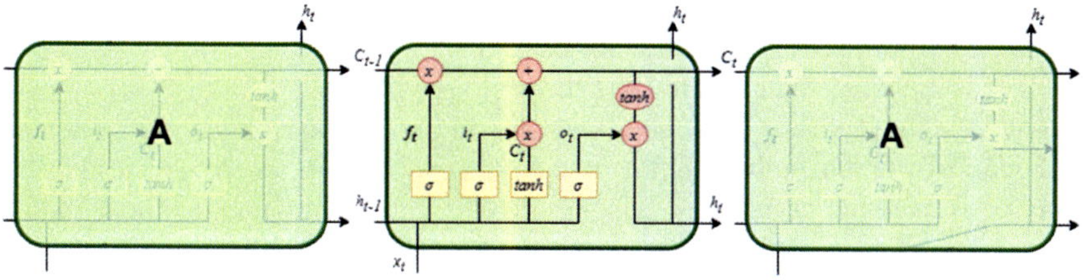

Fig. 2. Long Short-Term Memory (LSTM) structure

4.2 Feature Extraction and Dataset Pre-Processing

Our data set was collected from a real KIA SOUL machine and includes attacks such as DoS, Fuzz, Replay, Malfunction and Attack Free legitimate data [17]. At the beginning of our experiment, we had to clean and check the data for the content of incorrect data. After cleaning the data, we had to separate the attacked data from the legitimate data. Our set included 15 columns showing different information that was transferred between different devices. These columns contained information such as time since start, CAN ID, Time_gap which shows the time interval between CAN messages, Data Length Code (DLC), Payload in hex format, and 8 columns that show one each of Payload 1–8. In addition, in our dataset there is a marking of messages "R" and "T" which denote messages attacked or legitimate. We have added a new column to our dataset to convert CAN ID to categorical data to avoid the issue of huge CAN ID numbers during model training. However, many researchers have done research and they used all the columns to train their model. In our experiment, we only focused on two columns with which to classify attacks and detect them. We only used Categorical_CAN_ID and Time_gap. After fetching the desired columns, we connected all the datasets that we had, which is, DoS, Fuzz, Malfunction, Replay, Attack Free and then we labeled the data differently to distinguish them by the number we assigned them. We marked the data with numbers as follows: legitimate data was assigned the number 0, DoS - 1, Fuzz -2, Malfunction - 3 and Replay - 4. After combining all the data sets into one, we started pre-processing, which consisted of two stages: Data normalization and Data scaling. To make sure that all data points are on the same scale, we used "Mean Normalization", this method scales the data depending on the average and standard deviation. Our data preprocessing determines the mean and standard deviation for each dataset. We have changed the numbers to fit a certain range. Our function is used as follows (x_normalized = $(x - \mu)/\sigma$). x_normalized shows the normalized value of x, where x represents the original number. This way you can get the average of the data. In the second step, we used data scaling, which is done by transforming the values so that they are in a certain range. To do this, we used the formula (array-mean)/std to scale the data. After that, all the data that has passed through this function becomes scaled and zero-aligned. After pre-processing, we divided the datasets into small chunks in order to effectively regulate them. After that, we mix these segments and combine them again, after which we get the moved dataset. This method ensures that the data is evenly distributed and moved using this method, you can avoid overfitting the model and avoid any dependencies. Lastly, we used a time series that was organized with time step values of 10. To predict the next value, we developed a sliding eye method that takes as input data set equal to 10 of the time steps. The goal was to make a window of 10 and the model based on 10 messages was to predict the next value. The window is then moved one step down after the prediction and this cycle is repeated until the end of the data set. Then our dataset was split into two parts for training the model and for testing the LSTM. Each component has been converted to a NumPy array, and because our model is working with 3D data, we have changed the format of the data.

5 Evaluation Performance

We created an LSTM model using the TensorFlow library (version 2.11.0). Our model's architecture consists of four LSTM layers, each with 128 units. Each layer of the model was coupled to a batch normalization layer and a dropout layer with a dropout rate of 0.2. The hyperbolic tangent (tanh) is the activation function employed in the LSTM layers. A dense layer with five units and a Softmax activation function makes up the model's top layer. Using the above parameters, we trained the model using the backpropagation approach and the RMSprop optimizer. We shuffled the training data before to each epoch to improve the generalizability of the model. We used a variety of metrics and measurements to assess the model's efficacy. The percentage of samples that were properly categorized served as our main criterion for accuracy. The difference between the predicted and actual class labels was also measured using a loss estimator. We employed Early Stopping, a standard model training strategy, to reduce overfitting and enhance learning. When the chosen measure stops improving after a predetermined number of epochs, Early Stopping aims to end model training. It also stores and retrieves the model that produced the greatest outcomes with the least amount of loss. With a patience of 10 epochs and the recovery of the best weights, we used Early Stopping in our experiment. In order to constitute an improvement, a minimum change (min delta) of 0.0001 has to occur. In Table 1 and Fig. 3. The experiment's results showed that we were able to attain an accuracy of 98.87% and a loss coefficient of 0.0300 on the training sample. At epoch 97 out of 150, when this conclusion was reached, the early stopping parameters stopped the training. On the test sample, we had 99% accuracy. These findings show that the model is highly generalizable and capable of accurately classifying fresh data. We created two models for data categorization in this experiment. Data might be divided into five categories using the basic model: normal, dos, fuzzing, replay, and malfunction. For the second model, just the normal and attacked data classes were created, using BinaryCrossEntropy. The results of the second experiment showed that our binary model maintained the same high accuracy of 99% after 50 epochs with a loss of 0.0298 as shown in Table 2.

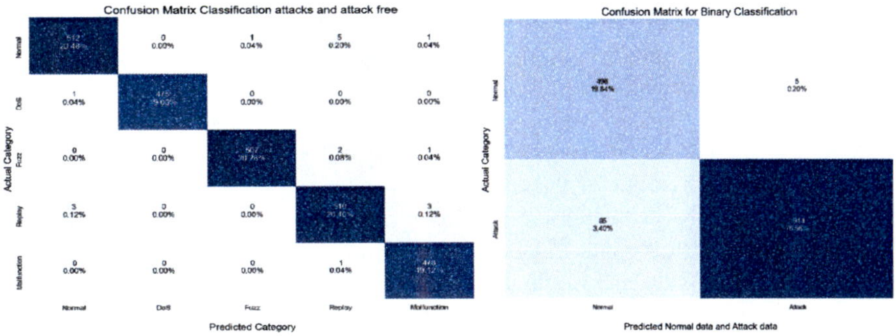

Fig. 3. Confusion Matrix for Multi-class and Binary Classification

Table 1. LSTM model report for Multi-class Classification.

	precision	recall	F1-score	support
Normal	0.99	0.99	0.99	519
DoS	1.00	1.00	1.00	476
Fuzz	1.00	1.00	1.00	510
Replay	0.98	0.99	0.99	516
Malfunction	0.99	0.99	0.99	479
Accuracy			0.99	2500
Macro avg	0.99	0.99	0.99	2500
Weighted avg	0.99	0.99	0.99	2500

Table 2. LSTM model report for Binary Classification.

	precision	recall	F1-score	support
Normal	0.97	0.99	0.98	484
Attack	1.00	0.99	1.00	2016
Accuracy			0.99	2500
Macro avg	0.99	0.99	0.99	2500
Weighted avg	0.99	0.99	0.99	2500

6 Conclusion

In the current experimental work, two LSTM models that can classify attacks and legitimate data were built in order to implement LSTM models on CAN bus Intrusion Detection Systems (IDS). The goal was to develop an effective intrusion detection system that can detect and classify attacks with high accuracy. In the initial experiment, we built an

LSTM model that can classify 4 types of attacks (DoS, Fuzz, Malfunction, Replay) and recognize legitimate data. Our initial experiment results showed that our model was able to classify attacks and anomaly detections with an accuracy of 99%. The results demonstrate that our model can be used as an effective intrusion detection system (IDS) for a CAN system. In the second experiment, we built an LSTM with binary classification, that is, to distinguish only two classes of attacks or legitimate data. The results of this experiment also showed much more promising accuracy with 99%. The results show that both models perform well in attack detection and classification tasks. Our future research will focus on extending and improving current models using different data preprocessing techniques and additional approaches. We expect that using these methods you can get much better results and reduce the risk that may arise from attackers. We think our results can help researchers create effective protection for vehicles.

Acknowledgments. This work was supported by the National Research Foundation of Korea (NRF) grant funded by the Korean government (MSIT) (No. 2021R1A4A2001810) and Institute for Information & communications Technology Planning & Evaluation(IITP) grant funded by the Korea government(MSIT) (No. 2022-0-01197, Convergence security core talent training business(SoonChunHyangUniversity).

References

1. ECU: https://www.linkedin.com/pulse/20140626152045-3625632-car-software-100m-lines-of-code-and-counting
2. Bozdal, M., Samie, M., Aslam, S., Jennions, I.: Evaluation of can bus security challenges. Sensors **20**(8), 2364 (2020)
3. Hossain, M.D., Inoue, H., Ochiai, H., Fall, D., Kadobayashi, Y.: An effective in-vehicle CAN bus intrusion detection system using CNN deep learning approach. In: GLOBECOM 2020–2020 IEEE Global Communications Conference, pp. 1–6. IEEE (2020)
4. Hanselmann, M., Strauss, T., Dormann, K., Ulmer, H.: CANet: an unsupervised intrusion detection system for high dimensional CAN bus data. IEEE Access **8**, 58194–58205 (2020)
5. Javed, A.R., Ur Rehman, S., Khan, M.U., Alazab, M., Reddy, T.: CANintelliIDS: Detecting in-vehicle intrusion attacks on a controller area network using CNN and attention-based GRU. IEEE Trans. Netw. Sci. Eng. **8**(2), 1456–1466 (2021)
6. Song, H.M., Kim, H.R., Kim, H.K.: Intrusion detection system based on the analysis of time intervals of CAN messages for in-vehicle network. In: 2016 International Conference on Information Networking (ICOIN), pp. 63–68. IEEE (2016)
7. Kang, M.J., Kang, J.W.: Intrusion detection system using deep neural network for in-vehicle network security. PLoS ONE **11**(6), e0155781 (2016)
8. Hossain, M.D., Inoue, H., Ochiai, H., Fall, D., Kadobayashi, Y.: LSTM-based intrusion detection system for in-vehicle can bus communications. IEEE Access **8**, 185489–185502 (2020)
9. Narasimhan, H., Vinayakumar, R., Mohammad, N.: Unsupervised deep learning approach for in-vehicle intrusion detection system. IEEE Consum. Electron. Mag. **12**, 103–108 (2021)
10. Hoang, T.N., Islam, M.R., Yim, K., Kim, D.: CANPerFL: improve in-vehicle intrusion detection performance by sharing knowledge. Appl. Sci. **13**(11), 6369 (2023)

11. Islam, M.R., Oh, I., Yim, K.: Universal intrusion detection system on in-vehicle network. In: Barolli, L. (ed.) Innovative Mobile and Internet Services in Ubiquitous Computing: Proceedings of the 17th International Conference on Innovative Mobile and Internet Services in Ubiquitous Computing (IMIS-2023), pp. 78–85. Springer Nature Switzerland, Cham (2023). https://doi.org/10.1007/978-3-031-35836-4_9
12. Hpl, S.: Introduction to the controller area network (CAN). Application Report SLOA101, pp. 1–17 (2002)
13. Lee, H., Jeong, S.H., Kim, H.K.: OTIDS: A novel intrusion detection system for in-vehicle network by using remote frame. In: 2017 15th Annual Conference on Privacy, Security and Trust (PST), pp. 57–5709. IEEE (2017)
14. Kang, H., Kwak, B. I., Lee, Y. H., Lee, H., Lee, H., Kim, H. K.: Car hacking and defense competition on in-vehicle network. In: Workshop on Automotive and Autonomous Vehicle Security (AutoSec), vol. 2021, p. 25 (2021)
15. Han, M.L., Kwak, B.I., Kim, H.K.: Anomaly intrusion detection method for vehicular networks based on survival analysis. Veh. Commun. **14**, 52–63 (2018)
16. Van Houdt, G., Mosquera, C., Nápoles, G.: A review on the long short-term memory model. Artif. Intell. Rev. **53**, 5929–5955 (2020)
17. Datasets: https://ocslab.hksecurity.net

Wireless Visual Sensor Node Placement Optimization Considering Different Distributions of Events

Yuki Nagai[1], Tetsuya Oda[2(✉)], Chihiro Yukawa[1], Kyohei Toyoshima[1], Kei Tabuchi[3], and Leonard Barolli[4]

[1] Graduate School of Engineering, Okayama University of Science (OUS), 1-1 Ridaicho, Kita-ku, Okayama 700-0005, Japan
{t22jm23rv,t22jm19st,t22jm24jd}@ous.jp

[2] Department of Information and Computer Engineering, Okayama University of Science (OUS), 1-1 Ridaicho, Kita-ku, Okayama 700-0005, Japan
oda@ous.ac.jp

[3] Graduate School of Science and Engineering, Okayama University of Science (OUS), 1-1 Ridaicho, Kita-ku, Okayama 700-0005, Japan
r23sm11og@ous.jp

[4] Department of Information and Communication Engineering, Fukuoka Institute of Technology, 3-30-1 Wajiro-Higashi-ku, Fukuoka 811-0295, Japan
barolli@fit.ac.jp

Abstract. Wireless Visual Sensor Networks (WVSNs) have a mesh mechanism, which enables a wide area imaging and image collection is done through multi-hop communication between multiple sensor nodes equipped with visual sensors. Therefore, they can be used for monitoring infrastructure facilities and rivers. However, the placement of sensor nodes has a significant impact on the connectivity and transmission loss of wireless communication in WVSNs and the imaging range of visual sensors is limited. Thus, it is necessary to find the optimum sensor node arrangement and imaging direction of the visual sensors in order to cover all events within the monitoring area. In this paper, we propose an intelligent system for optimizing sensor node placement in WVSNs. The proposed system integrates two methods for sensor node placement by considering the number of events within the imaging range of the visual sensor and the imaging direction of the visual sensor. To evaluate the proposed method, computer simulations are performed for two scenarios with uniform and normal distribution of events. The simulatione results show that the proposed method maximizes the Size of Giant Compponent (SGC) and all sensor nodes are connected. From the visualization results, we found that in CCM-based SA, the nodes are spread over a wider area and all events are covered within the imaging range of the visual sensor.

1 Introduction

A Wireless Visual Sensor Network (WVSN) is a network of sensor nodes equipped with a mesh mechanism, where sensor nodes with visual sensors perform multi-hop communication between multiple sensor nodes [1–5]. Therefore, in WVSNs the communication can be done through other sensor nodes even if some network links fail.

L. Barolli (Ed.): BWCCA 2023, LNDECT 186, pp. 312–322, 2024.
https://doi.org/10.1007/978-3-031-46784-4_30

They can provide stable wireless connectivity and the image collection can be done over a wide area. Thus, they can be used for monitoring infrastructure facilities and rivers.

In a previous study, we implemented a WVSN with stereo camera for monitoring water levels in rivers [6]. We found that multiple sensor nodes need to be deployed around the monitoring target. However, the placement of sensor nodes has a significant impact on the connectivity and transmission loss of wireless communication [7–12]. Therefore, the placement optimization of mesh routers is a very important problem [9, 13–16]. For this reason, we proposed the Coverage Construction Method (CCM) [17], CCM-based Hill Climbing (HC) [18] and CCM-based Simulated Annealing (SA) [19, 20] to solve this problem. In addition, in WVSNs the imaging range of visual sensors is limited because the lens and sensor size of the visual sensor are used to decide the distance and viewing angle captured by the visual sensor. Therefore, in order to take the image all events at different locations in the target space with multiple visual sensors, it is essential to find the optimal arrangement of the sensor nodes and the imaging direction considering the viewing angle of the visual sensors.

In this paper, an intelligent system for optimizing sensor node placement in WVSNs is proposed. The proposed system integrates two methods (CCM and CCM-based SA) for sensor node placement, considering the Number of Covered Events (NCE) by the imaging range of the visual sensor and the imaging direction of the visual sensor.

To evaluate the proposed method, computer simulations are performed for two scenarios with uniform and normal distribution of events. The Size of Giant Component (SGC) of both methods is maximized, so all sensor nodes are connected. In the case of CCM, not all events are covered due to the dense deployment of sensor nodes. While in CCM-based SA, sensor nodes are deployed over a larger area and in both scenarios NCE is maximized and all events are covered by the imaging range of the visual sensor.

The paper is organized as follows. In Sect. 2, we present an overview of the proposed system. In Sect. 3, we show the simulation results. Finally, in Sect. 4, we conclude the paper.

2 Proposed Intelligent System

We show the structure of the proposed system in Fig. 1, while Fig. 2 shows the flowchart of the proposed system. The proposed system constructs a WVSN by deciding the optimal placement of sensor nodes and the imaging direction of visual sensors based on CCM and CCM-based SA in order to maximize the number of events within the imaging range of the visual sensors. Therefore, the objective function considers two factors which are the SGC which indicates network connectivity and the NCE which indicates the number of events within the imaging range of the visual sensor.

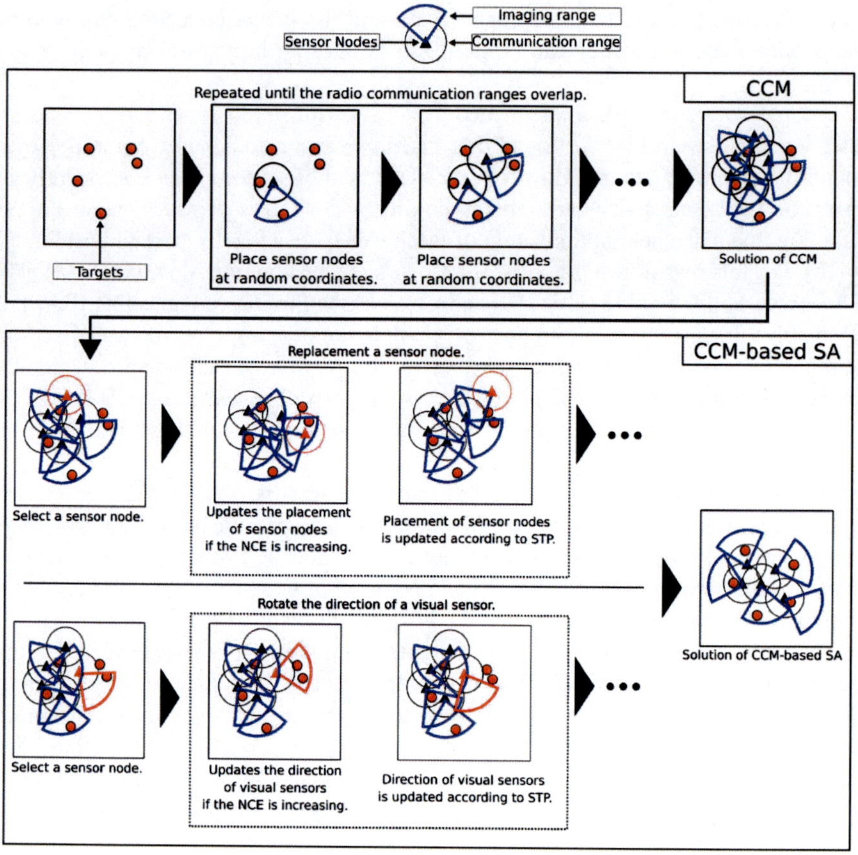

Fig. 1. Structure of proposed system.

2.1 CCM

CCM is a method for maximizing SGC and NCE for mesh router optimization problem. In CCM, the coordinates of sensor node placements and the imaging direction of the visual sensors are placed randomly in the target area. The CCM consider the wireless communication range of the sensor nodes as a circle and performs collision detection between the wireless communication ranges of each sensor node. The placement and collision detection process is repeated until the wireless communication ranges overlap between the already placed sensor nodes and the other sensor nodes. If the wireless communication range of a sensor node and the already placed sensor nodes do not overlap, the sensor node is deleted and replaces with a new one. Otherwise, if the wireless communication range of a sensor nodes and the already placed sensor node overlap, the seson node is added to the already placed sensor nodes. This process continues until

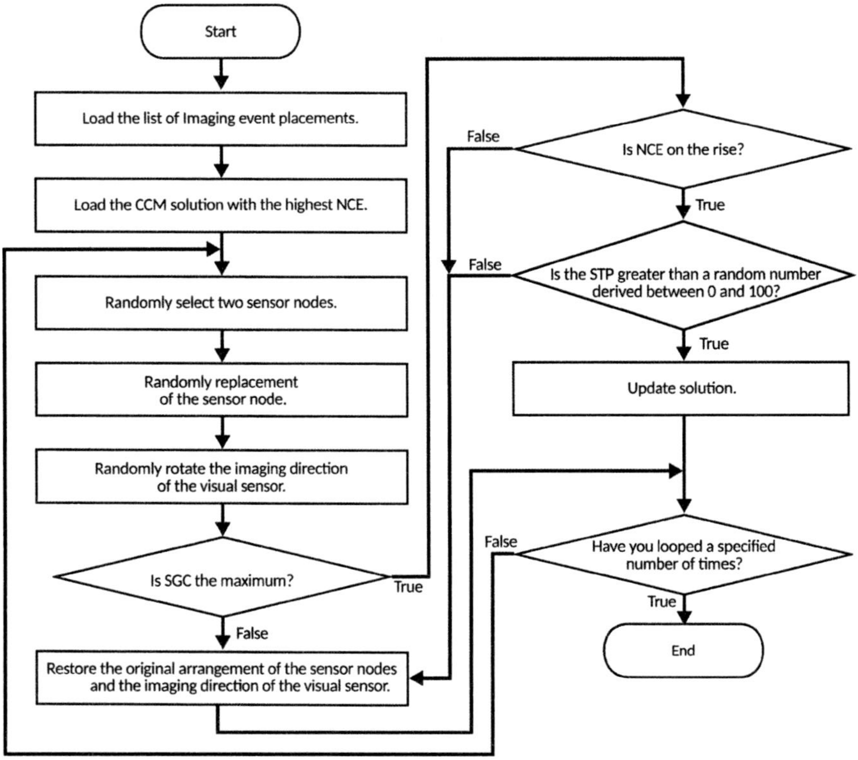

Fig. 2. Flowchart of proposed system.

a determined number of repetitions or the SGC reaches the defined number of sensor nodes. Then, these processes are repeated and the solution with the highest NCE among the generated solutions is the final solution of the CCM.

2.2 CCM-Based SA

CCM-based SA method derives the optimal solution by SA based on the placement of sensor nodes and the imaging direction of the visual sensor decided by CCM. CCM-based SA is a nearest neighbor search method, which allows escaping from local solutions by neighborhood search and state transitions by random selection. Also, SA updates the solution according to the State Transition Probability (STP), which consists of the solution evaluation and the temperature.

The CCM-based SA randomly selects two sensor nodes. One selected sensor node is relocated to random coordinates in the simulation area and the imaging direction of the visual sensor is not changed. Another sensor node rotates the imaging direction of the visual sensor in an arbitrary imaging range and changes the imaging direction, while the coordinates of the sensor node remain unchanged. A collision detection of the wireless communication range of the sensor nodes is performed and if the SGC is maximized, then the NCE is calculated. If the NCE is increased, the placement of the sensor node and the imaging direction of the visual sensor are updated. Otherwise, if the NCE is decreased or the NCE is not updated, the placement of the sensor nodes and the imaging direction of the visual sensor are updated according to the STP. This process is repeated until a determined number of times to derive the optimal sensor node placement and visual sensor imaging direction.

3 Simulation Results

We evaluated the performance of the proposed system by computer simulations. Table 1 shows the simulation parameters. The simulation are performed for two scenarios where the imaging targets of the visual sensor are randomly placed with uniform and normal distribution in the simulation area. The moving distance of a sensor node is set in the range of ±5 [m] from the coordinates of the sensor node in the X and Y coordinate direction. While, the rotation angle of the visual sensor for the imaging direction is set in the range of ±90 [°] from the current angle. The view angle of visual sensor is set at 90 [°] and the maximum imaging range is 20 [m] considering the standard lens specifications.

Figure 3 shows the simulation results for uniform distribution. While, Fig. 3(a) and Fig. 3(b) show simulation results of CCM and the CCM-based SA, respectively. Figure 4 shows the simulation results for normal distribution. While, Fig. 4(a) and Fig. 4(b) show simulation results of CCM and the CCM-based SA, respectively. It can be seen that the SGC is 16 [*unit*], so this is the maximum number of sensor nodes. For CCM simulation and uniform distribution, the NCE at the end of the iterations covers 31 [*unit*] events in the imaging range of the visual sensor. While for CCM and normal distribution, the NCE at the end of the iterations covers 39 [*unit*] events in the imaging range of the visual sensor. For both scenarios, the CCM-based SA is able to cover all 48 [*unit*] events.

Figure 5 shows visualisation results for uniform distribution, while in Fig. 5(a) and Fig. 5(b) are shown visualisation results of CCM and CCM-based SA, respectively. Figure 6 shows visualisation results for normal distribution, while in Fig. 6(a) and Fig. 6(b) are shown visualisation results of CCM and CCM-based SA, respectively. From the simulation results, it can be seen that in both scenarios, the proposed system has a wider range of sensor nodes, an increased communication range and can cover all imaging events by the imaging range of the visual sensor.

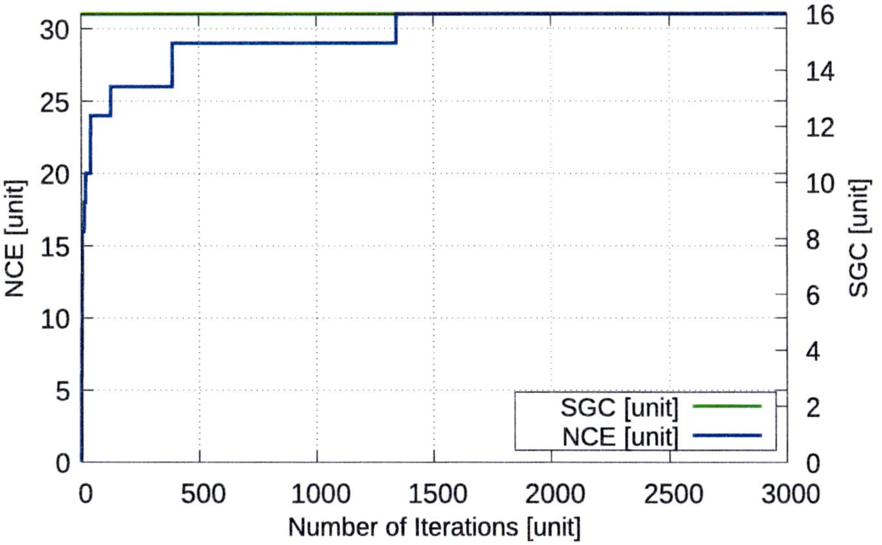

(a) Results of CCM.

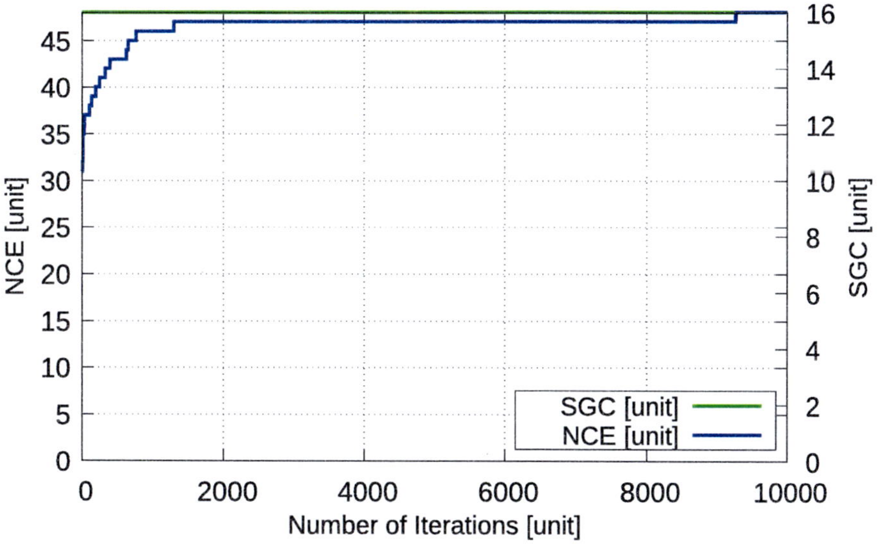

(b) Results of CCM-based SA.

Fig. 3. Simulation results for uniform distribution.

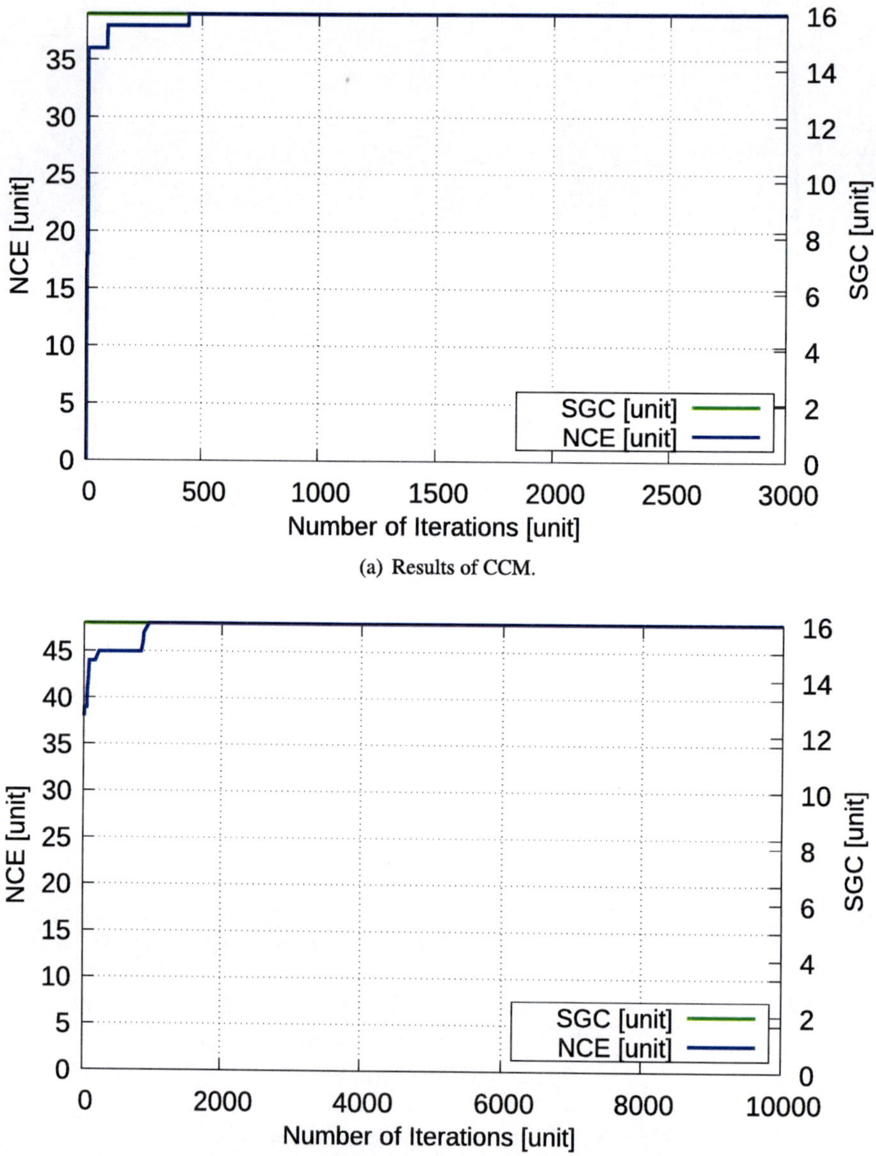

(a) Results of CCM.

(b) Results of CCM-based SA.

Fig. 4. Simulation results for normal distribution.

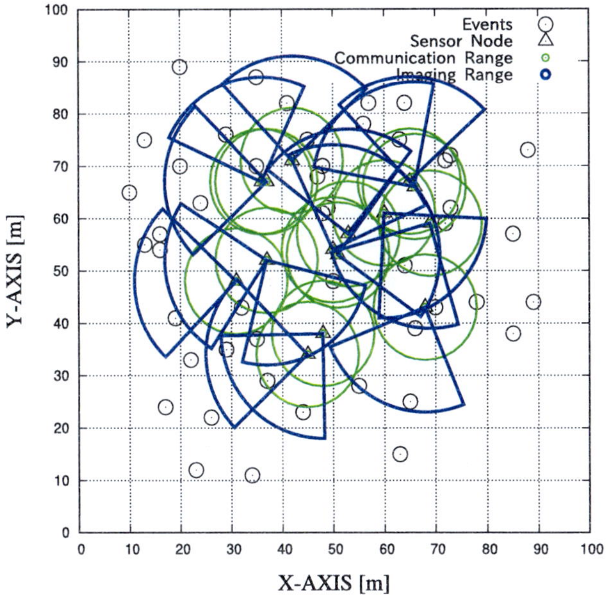

(a) Results of CCM.

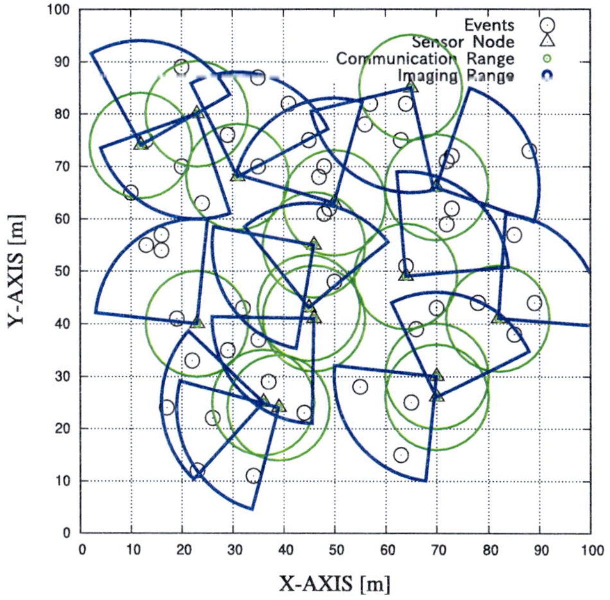

(b) Results of CCM-based SA.

Fig. 5. Visualization results for uniform distribution.

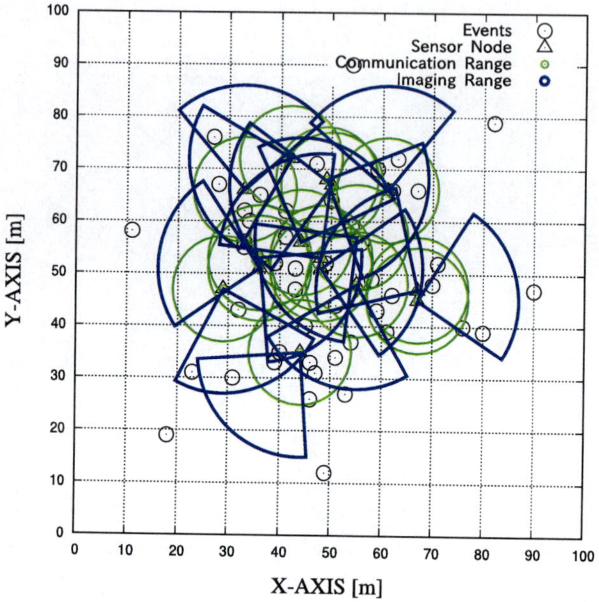

(a) Results of CCM.

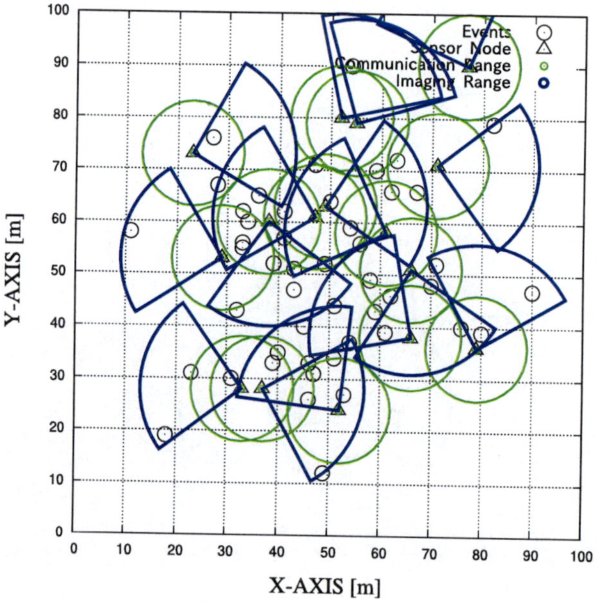

(b) Results of CCM-based SA.

Fig. 6. Visualization results for normal distribution.

Table 1. Parameters and values for simulations.

Functions	Values
Width of Simulation Area	100 [m]
Hight of Simulation Area	100 [m]
Number of Mesh Routers	16
Radius of Radio Communication Range of Mesh Routers	10 [m]
Number of Events	48
Number of Loops for CCM	3000 [*unit*]
Number of Loops for CCM-based SA	10000 [*unit*]
Imaging Range	20 [m]
View Angle	90 [°]

4 Conclusions

In this paper, we proposed an intelligent system for optimization of sensor node placement in WVSNs.

We evaluated the CCM and CCM-based SA methods by computer simulations under two different scenarios, one with a uniform distribution and the other with a normal distribution of events. From the simulation results, we conclude as follows.

- The SGC for both methods is maximized, so all sensor nodes are connected.
- For the uniform distribution of events, the NCE for CCM at the end of iterations covers 31 [*unit*] events, while for CCM-based SA all 48 [*unit*] events.
- For the normal distribution of events, the NCE for CCM at the end of iterations covers 39 [*unit*] events, while for CCM-based SA all 48 [*unit*] events.
- For CCM-Based SA the nodes are spread over a wider area and all events are covered by the imaging range of visual sensors.

In the future, we would like to carry out additional simulations for different scenarios as well as considering large-scale simulations and the placement of events in three dimensions.

Acknowledgement. This work was supported by JSPS KAKENHI Grant Number JP20K19793.

References

1. Idoudi, M., et al.: Wireless visual sensor network platform for indoor localization and tracking of a patient for rehabilitation task. IEEE Sens. J. **18**(14), 5915–5928 (2018)
2. Peng-Fei, W., et al.: Node scheduling strategies for achieving full-view area coverage in camera sensor networks. Sensors **17**(6), 1303–1321 (2017)
3. Akyildiz, I.F., et al.: Wireless mesh networks: a survey. Comput. Netw. **47**(4), 445–487 (2005)
4. Jun, J., et al.: The nominal capacity of wireless mesh networks. IEEE Wirel. Commun. **10**(5), 8–15 (2003)

5. Oyman, O., et al.: Multihop relaying for broadband wireless mesh networks: from theory to practice. IEEE Commun. Mag. **45**(11), 116–122 (2007)
6. Nagai, Y., et al.: A triangulation based water level measuring system for a water reservoir tank. In: Barolli, L. (ed.) AINA 2023. LNNS, vol. 655, pp. 376–383. Springer, Cham (2023). https://doi.org/10.1007/978-3-031-28694-0_36
7. Oda, T., et al.: A GA-based simulation system for WMNs: performance analysis of WMN-GA system for different WMN architectures considering DCF and EDCA. In: Proceedings of the 7-th International Conference on Intelligent Networking and Collaborative Systems (INCoS-2015), pp. 232–238 (2015)
8. Oda, T., et al.: Analysis of node placement in wireless mesh networks using friedman test: a comparison study for tabu search and hill climbing. In: Proceedings of the 9-th International Conference on Innovative Mobile and Internet Services in Ubiquitous Computing (IMIS-2015), pp. 133–140 (2015)
9. Oda, T., et al.: Analysis of mesh router placement in wireless mesh networks using Friedman test considering different meta-heuristics. Int. J. Commun. Netw. Distrib. Syst. **15**(1), 84–106 (2015)
10. Oda, T., et al.: Effects of population size for location-aware node placement in WMNs: evaluation by a genetic algorithm-based approach. Pers. Ubiquit. Comput. **18**, 261–269 (2014)
11. Oda, T., et al.: Analysis of mesh router node placement using WMN-GA system considering different architectures of WMNs. In: Proceedings of the 17-th International Conference on Network-Based Information Systems (NBiS-2014), pp. 39–44 (2014)
12. Sakamoto, S., et al.: Performance analysis of two wireless mesh network architectures by WMN-SA and WMN-TS simulation systems. J. High Speed Netw. **23**(4), 311–322 (2017)
13. Oda, T., et al.: WMN-GA: a simulation system for WMNs and its evaluation considering selection operators. J. Ambient. Intell. Humaniz. Comput. **4**(3), 323–330 (2013)
14. Ikeda, M., et al.: Analysis of WMN-GA simulation results: WMN performance considering stationary and mobile scenarios. In: Proceedings of the 28-th IEEE International Conference on Advanced Information Networking and Applications (IEEE AINA-2014), pp. 337–342 (2014)
15. Oda, T., et al.: A genetic algorithm-based system for wireless mesh networks: analysis of system data considering different routing protocols and architectures. Soft. Comput. **20**(7), 2627–2640 (2016)
16. Sakamoto, S., Ozera, K., Oda, T., Ikeda, M., Barolli, L.: Performance evaluation of intelligent hybrid systems for node placement in wireless mesh networks: a comparison study of WMN-PSOHC and WMN-PSOSA. In: Barolli, L., Enokido, T. (eds.) IMIS 2017. AISC, vol. 612, pp. 16–26. Springer, Cham (2018). https://doi.org/10.1007/978-3-319-61542-4_2
17. Hirata, A., et al.: Approach of a solution construction method for mesh router placement optimization problem. In: Proceedings of the IEEE 9-th Global Conference on Consumer Electronics (IEEE GCCE-2020), pp. 1–2 (2020)
18. Hirata, A., Oda, T., Saito, N., Hirota, M., Katayama, K.: A coverage construction method based hill climbing approach for mesh router placement optimization. In: Barolli, L., Takizawa, M., Enokido, T., Chen, H.-C., Matsuo, K. (eds.) BWCCA 2020. LNNS, vol. 159, pp. 355–364. Springer, Cham (2021). https://doi.org/10.1007/978-3-030-61108-8_35
19. Hirata, A., et al.: A voronoi edge and CCM-based SA approach for mesh router placement optimization in WMNs: a comparison study for different edges. In: Barolli, L., Hussain, F., Enokido, T. (eds.) AINA 2022. LNNS, vol. 451, pp. 220–231. Springer, Cham (2022). https://doi.org/10.1007/978-3-030-99619-2_22
20. Oda, T.: A delaunay edges and simulated annealing-based integrated approach for mesh router placement optimization in wireless mesh networks. Sensors **23**(3), 1050 (2023)

A Music Therapy Model Based on Consortium Blockchain Platform with Evidence and Tracking Efficacy of Hypertension Treatment

Hsing-Chung Chen[1,2](✉), Pei-Yu Hsu[1], and Jhih-Sheng Su[1]

[1] Department of Computer Science and Information Engineering, Asia University, Taichung 413305, Taiwan
cdma2000@asia.edu.tw, shin8409@ms6.hinet.net, hcliebe2019@gmail.com, a50b50a50@yahoo.com.tw

[2] Department of Medical Research, China Medical University Hospital, China Medical University, Taichung 404327, Taiwan

Abstract. This study innovatively designs a music therapy music selection recommendation system based on Case-based Spiral Diagnosis Reasoning with Consortium Blockchain Knowledge Base (CBSDR-CBKB) model. The purpose of this study is that hypertensive patients could effectively improve their physical and mental health by normalizing blood pressure through music therapy provided by CBSDR-CBKB system. In view of the lack of a music therapy recommendation system, this paper adopts the combination of case-based reasoning with spiral diagnosis procedure to recommend the music selection for the music ther apist who provides to patients the therapist to listen to different types of music to make an integrated music selection recommendation application system. It not only makes it easy for hypertensive patients and their families to find music styles and selection principles suitable for music therapy, but also retains successful cases and failed cases and can screen cases more efficiently. Furthermore, this study is first design with a music therapy selection recommendation system. In CBSDR-CBKB. The steps of this study are to input the characteristic attributes of hypertensive blood pressure subjects, use the spiral theoretical model based on case reasoning to search for initial similar cases, and then perform periodic music therapy and recording after modification according to the case attributes. Next, the results output a music therapy record of the case according to the assessment of the Taiwan Hypertension Society and Taiwan Society of Cardiology, which includes a set of relative or modified music therapy strategies and a set of corresponding treatment effects, providing a music therapy recommendation. Each record will then be stored in Consortium Blockchain Platform. The system enables patients with hypertension caused by autonomic nervous system disorders to find suitable music styles to assist in the treatment or control of hypertension.

Keywords: music therapy · hypertension · case-based reasoning mechanism · spiral diagnosis procedure · consortium Blockchain platform

L. Barolli (Ed.): BWCCA 2023, LNDECT 186, pp. 323–334, 2024.
https://doi.org/10.1007/978-3-031-46784-4_31

1 Introduction

Due to the high pressure of life and the fast pace of life, people are often under tension. As a result, the blood circulation of these people gradually deteriorates. Next, hypertension is the most prevalent chronic disease. Hypertension is a common global health issue and the most common chronic disease, known as the "silent killer"[1]. The accumulation of high blood pressure will cause systemic complications, which it will be posed some major threats to human health, such as myocardial infarction, stroke, *etc.* It is the autonomic nerve that regulates blood pressure [1–8]. Patients with high blood pressure easily increase the risk of stroke, heart disease, and kidney disease. The function of regulating blood pressure is to relax the autonomic nerves. For example, when exercising, the sympathetic nerves of people's bodies become excited, and blood pressure rises. On the contrary, if people are in a state of rest, the parasympathetic nerves of people's bodies will make blood pressure drop [1–8]. Furthermore, the changes in mood will also be reflected in blood pressure, e.g., the excited or excited mood will increase blood pressure. Otherwise, the calm mood will lower blood pressure [1–8]. However, a recommendation system for music therapy is lacking.

First, the proposes of this study are establishing a procedure that combines case-based reasoning with spiral diagnosis [9, 10]. It will recommend song selections and treatment procedures for the test subjects. It could provide an application system for therapists to listen to different types of music to make integrated music selection recommendations. Finally, the system could be recorded in the consortium blockchain. These successful music therapy process cases include basic information of the subjects, song selection, treatment time and cycle, *etc.* based on privacy protection.

The Hyperledger Fabric blockchain platform [11–16] of this system is designed as a consortium chain platform with point economic operation. When a successful case is generated, the music therapist can remove the private or confidential information in the personal data of the treated person and store it in the alliance chain. It can form the knowledge base based on blockchain technology for this study. This system will design a decentralized application (DApp) [11–16]. It can be used to connect cases where music therapists diagnose cases with the real blockchain world that provides payment mechanisms. In this system, subscribing customers can access successful cases by using smart contracts [11–16].

Based on the above, this study innovatively designs a music therapy music selection recommendation system based on the case-based spiral diagnosis reasoning with consortium blockchain knowledge base (CBSDR-CBKB) model. In other words, this article provides a business service model for accessing music therapy cases based on the blockchain alliance chain.

The remainder of this paper is organized as follows: In Sect. 2, we introduce the related works. In Sect. 3, we first present CBSDR-CBKB model. Then, the experimental results and discusses are described in Sect. 4. Finally, we draw our conclusions in Sect. 5.

2 Related Works

Music therapy is a tool used to treat patients with physical or mental illnesses through music, singing, and rhythm [1–8]. For cases that require treatment in terms of body and mind, Music therapy provides musical elements (timbre, rhythm, melody, and harmony) in order to meet the emotional, psychosocial, cognitive, and social needs of the healer [1–8]. Music therapy is a comprehensive science that integrates medicine, psychology, and musicology. Today, there are many successful cases of research and treatment. Therefore, many medical institutions have also adopted music therapy as a formal diagnostic department [1]. For cases in need of physical and mental treatment, the CBSDR-BKB model is proposed in this article to recommend appropriate music and treatment processes [1]. It can help the treated person quickly find the appropriate music and treatment process.

Next, the case-based reasoning with spiral diagnosis (CBR-SD) [10] is an extended by Case-based reasoning (CBR) mechanism. As we known, CBR technique was proposed by Aamodt and Plaza [9] in (1994) is the method of solving the new problems through the solutions of the similar previous problems [9]. CBR has been formalized as case retrieve, case reuse, case revise, and case retain techniques [9]. In 2017, the RBAC (Role-based Access Control) with CBR (case-based reasoning) based negotiation technique is proposed by S.-S. Tseng and H.-C. Chen et al. [10], which is the CBR-N (CBR-based Negotiation) RBAC model by using spiral negotiating process. It could meet the requirements of negotiating role assignment (NRA) task and cooperative role assignment (CRA) task in any platform. Their novel RBAC scheme [10] could efficiently manage not only one application server, but also multiple application servers on the cooperation domain through supporting both negotiating role assignment (NRA) task and cooperative role assignment (CRA) task in any platform by using a spiral negotiating process [10]. In the platform, the NRA task using CBR-N model with spiral negotiating process consists of four steps: similar case inquiry, negotiating, fitting, and determining, where the CBR-N technique including similar scenarios and corresponding case inquiry, negotiating for REUSE or REVISE of initial case, fitting a local role or a virtual role, and case determining and retaining technique is used to find a similar case assigning an appropriate role(s) to user[10]. Thus, the basic CBR-N technique [10] consists of four steps: similar case inquiry, negotiating, fitting, and determining applying into the URM-platform, where the CBR-N technique including *similar scenarios and corresponding case inquiry*, *negotiating for REUSE or REVISE of initial case*, *fitting a local role or a virtual role*, and *case determining and retaining technique* is used to find the similar case in order to re-assign the appropriate role(s) to user [10]. The basic CBR-based Negotiation (CBR-N) model with spiral negotiating process is described below [10].

Step 1. *Inquiring similar case*: According to the characteristics of the model, CBR-based Negotiation technique is used to inquire the most similar case with the corresponding similarity scenario to be an initial case in the negotiation process. This step corresponds to NRA inquiring step [10].

Step 2. *Negotiating for REUSE or REVISE of initial case*: According to the CBR-based negotiating processing, CBR case reuse or case revise technique is used to assign appropriate role(s) consisting of a local role and virtual role if it is needed in the negotiating process [10].

Step 3. *Fitting a local role or a virtual role*: Fitting a local role by the serving server or a virtual role by the cooperative server via NRA process [3]. This step corresponds to NRA platform presentation step [10].

Step 4. *Case determining and retaining*: CBR assigning role case retain technique is used to clarify the negotiating project and assist user to get appropriate role and meet their requirements in the negotiating process [10]. This step corresponds to NRA analysis and conclusion steps [10].

Subsequently, CBR-based Negotiation (CBR-N) procedures with spiral negotiating process [10] are proposed to assist RBAC servers to reuse the previous appropriate cases and reserve to user in the system, as shown in Fig. 1.

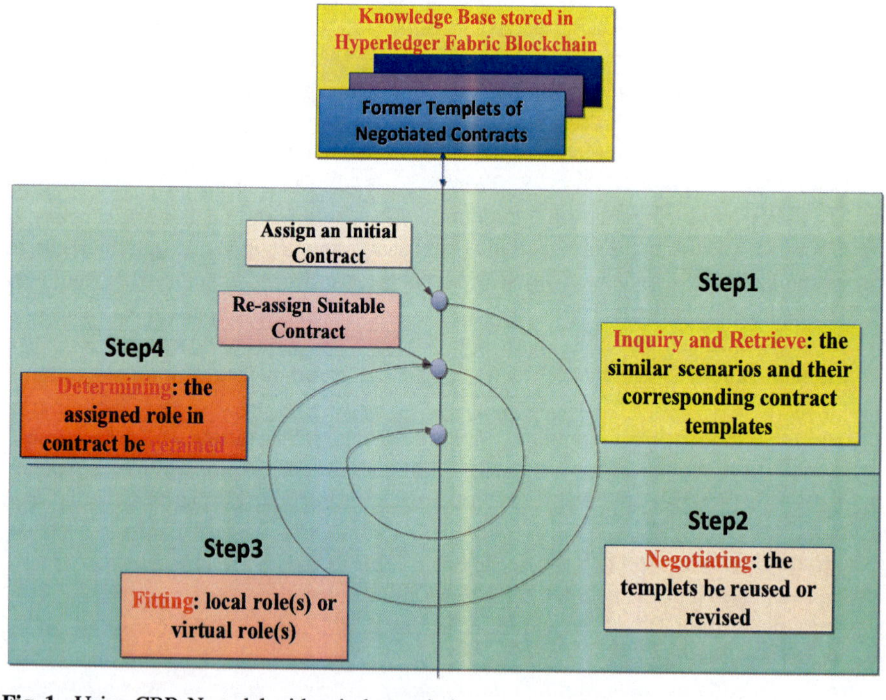

Fig. 1. Using CBR-N model with spiral negotiating process to build the Knowledge Base stored in Hyperledger Fabric Blockchain platform [10]

On the other hands, the blockchain technology benefit in tackling online shopping transaction revocation issue is presented by H.-C. Chen *et al.* [11]. They explained how blockchain technology might help solving the problem through four aspects: (1) digital access rules, (2) data immutability, (3) package tracking and update, and (4) easy payment management. They then propose payment distribution formulas to address the revocation issue [11]. According to the basic blockchain technologies consisting of the ledger, smart contracts, consensus, and cryptography, the innovation trust chain mechanism [12,

17] was first proposed and implemented by H.-C. Chen *et al.*, which could be provided the system-level security for any IoT application with System-Module-IC chips architecture toward Industrial 4.0 in order to meet the requirements of both manufacturing innovation and product innovation in Sustainable Development Goals (SDGs) [12, 17]. The reconfigurable PM2.5 sensor green deployment mechanism based on blockchain technology was first proposed by H.-C. Chen *et al.* [13]. Their results shown that the modules which are measured the similar data will be distributed into the candidate group instead of the current working group, where the total corresponding energy consumption will be reduced for the Blockchain transaction operations in their blockchain network [13]. Moreover, the relevant data storages will be also avoided to store the duplication data as well as redundant data [13]. Finally, they want to improve the performances not only duplication or similar data reducing but also the relevant Blockchain transaction operations [6]. H.-C. Chen *et al.* [14] presented the trust evaluation gateway for distributed blockchain IoT network, where the idea of the trust evaluation gateway is first proposed for distributed BC IoT network. In this paper, the three types of trust evaluation function for Machine to Individual (M2I), Machine to Machine (M2M), and Individual to Individual (I2I), selectively, are provided the appropriate solution for solving the online authentication problem operated among the general Internet and distributed Blockchain networks [14]. H.-C. Chen *et al.* [15] illustrated how IoT devices could be evaluated by the sink nodes acted as a blockchain nodes in order to give the contribution for cooperative evaluation in the blockchain for the integration IoT application. Their cooperative evaluation method is required while executing transaction process in Blockchain network, which the designed smart contract could validate IoT devices by these collaboration blockchain agent nodes [15]. The innovative secure private Over-The-Top (*p*-OTT) platform is proposed by H.-C. Chen *et al.* [16], which the transaction process could deal with the credible and customizable services by adopting blockchain technology. Each product of digital media content (DMC) could be pre-defined a template of smart contract. After negotiation process by using this smart contract, the real data contract (RDC) will be created and outputted by smart contract [16]. According to the predefined template of smart contract, the content producer or provider could easily implement his transaction policy [16]. In addition, they designed the online authentication key for each member and a shadow key pair for each booked DMC during a transaction period [16]. Therefore, they presented the secure OTT model with blockchain technology. Finally, the results show that it is not only a secure OTT platform but also a flexible transaction via preforming smart contract [16].

3 System Design

The CBSDR-CBKB model proposed in this paper is a new modified model based on the spiral negotiation (CBR-SN) model of case reasoning proposed by the scholars S.-S. Tseng and H.-C. Chen [10] in 2017. This study proposes a music therapy recommendation system based on the CBSDR-CBKB model. The system design combines case-based reasoning with spiral diagnostic procedures. It could recommend music tracks to test subjects and provide different music types to hypertensive patients with autonomic nervous system disorders. This study explored the effects of different types of music on the treatment of hypertensive patients with dysautonomia. Learn about the best suitability and accuracy of a customized music therapy selection recommendation system. The CBSDR-CBKB model could be used to provide the high blood pressure self-measurement record form with efficiently and traceability. The music therapy records of the case can be stored into the Hyperledger Fabric blockchain platform. System architecture diagram is shown in Fig. 2.

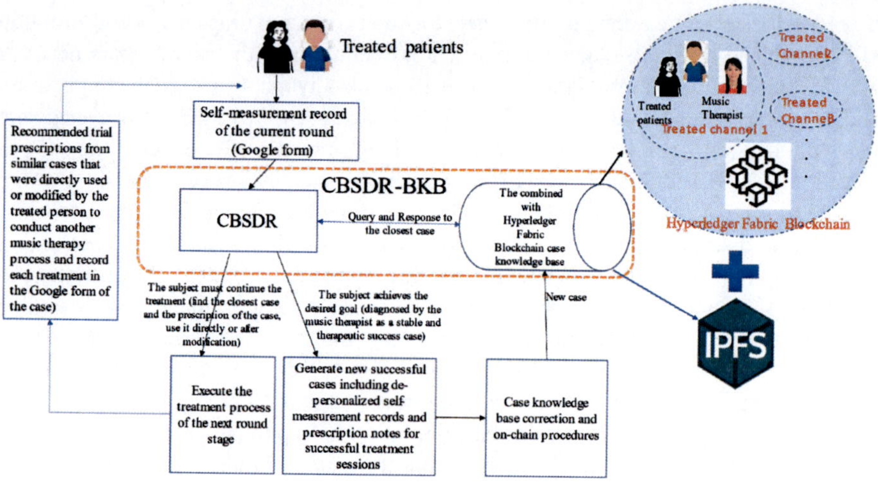

Fig. 2. The flow chart of CBSDR-CBKB model

According to A. Aamodt (1994) *et al.* [9], each case is stored in a network structure with categories, cases, and index pointers. Each case is associated with a category. Index pointers can point to categories or cases.

The design of this system is recommended for music therapists in music therapy clinics. To promote and apply the results of this research, the Hyperledger Fabric blockchain platform [17] of this system is deployed as a consortium chain platform with point economic operation. When a successful case is generated, the music therapist can remove the private or confidential information in the personal data of the treated person and store it in the alliance chain. It can form the knowledge base based on blockchain technology for this study. This system will design a DApp [17]. It can be used to connect cases where music therapists diagnose cases with the real blockchain world that provides payment mechanisms. In this system, subscribing customers can access successful cases by using smart contracts [17]. Importantly, the records of all successful cases will be stored in the InterPlanetary File System (IPFS) [17], and the content identifiers (CIDs) which are hash values will be obtained. These depersonalized treated records together with the corresponding CID values will be stored into the customized channel of Hyperledger Fabric blockchain platform, where the customized channel is organized by the authorized music therapist and treated patients. Each depersonalized treated record includes de-characterized data of successful cases that are basically related to hypertension symptoms, a set of corresponding or modified music therapy strategies and a corresponding set of treatment effectiveness, as well as a music therapy selection and listening sequence, number of treatment days per week and length of each treatment, *etc.*

Furthermore, when the patient fills out the Google form for self-measurement records at this phase, the music therapist could use the spiral diagnosis process of CBSDR-CBKB to find the prescription that is closest to this patient. The music therapist could directly use or modify the prescription and carry out the treatment processes of the next round. Finally, the treated person achieves the expected goal (a stable and successful case diagnosed by the music therapist). Successful new cases will be stored in the case knowledge base, IPFS as well as Hyperledger Fabric. As a choice of treatment mode that music therapists could refer to. The architecture diagram of this system is shown in Fig. 2.

4 Experimental Results and Discussion

This session mainly uses CBSDR-CBKB for case simulation and inference and analyzes and discusses the inferred results. In other words, this session uses the designed CBSDR-CBKB model to analyze and discuss a hypertension self-measurement record form, and validates this model based on the self-measurement records obtained by the treated patients filling out the Google form during treatment. The self-measurement record sheet could help the treated person check his or her own blood pressure and health status and could assist the music therapist as a reference for music prescriptions. Based on the results of the pre-test and post-test self-measurement record forms filled by the patients, the music therapist could understand the treatment records of the patients to evaluate the effectiveness of the case. The CBSDR-CBKB model is used for correction and adjustment, with the cycle on a prescription pad as the unit or called a treatment cycle. In this session, the system presents the results of the case study. The data in the self-measurement record sheet are counted and analyzed.

Case Study

Case Feature 1: the basic case information without personal characteristics. This study deals with the case of a 36-year-old woman receiving music therapy. Her occupation is a senior executive. She is a prehypertensive patient, who is receiving music therapy four months.
Case Feature 2: the attribute classification set is listed below, where all notations are described in Table 1. HP-2 (Prehypertension), HB-2 (Heartbeat rate: normal), SP-1 (Blood oxygen concentration: normal), TM-3 (Whether taking medication: no medication), BMI-2 (Body mass index: normal weight), LP-1 (Adjustment to life stress: sometimes (3 points), WP-2 (Adjustment to work pressure: sometimes (3 points), EH-1 (Eating habits: never (1 point), WLE-4 (Working and living environment: often comfortable (4 points)), LH-1 (Lifestyle Habits: Never (1 point), EXH-1 (Exercise Habits: Never (1 point).
Treatment Feature 1: the prescription notes are listed below, where all notations are described in Table 1: LM-4 (The type of music listening is nature), MT-5 (The blood pressure measured during music listening period is 6 a.m. to 9 a.m. in the morning),

Table 1. Notations

Symbol	Description
HP-*j*	Blood pressure standard score scale: $j = 1$ means normal blood pressure (systolic blood pressure < 120 (mmHg) and diastolic blood pressure < 80 (mmHg)); $j = 2$ means prehypertension (systolic blood pressure = 120–129 (mmHg) and diastolic blood pressure) < 80 (mmHg)); $j = 3$ indicates the first stage of hypertension (systolic blood pressure = 130 ~ 139 (mmHg) and diastolic blood pressure = 80 ~ 89 (mmHg)); $j = 4$ indicates the second stage of hypertension (systolic blood pressure > 140 (mmHg) and diastolic blood pressure > 90 (mmHg))
HB-*j*	Heart rate standard score scale: $j = 1$ means slow heartbeat (beats per minute < 60 beats); $j = 2$ means normal standard value (beats per minute = 60 ~ 100 beats); $j =$ 3 means the first period High blood pressure (beats above 100 beats per minute)
SP-*j*	Blood oxygen concentration standard score scale: $j = 1$ means the blood oxygen concentration is in the normal range (SpO2 = 95% ~ 100%); $j = 2$ means the blood oxygen concentration is in the abnormal range (SpO2 = 91% ~ 94%); $j = 3$ indicates that the blood oxygen concentration is in the severely hypoxic range (SpO2 < 90%)
BMI-j	Body mass index score scale: $j = 1$ means underweight (BMI < 18.5); $j = 2$ means normal range (18.5≦BMI < 24); $j = 3$ means overweight (first-grade obesity) (24≦BMI < 27); $j = 4$ indicates mild obesity (level 2 obesity) (27≦BMI < 30); $j =$ 5 indicates moderate/severe obesity (level 3 obesity) (BMI≧30)
LP-j	Adjustment score scale of life stress: $j = 1$ means never having the ability to self-adjust; $j = 2$ means occasionally have the ability to self-adjust; $j = 3$ means sometimes have the ability to self-adjust; $j = 4$ means often have the ability to self-adjust; $j = 5$ means always have self-adjustment ability
WP-j	Adjustment score scale of work stress: $j = 1$ means never have work pressure; $j = 2$ means occasionally work pressure; $j = 3$ means sometimes work pressure; $j = 4$ means often work pressure; $j = 5$ means always There is work pressure
EH-j	Dietary Habit Score Scale: $j = 1$ means never having a low-fat, low-sodium, and low-sugar diet, increasing vegetables and fruits, choosing whole grains, replacing red meat with white meat, reducing irritating foods, etc. [68]; $j = 2$ means occasionally do it; $j = 3$ means sometimes do it; $j = 4$ means do it often; $j = 5$ means always do it
WLE-j	Work and living environment comfort score scale: $j = 1$ means never comfortable; $j = 2$ means occasionally comfortable; $j = 3$ means sometimes comfortable; $j = 4$ means often comfortable; $j = 5$ means always comfortable
LH-*j*	Life Habit Score Scale: $j = 1$ ~ 5 means never do it, do it occasionally, do it sometimes, do it often, and always do it, respectively
EXH-*i*	Exercise habit score scale: $i = 1$ ~ 5 means never do it, occasionally do it (1–2 times a month), sometimes do it (2–3 times a month), often do it (at least 3 times a week)), and always do (almost every day), respectively

(*continued*)

Table 1. *(continued)*

Symbol	Description
LM*j*	Scale of music genres and repertoire scores for listening to music: $j = 1$ means classical music, *e.g.,* Mozart's "Sonata for Two Pianos in D Major, Op. 448"-three movements in total; $j = 2$, Taiwanese aboriginal music, *e.g.,* "Walk Along the River" album-10 tracks; $j = 3$, nature music, e.g., Wu Jintai's natural music collection-20 tracks; $j = 4$, light music, *e.g.*, Kobayashi Hiroyuki, Oya Kenharu (Famous Doctor's Autonomic Nervous Music Therapy) CD Book - 9 Tracks
MT*i*	The description of the time period for measuring blood pressure and listening to music[71] Score scale: $i = 1$ means morning (6 o'clock-9 o'clock); $i = 2$ means early noon (9 o'clock-12 o'clock); $i = 3$ means afternoon (12 o'clock-12 o'clock) 15 o'clock); $i = 4$ means evening (15 o'clock-18 o'clock); $i = 5$ means after dinner (18 o'clock-21 o'clock); and $i = 6$ means before going to bed (21 o'clock-24 o'clock)
LMF*i*	The emotional trait score scale of listening to music: $i = 1$ means the spirit is comforted; $i = 2$ means reducing stress and anxiety; $i = 3$ means calm and comfortable; $i = 4$ means feeling relaxed and happy; $i = 5$ means no special feeling and $i = 6$ means feeling down
TC*i*	Treatment cycle score scale: $i = 1$ means once a week; $i = 2$ means once every 2 weeks; $i = 3$ means once every 4 weeks; $i = 4$ means once every 4 to 12 weeks; $i = 5$ means > three months ≤ six months; $i = 6$ means more than six months
NT_i	Score scale of treatment times: $i = 1$ means once a week; $i = 2$ means 2 ~ 4 times a week; $i = 3$ means 5 ~ 7 times a week; $i = 4$ means 8 ~ 10 times a week; $i = 5$ means more than 10 times per week
TT_i	Scale of treatment time points: $i = 1$ means 10 min each time; $i = 2$ means 30 min each time; $i = 3$ means 60 min each time; $i = 4$ means 90 min each time; $i = 5$ means 120 min each time
R*i*	The treatment results after adjusting and revising the prescription: $i = 1$ means there is a significant effect and $i = 2$ means there is no obvious effect

LMF-1 (the emotional qualities after listening to music soothe the spirit), TC- 4 (the treatment cycle is large than three months and less than six months), NT-4 (the number of treatments is four to eight times per week), TT-4 (the treatment time is 60 min for each time).

Treatment Feature 2: Treatment effectiveness. The treatment effect was R-1 (Treatment results: significant effect shown in Table 1), where the treatment sessions for this case study are listed in Table 2.

Table 2. Treatment session for this case study

The next treatment time for the patient	The treatment cycle	The number of treatments	The treatment time	The evaluation of the treatment results
July 1, 2022	One week	Two to four times per week	10 min each time	Invalid
July 8, 2022	One week	Four to eight times per week	30 min each time	Invalid
July 15, 2022	Half a month	Four to eight times per week	60 min each time	Valid
July 29, 2022	One month	Four to eight times per week	60 min each time	Valid
September 2, 2022	Two months	Four to eight times per week	60 min each time	Valid
November 4, 2022	-	-	-	Store successful cases into Hyperledger Fabric blockchain platform

5 Conclusion and Future Work

This study proposes the CBSDR-CBKB model as a music selection and treatment plan recommendation system for music therapy. For music therapists, it can provide an efficient auxiliary system that can store treatment evidence and track results; for the treated, it can provide music selection and recommendation. By listening to different types of music and music therapy procedures, it allows the treated patients to stabilize their autonomic nerves to achieve the purpose of controlling blood pressure. The main contribution of this recommendation system will be to assist the music therapist in the choice of treatment model for the person being treated. It can also track the records of patients undergoing music therapy and evaluate the effectiveness of treatment.

Furthermore, the CBSDR-CBKB model proposed in this paper. This mode can provide music therapists with a music therapy auxiliary system mode, which can help music therapists quickly find similar cases and corresponding prescriptions as a timely diagnosis reference. It can improve the diagnostic efficiency of music therapists and enable non-repudiation of case documentation, tracking and analysis.

Finally, further research work in the future of this study will extend the contribution of this paper. This research will continue to realize an economic model of music therapy services based on point services. The economic model of the point service is to provide an incentive and crowdfunding mechanism. The cryptocurrency is called "Music Therapy Coin" or "Music Therapy Points". It is a mechanism for crowdfunding funds by selling service points. In order to collect feedback evaluations of the service, it also provides a points feedback mechanism.

Acknowledgment. This work was supported by the Chelpis Quantum Tech Co., Ltd., Taiwan, under the Grant number of Asia University: I112IB120. This work was supported by the National Science and Technology Council (NSTC), Taiwan, under NSTC Grant numbers: 111-2218-E-468-001-MBK, 110-2218-E-468-001-MBK, 110-2221-E-468-007, 111-2218-E-002-037 and 110-2218-E-002-044. This work was also supported in part by Asia University, Taiwan, and China Medical University Hospital, China Medical University, Taiwan, under grant numbers below. ASIA-111-CMUH-16, ASIA-110-CMUH-22, ASIA108-CMUH-05, ASIA-106-CMUH-04, and ASIA-105-CMUH-04.

References

1. Hsu, P.-Y.: The Study of Music Therapy Model Design Based on Blockchain Platform and Case-based Reasoning Mechanism. Master thesis, Department of Computer Science and Information Engineering, Asia University, Taiwan. National Digital Library Theses and Dissertations in Taiwan, National Central Library, R.O.C, (2023)
2. Xu, W.: Response of music therapy to body mechanism and automatic efficacy evaluation system based on artificial intelligence technology. In: 2021 2nd International Seminar on Artificial Intelligence, Networking and Information Technology (AINIT), INSPEC Accession Number: 21665597 (2021). https://doi.org/10.1109/AINIT54228.2021.00020
3. Zhao, N.: Intelligent system of somatosensory music therapy information feedback in deep learning environment. Complexity **2021**, 1–11 (2021). https://doi.org/10.1155/2021/9935504
4. Dutta, E., Bothra, A., Chaspari, T., Ioerger, T., Mortazavi, B.J.: Reinforcement Learning using EEG signals for Therapeutic Use of Music in Emotion Management. In: 2020 42nd Annual International Conference of the IEEE Engineering in Medicine & Biology Society (EMBC), INSPEC Accession Number: 19963922, pp 1–4 (2020)
5. Xu, C., He, Z., Shen, Z., Huang, F.: Potential benefits of music therapy on stroke rehabilitation. Oxidative Med. Cell. Longevity **2022**, 1–11 (2022). https://doi.org/10.1155/2022/9386095
6. Krout, R.E.: Music listening to facilitate relaxation and promote wellness: Integrated aspects of our neurophysiological responses to music. Arts Psychother. **34**(2), 134–141 (2007). https://doi.org/10.1016/j.aip.2006.11.001
7. Lim, M.S., et al.: Leisure sedentary time is differentially associated with hypertension, diabetes mellitus, and hyperlipidemia depending on occupation. BMC Public Health **17**, 278 (2017). https://doi.org/10.1186/s12889-017-4192-0
8. Gupta, N., Korshøj, M., Dumuid, D., Coenen, P., Allesøe, K., Holtermann, A.: Daily domain-specific time-use composition of physical behaviors and blood pressure. Int. J. Behav. Nutr. Phys. Act. **16**, 4 (2019). https://doi.org/10.1186/s12966-018-0766-1
9. Aamodt, A., Plaza, E.: Case-based reasoning: foundational issues, methodological variations, and system approaches. AI Commun. **7**(1), 39–59 (1994). https://doi.org/10.3233/AIC-1994-7104
10. Tseng, S.-S., Chen, H.-C., Hu, L.-L., Lin, Y.-T.: CBR-based negotiation RBAC model for enhancing ubiquitous resources management. Int. J. Inform. Manag. **37**(1), 1539–1550 (2017). https://doi.org/10.1016/j.ijinfomgt.2016.05.009
11. Chen, H.-C., Damarjati, C., Prasetyo, E., Arofiati, F., Sugiyo, D.: Blockchain technology benefit in tackling online shopping transaction revocation issue. In: Proceedings of the 2020 The 6th International Conference on Frontiers of Educational Technologies (ICFET 2020), pp. 191–195. Tokyo, Japan (2020). https://doi.org/10.1145/3404709.3404757
12. Chen, H.-C., et al.: An implementation of trust chain framework with hierarchical content identifier mechanism by using Blockchain technology. Sensors **22**(13), 4831 (2022). https://doi.org/10.3390/s22134831

13. Chen, H.-C., Liang, Y.-H., Hsu, P.-Y.: Reconfigurable PM2.5 sensor green deployment mechanism based on Blockchain technology. In: Tsihrintzis, G.A., Wang, S.-J., Lin, I.-C. (eds.) 2021 International Conference on Security and Information Technologies with AI, Internet Computing and Big-data Applications, pp. 368–378. Springer International Publishing, Cham (2023). https://doi.org/10.1007/978-3-031-05491-4_37
14. Chen, H.-C.: A trust evaluation gateway for distributed Blockchain IoT network. In: Chen, J.-L., Pang, A.-C., Deng, D.-J., Lin, C.-C. (eds.) Wireless Internet: 11th EAI International Conference, WiCON 2018, Taipei, Taiwan, 15–16 Oct 2018, Proceedings, pp. 156–162. Springer International Publishing, Cham (2019). https://doi.org/10.1007/978-3-030-06158-6_16
15. Chen, H.-C., Irawan, B., Shae, Z.-Y.: A cooperative evaluation approach based on Blockchain technology for IoT application. In: Barolli, L., Xhafa, F., Javaid, N., Enokido, T. (eds.) IMIS 2018. AISC, vol. 773, pp. 913–921. Springer, Cham (2019). https://doi.org/10.1007/978-3-319-93554-6_92
16. Chen, H.-C., Kuo, S.-S., Chen, H.-M.: Secure OTT service scheme based on Blockchain technology. In: The 32nd IEEE International Conference on. Advanced Information Networking and Applications (IEEE AINA-2018), pp. 645–650. Pedagogical University of Cracow, Poland (2018). https://doi.org/10.1109/WAINA.2018.00161
17. Chen, H.-C., Liang, Y.-H., Su, J.-S., Tsai, K.-Y., Song, Y.-L., Hsu, P.-Y., Cai, J.-S.: The implement of a reconfigurable intelligence trust chain platform with anti-counterfeit traceable version function for the customized system-module-IC. In: Barolli, L. (ed.) Innovative Mobile and Internet Services in Ubiquitous Computing: Proceedings of the 17th International Conference on Innovative Mobile and Internet Services in Ubiquitous Computing (IMIS-2023), pp. 261–272. Springer Nature Switzerland, Cham (2023). https://doi.org/10.1007/978-3-031-35836-4_28

Effect of Multiple Unmanned Aerial Vehicles on Data Transmission Considering DTN-Based V2V Communication in Urban Area

Shura Tachibana[1], Ryuki Shiromoto[2], Makoto Ikeda[2](✉), and Leonard Barolli[2]

[1] Graduate School of Engineering, Fukuoka Institute of Technology, 3-30-1 Wajiro-Higashi, Higashi-Ku, Fukuoka 811-0295, Japan
mgm23106@bene.fit.ac.jp

[2] Department of Information and Communication Engineering, Fukuoka Institute of Technology, 3-30-1 Wajiro-Higashi, Higashi-Ku, Fukuoka 811-0295, Japan
makoto.ikd@acm.org, barolli@fit.ac.jp

Abstract. In this paper, we present the performance of multiple Unmanned Aerial Vehicles (UAVs) and Vehicle-to-Vehicle (V2V) communication in an urban environment. We employ the Epidemic protocol as the communication method to evaluate the dissemination of bundle messages by multiple UAVs and regular vehicles. We consider the urban grid and Tenjin area in Fukuoka City, where UAVs are assumed to move in a vertical direction randomly. From the simulation results, we found that the presence or absence of a UAV had a significant impact on the message delivery probability, especially when the number of vehicles was less than 90. Also, the use of UAVs improves message reachability and storage consumption on complex roads.

Keywords: Epidemic Protocol · UAV · V2V · DTN

1 Introduction

In many crises and public gatherings, the utilization of aerial drones and conventional vehicles for the Internet and applied services has proven to be highly beneficial [5]. Nevertheless, the utilization of ad-hoc networks might present certain challenges, particularly in situations where the network becomes overwhelmed due to factors such as natural disasters or social gatherings. Hence, Delay/Disruption/Disconnection-Tolerant Networking (DTN) emerges as a viable option due to its ability to facilitate message transmission in challenging circumstances [2,12,19].

In [24,25], Zhao, *et al.* introduced the concept of message ferry system, which ascertain the needed quantity of ferries and the performance is improved. However, the authors did not consider the impact of mobile networks or the variability in traffic patterns.

L. Barolli (Ed.): BWCCA 2023, LNDECT 186, pp. 335–344, 2024.
https://doi.org/10.1007/978-3-031-46784-4_32

In [15], the authors evaluated the encounter rate between two groups of agents in a mutual search by Lévy walk on a unit-disk graph. They focused on the inter-group encounters among agents within a unit-disk graph, where each mobile agent operates under action constraints that limit its movement to the graph's links. In our previous work, we investigated the utilization of UAVs and vehicles in grid-based scenarios during crisis scenarios [3,4]. However, there are not specific models to incorporate the real-map scenario and assess the effects of changing UAV altitudes.

In this paper, we present the effect of multiple UAVs in data trasmission taking into account fluctuations in UAV altitude and V2V communication. We use the Epidemic Protocol for transmission of bundle messages. From the simulation results, we confirmed that the presence or absence of a UAV had a significant impact on the message delivery probability, especially when the number of vehicles was less than 90. Also, the use of UAVs improves message reachability and storage consumption on complex roads.

The rest of the paper is organized as follows. In Sect. 2, we provide the overview of DTN scheme and Epidemic protocol. In Sect. 3, we describe DTN-based message delivery platform and considered scenarios. In Sect. 4, we present the evaluation results. Finally, we conclude the paper in Sect. 5.

2 Background of Delay-Tolerant Networking

In this section, we present the overview of DTN and Epidemic protocol.

2.1 DTN Overview

DTN was initially proposed as a primary communication mechanism for inter-satellite networks and various delivery mechanisms were suggested and proposed in literature [9–11,13,16,20–22]. Some research work discuss the benefits of employing longer latency and recovery strategies [7,14,17]. Also, the message replication system has been identified as a reliable and flexible method for creating internet connectivity across various operations [23].

During crisis situations, delays and frequent disruptions may be extended. For next generation wireless networks, the velocity of vehicles and UAVs is improved, so making traditional end-to-end routing strategies inappropriate. As a result, the DTN technique, which was initially developed for space missions, has been examined for its potential suitability in terrestrial, aerial, and underwater settings [8]. In DTN, the message replication strategy utilizes bundle messages to facilitate the transmission of a message to an unidentified destination [6].

2.2 Epidemic Protocol

The Epidemic protocol [16,22] has potential mechanisms for message replication in disaster situations, but there are discussions and suggestions regarding the integration of recovery functionalities. The algorithm presented in Algorithm 1 demonstrates the processing flow for each incoming message that is handled by the Epidemic protocol.

Algorithm 1. Epidemic protocol.

```
Start
Listen {Depending on the type of message, the following processes are performed}

while True do
  if receive SV then
    if find unknown contents then
      Send REQUEST message
    end if
  else if receive REQUEST then
    Send bundle message
  else if receive bundle message then
    Store the bundle message in the storage
  else if receives anti-packet then
    if end-point then
      end-point generates anti-packet to the network
    else
      Intermediate node deletes the corresponding bundle message
      Send anti-packet to other nodes
    end if
  end if
end while
Stop
```

The Epidemic protocol manages the reproduction of a bundle message by employing two control messages. During epidemic phase, each node is responsible for transmitting a Summary Vector (SV) and maintain a record of bundles stored in each node. After the nodes receive SVs, they initiate a process of comparing them with their own SV. When the received SV includes a bundle with unknown contents, the node transmits a REQUEST message. However, the duplication of bundles to neighboring nodes decreases the storage space. The received bundles are subsequently stored in the storage and undergo ongoing duplication, regardless of whether the end-point receives the bundles. In the context of an anti-packet, the end-point is responsible for sending an anti-packet that contains a list of bundles that have been received at the end-point.

3 DTN-Based Delivery Platform Using UAV and V2V

In this section, we introduce the DTN-based delivery platform using UAVs and V2V communication. We implemented two scenarios considering urban grid and real-map scenarios on the Scenargie simulator [18]. In Fig. 1(a) is shown the urban grid model. While, the model of Tenjin area in Fukuoka City is displayed in Fig. 1(b).

The altitudes of the UAVs are varying between 20 m (referred to as Z_{min}) and 40 m (referred to as Z_{max}). The Algorithm 2 is used to produce a maximum of nine random trip paths.

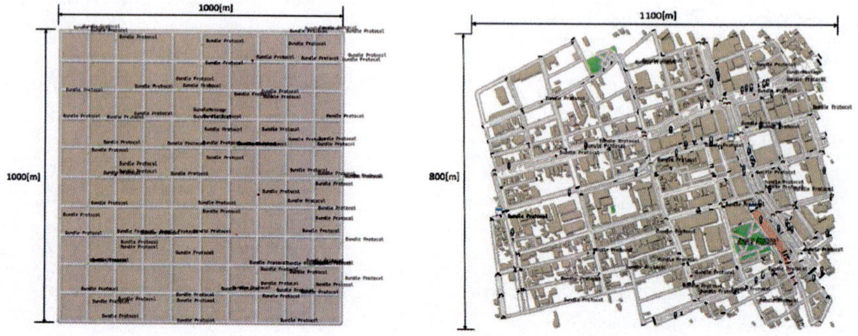

(a) Urban grid scenario

(b) Tenjin scenario

Fig. 1. Evaluation scenarios considering urban grid (a) and real map (b) situations.

Algorithm 2. Mobility pattern generator of UAV.

```
Start
srand((unsigned int)time(NULL)) {Initialize}
for int t = START_TIME; t <= END_TIME; t += INTERVAL do
    print ("UAV_ID and t");
    print ("%d", -X_r/2+ rand() %X_r + 1); {X_r: range of X-axis}
    print ("%d", -Y_r/2+ rand() %Y_r + 1); {Y_r: range of Y-axis}
    print ("%d", Z_min+ rand() %Z_max − Z_min + 1); {Z-axis}
    print ("additional information and EOL");
end for
Stop
```

The simulation conditions from physical layer to application layer are presented in Table 1. Both UAVs and general vehicles utilize the Epidemic protocol. The start-point and end-point are randomly moved based on a GIS-based random way-point mobility model. UAVs periodically move randomly and mobility generator is displayed in Algorithm 2. Then, the UAVs proceed to the next point in accordance with a predetermined schedule based on trace files.

The start-point sends a set of bundle messages for intermediate vehicles or UAVs, with a time interval of 0 to 400 seconds. The radio frequency utilized in this study is 5.9 GHz, as specified by the ITU-R P.1411 [1], which serves as the propagation model. We evaluate the impact of various vehicles and UAVs on delivery ratio, delay, overhead, and storage usage.

- The delivery ratio is an indicator to assess the percentage of bundle messages that are successfully delivered to the selected end-point, relative to the total number of bundle messages generated.
- The delay denotes the amount of time needed for the bundle to reach its final destination (end-point).
- The overhead denotes the quantity of duplicated bundles in the network.
- The storage usage refers to the aggregate amount of storage consumption, measured in bytes. The assessment does not consider recovery schemes.

Table 1. Simulation parameters.

Items	Setup Values
Simulation Duration	500 [s]
Area Size: Urban Grid	$1,000 \times 1,000$ [m^2] (10×10 grids)
Area Size: Tenjin in Fukuoka City	Selected area: $1,100 \times 800$ [m^2]
Scenario: Building Height	15 [m]
DTN: Protocol	Epidemic
DTN: Recovery Function	Disabled
DTN: Message Type	Bundle
DTN: Bundle Size	500 [bytes]
DTN: Bundle Generation	0 to 400 [s]
Number of UAVs	0, 3, 6 and 9 [nodes]
UAV: Mobility Pattern	Random
UAV: Altitude Pattern	Randomly move from 20 to 40 [m]
Number of General Vehicles	30, 60, 90, 120 and 200 [vehicles]
General Vehicle: Mobility Model	GIS-based Random Way-Point
General Vehicle: Minimum Speed	8.333 [m/s]
General Vehicle: Maximum Speed	16.666 [m/s]
Antenna Model	Omni-directional
Physical Model	IEEE 802.11p

4 Evaluation Results

The delivery ratio results for different cases are shown in Fig. 2. In Fig. 2(a), the delivery ratio reaches 100% in all cases, and all bundle messages reach their end-point in time. In the Tenjin area (see Fig. 2(b)), we observed that the message delivery ratio reaches a maximum value of 100% when the number of deployed UAVs is equal or higher than six. On the other hand, in scenarios where UAVs are not used or are limited to three, a decrease in delivery performance is observed. In the case when UAVs are not used, the delivery ratio increases with increasing vehicle densities. After 150 vehicles, all bundle messages reach the end-point.

The delay results for different cases are shown in Fig. 3. For both scenarios, the long delay is for 30 vehicles. Then it decreases with increasing the number of UAVs. However, there are some cases where the delay time does not improve even when the number of UAVs is increased. We think that this is related to the size of the simulation environment and the irregular movements of the vehicles and UAVs. In addition, the difference in delay time between the urban grid and the Tenjin area is big because of the complex roads in this area.

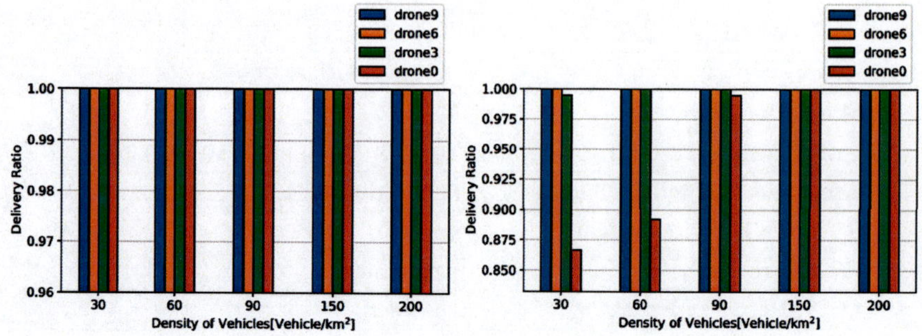

(a) Urban grid

(b) Tenjin

Fig. 2. Delivery ratio results for different UAVs and vehicles.

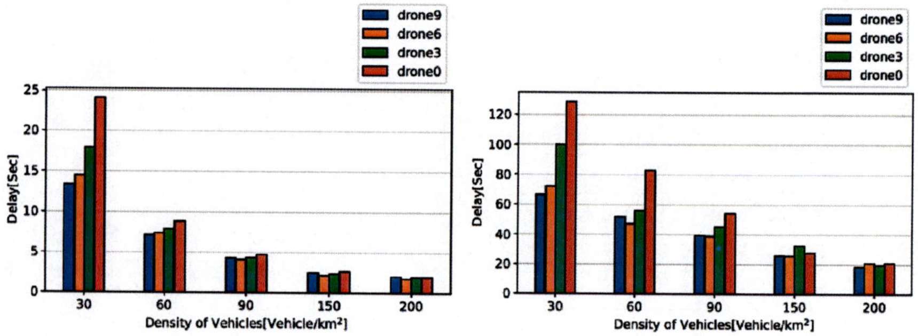

(a) Urban grid

(b) Tenjin

Fig. 3. Delay results for different UAVs and vehicles.

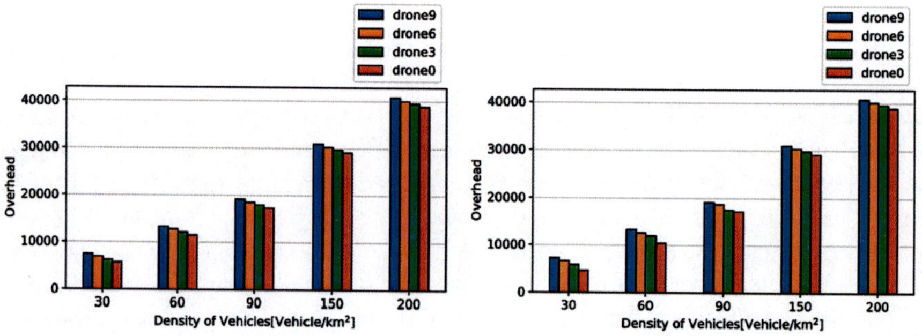

(a) Urban grid

(b) Tenjin

Fig. 4. Overhead results for different UAVs and vehicles.

The overhead results for different cases are shown in Fig. 4. We observe that the overhead increases with increasing the number of vehicles. In both scenarios, the overhead does not change significantly, and we confirmed that the overhead also increases with increasing the number of UAVs.

The storage usage results for different vehicles are shown in Fig. 5 and Fig. 6. In the case of urban grid scenario, when the number of vehicles is 30, the amount of storage usage increases slightly with the increasing number of UAVs. Furthermore, in the Tenjin area case, a higher quantity of UAVs leads to a decrease in storage usage. As shown in Fig. 6, when the number of vehicles increases, there is a small difference observed.

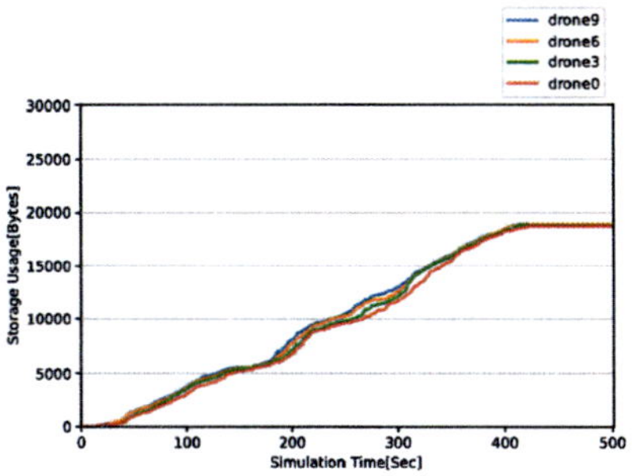

(a) Urban grid: 30 nodes

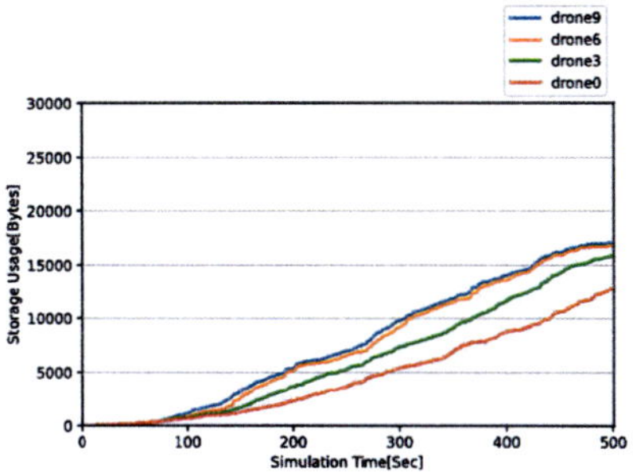

(b) Tenjin: 30 nodes

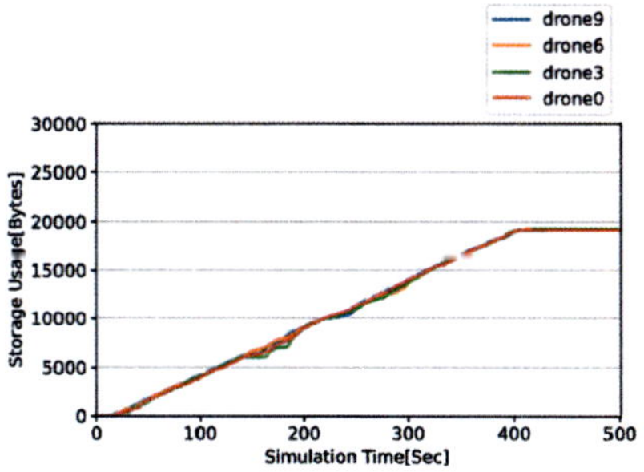

(c) Urban grid: 60 nodes

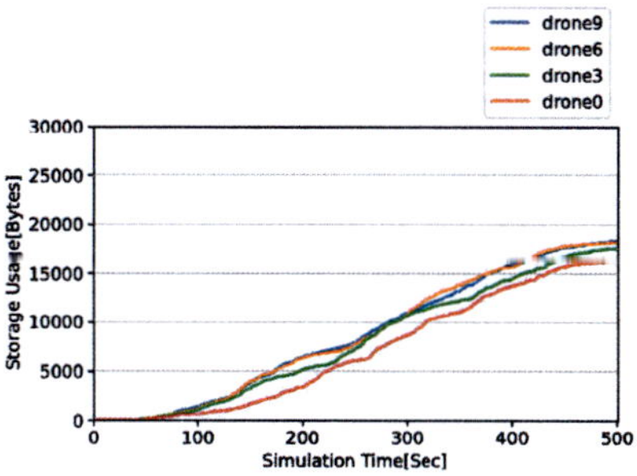

(d) Tenjin: 60 nodes

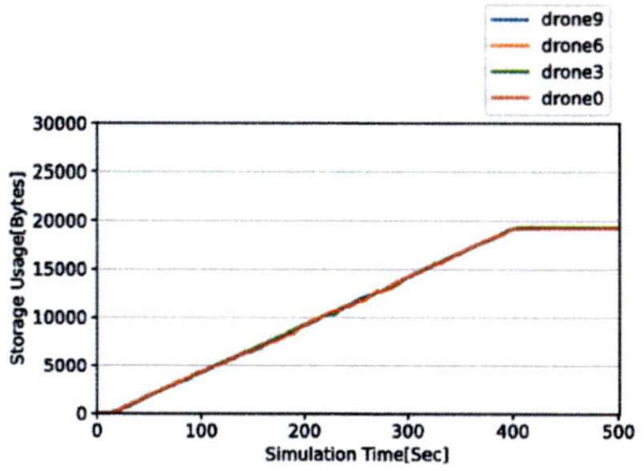

(e) Urban grid: 90 nodes

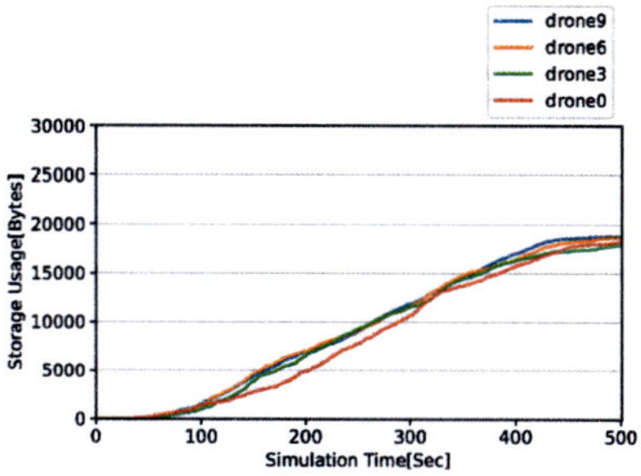

(f) Tenjin: 90 nodes

Fig. 5. Storage usage results from 30 to 90 vehicles.

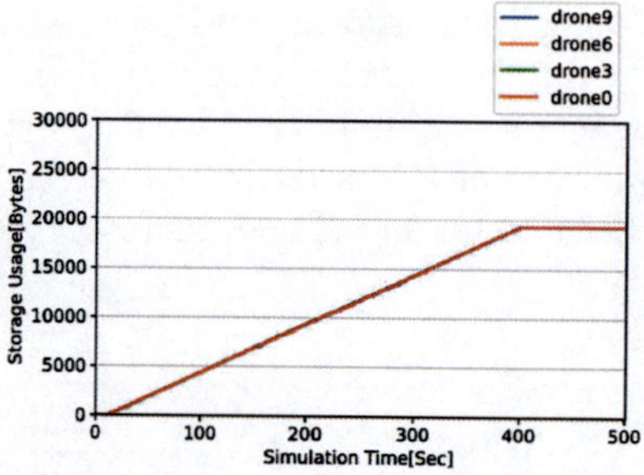

(a) Urban grid: 150 nodes

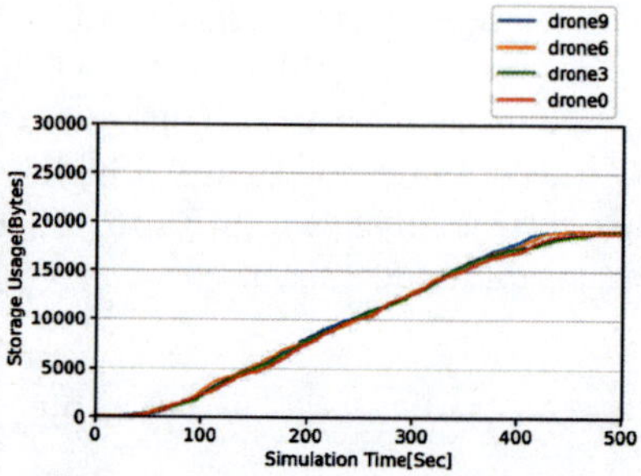

(b) Tenjin: 150 nodes

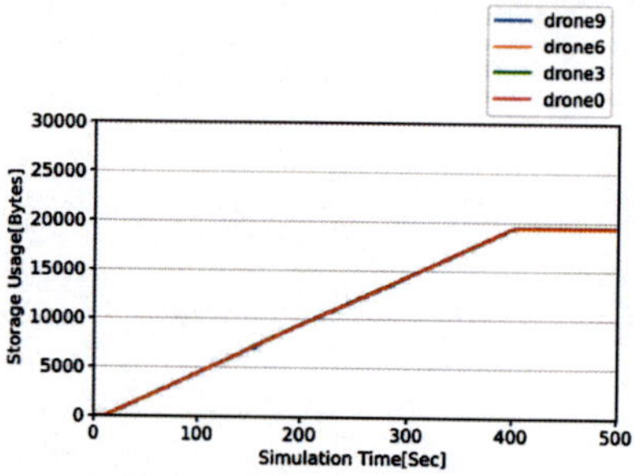

(c) Urban grid: 200 nodes

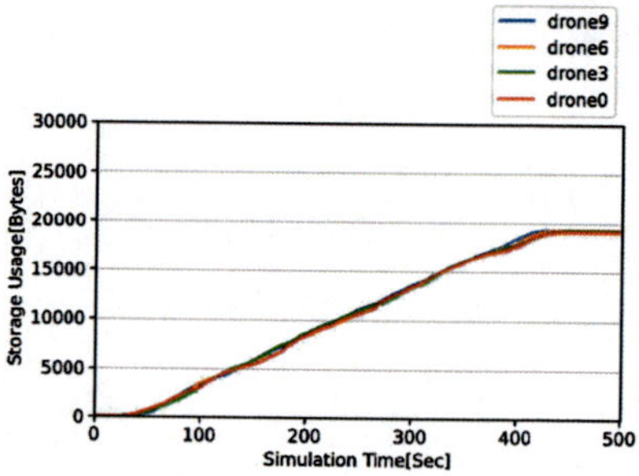

(d) Tenjin: 200 nodes

Fig. 6. Storage usage results from 150 to 200 vehicles.

5 Conclusions

In this paper, we investigated the impact of varying the number of UAVs on data transmission efficiency under DTN-based V2V communication. Four patterns of UAV quantities (0, 3, 6, and 9 units) were randomly deployed at altitudes ranging from 20 to 40 m. The objective was to explore the relationship between the number of UAVs and their effect on communication performance.

From the simulation results, we confirmed that the presence or absence of a UAV had a significant impact on the message delivery probability, especially when the number of vehicles was less than 90. This confirms that the use of UAVs improves message reachability and storage consumption on complex roads.

In the future work, we would like to consider recovery methods and other parameters.

References

1. Recommendation ITU-R P.1411-11: Propagation data and prediction methods for the planning of short-range outdoor radiocommunication systems and radio local area networks in the frequency range 300 MHz to 100 GHz. ITU (2019). https://www.itu.int/rec/R-REC-P.1411-11-202109-I/en

2. Arafat, M.Y., Moh, S.: Location-aided delay tolerant routing protocol in UAV networks for post-disaster operation. IEEE Access **6**, 59891–59906 (2018). https://doi.org/10.1109/ACCESS.2018.2875739
3. Azuma, M., Uchimura, S., Ikeda, M., Barolli, L.: AAR: an adaptive anti-packet recovery approach for improving delay tolerant networking data transfer using UAVs and vehicles. In: Barolli, L. (ed.) 3PGCIC 2022. LNNS, vol. 571, pp. 181–191. Springer, Cham (2022). https://doi.org/10.1007/978-3-031-19945-5_18
4. Azuma, M., Uchimura, S., Sako, S., Ikeda, M., Barolli, L.: Performance evaluation of an adaptive anti-packet recovery method considering UAVs and vehicles in an urban scenario. In: Barolli, L. (ed.) IMIS 2022. LNNS, vol. 496, pp. 230–237. Springer, Cham (2022). https://doi.org/10.1007/978-3-031-08819-3_23
5. Cao, Y., Jiang, T., Kaiwartya, O., Sun, H., Zhou, H., Wang, R.: Toward preempted EV charging recommendation through V2V-based reservation system. IEEE Trans. Syst. Man Cybern.: Syst. **51**(5), 3026–3039 (2021). https://doi.org/10.1109/TSMC.2019.2917149
6. Cerf, V., et al.: Delay-tolerant networking architecture. IETF RFC 4838 (Informational) (2007)
7. Cui, J., Cao, S., Chang, Y., Wu, L., Liu, D., Yang, Y.: An adaptive spray and wait routing algorithm based on quality of node in delay tolerant network. IEEE Access **7**, 35274–35286 (2019). https://doi.org/10.1109/ACCESS.2019.2904750
8. Dendy, R., Mortensen, D., Zeleznikar, D., Booth, S.: Flexible user radio for lunar missions. In: Proceedings of the IEEE Aerospace Conference 2023, pp. 1–11 (2023). https://doi.org/10.1109/AERO55745.2023.10115724
9. Henmi, K., Koyama, A.: Hybrid type DTN routing protocol considering storage capacity. In: Barolli, L., Okada, Y., Amato, F. (eds.) EIDWT 2020. LNDECT, vol. 47, pp. 491–502. Springer, Cham (2020). https://doi.org/10.1007/978-3-030-39746-3_50
10. Iranmanesh, S., Raad, R., Raheel, M.S., Tubbal, F., Jan, T.: Novel DTN mobility-driven routing in autonomous drone logistics networks. IEEE Access **8**, 13661–13673 (2020). https://doi.org/10.1109/ACCESS.2019.2959275
11. Ito, M., Nishiyama, H., Kato, N.: A novel routing method for improving message delivery delay in hybrid DTN-MANET networks. In: Proceedings of the IEEE Global Communications Conference (GLOBECOM-2013), pp. 72–77 (2013). https://doi.org/10.1109/GLOCOM.2013.6831050
12. Kawabata, N., Yamasaki, Y., Ohsaki, H.: Hybrid cellular-DTN for vehicle volume data collection in rural areas. In: Proceedings of the IEEE 43rd Annual Computer Software and Applications Conference (COMPSAC-2019), vol. 2, pp. 276–284 (2019). https://doi.org/10.1109/COMPSAC.2019.00048
13. Lent, R.: Enabling cognitive bundle routing in NASA's high rate DTN. In: 2022 International Wireless Communications and Mobile Computing (IWCMC), pp. 1323–1328 (2022). https://doi.org/10.1109/IWCMC55113.2022.9824158
14. Marchese, M., Patrone, F., Cello, M.: DTN-based nanosatellite architecture and hot spot selection algorithm for remote areas connection. IEEE Trans. Veh. Technol. **67**(1), 689–702 (2018). https://doi.org/10.1109/TVT.2017.2739298
15. Nishigami, C., Hayashibara, N.: The probability of encounters of mutual search using lévy walk on unit disk graphs. In: Barolli, L., Miwa, H., Enokido, T. (eds.) NBiS 2022. LNNS, vol. 526, pp. 90–101. Springer, Cham (2022). https://doi.org/10.1007/978-3-031-14314-4_9
16. Ramanathan, R., Hansen, R., Basu, P., Hain, R.R., Krishnan, R.: Prioritized epidemic routing for opportunistic networks. In: Proceedings of the 1st International

MobiSys Workshop on Mobile Opportunistic Networking (MobiOpp 2007), pp. 62–66 (2007). https://doi.org/10.1145/1247694.1247707
17. Sato, F., Kikuchi, R.: Hybrid routing scheme combining with geo-routing and DTN in VANET. In: Proceedings of the 10th International Conference on Innovative Mobile and Internet Services in Ubiquitous Computing (IMIS-2016), pp. 250–255 (2016). https://doi.org/10.1109/IMIS.2016.107
18. Scenargie: Space-time engineering, LLC. http://www.spacetime-eng.com/
19. Solpico, D., et al.: Application of the V-HUB standard using LoRa beacons, mobile cloud, UAVs, and DTN for disaster-resilient communications. In: Proceedings of the IEEE Global Humanitarian Technology Conference (GHTC-2019), pp. 1–8 (2019). https://doi.org/10.1109/GHTC46095.2019.9033139
20. Spyropoulos, T., Psounis, K., Raghavendra, C.S.: Spray and wait: an efficient routing scheme for intermittently connected mobile networks. In: Proceedings of the ACM SIGCOMM workshop on Delay-tolerant networking 2005 (WDTN 2005), pp. 252–259 (2005). https://doi.org/10.1145/1080139.1080143
21. Tornell, S.M., Calafate, C.T., Cano, J.C., Manzoni, P.: DTN protocols for vehicular networks: an application oriented overview. IEEE Commun. Surv. Tutor. **17**(2), 868–887 (2015). https://doi.org/10.1109/COMST.2014.2375340
22. Vahdat, A., Becker, D.: Epidemic routing for partially-connected ad hoc networks. Technical report, Duke University (2000)
23. Xu, Y., Zhang, H., Pang, Z., Kang, Y., Su, Y.: Design and implementation of a transport protocol with network coding for delay tolerant underwater acoustic sensor networks. In: OCEANS 2022 - Chennai, pp. 1–4 (2022). https://doi.org/10.1109/OCEANSChennai45887.2022.9775134
24. Zhao, W., Ammar, M., Zegura, E.: Controlling the mobility of multiple data transport ferries in a delay-tolerant network. In: Proceedings IEEE 24th Annual Joint Conference of the IEEE Computer and Communications Societies, vol. 2, pp. 1407–1418 (2005). https://doi.org/10.1109/INFCOM.2005.1498365
25. Zhao, W., Ammar, M.: Message ferrying: proactive routing in highly-partitioned wireless ad hoc networks. In: The Ninth IEEE Workshop on Future Trends of Distributed Computing Systems 2003, FTDCS 2003. Proceedings, pp. 308–314 (2003). https://doi.org/10.1109/FTDCS.2003.1204352

Moving Accuracy Measurement of Omnidirectional Robot for MOAP and Wheelchair Tennis

Keita Matsuo[1(✉)], Elis Kulla[2], and Leonard Barolli[1]

[1] Department of Information and Communication Engineering, Fukuoka Institute of Technology (FIT), 3-30-1 Wajiro-Higashi, Higashi-Ku, Fukuoka 811-0295, Japan
{kt-matsuo,barolli}@fit.ac.jp

[2] Department of System Management, Fukuoka Institute of Technology (FIT), 3-30-1 Wajiro-Higashi, Higashi-Ku, Fukuoka 811-0295, Japan
kulla@fit.ac.jp

Abstract. Recently, many kinds of robots are developed and they collaborate with various things to help humans. Over the past few decades, industrial robots have been used in factory operations. Also, these robots such as vacuum cleaner robot, security robot, therapy robot and wheelchair robot can support humans for many operations. Thus, using robots and related technologies are very meaningful to enrich our lives. In the context of welfare, in many countries in the world there are millions of people with disabilities who rely on wheelchairs. In this paper, we present an omnidirectional robot and introduce two applications: Moving Omnidirectional Access Point Robot (MOAP) and Wheelchair for tennis. These applications required a precise control of omnidirectional robot. Therefore, we measured the accuracy of moving omnidirectional robot using supersonic signals. The results shows that the omnidirectional robot was controlled correctly, because the error rate in x direction was within 1.00 [%]. However, when the robot moved in y direction, the error rate was 2.9 [%].

1 Introduction

Recently, there are different kinds of robots, which can support humans during their daily life. Over the past few decades, industrial robots have been used in factory operations. One of the industrial robot field is Mechatronics, used by Yaskawa Electric Company, Japan [13]. Also, microrobots are used in medical industry for non-invasive medical diagnostics and therapies within the human body, utilizing miniaturized sensors and actuators [10].

Connected cars can be considered as mobile robots, which has been used for collecting traffic data in a cloud system in order to enable life-critical use cases such as safety of driving [6]. Also, by the advancement of automated vehicle

L. Barolli (Ed.): BWCCA 2023, LNDECT 186, pp. 345–353, 2024.
https://doi.org/10.1007/978-3-031-46784-4_33

technologies many leading countries are carrying out trials of automated shuttle buses on public roads [5].

In our research work, we are exploring the possibilities of utilizing omnidirectional moving robots. So far, we have been focused on applying the omniderectional moving robot as a mobile router for network support [14–16], and for helping humans by using omnidirectional wheelchair [7–9].

Usage of robots and related technologies can enrich our lives. In the context of welfare, in 34 developed and 156 developing countries, approximately 132 million people with disabilities rely on wheelchairs, accounting for roughly 1.86% of the global population. Additionally, numerous individuals experience health conditions linked to motor disabilities, which hinder their capacity to control their limbs or even their head [3].

Therefore, maintaining a healthy and active lifestyle is very important for everyone. Sports can play a crucial role in improving the quality of life, and there are specific activities accessible to individuals who utilize wheelchairs.

In this paper, we present the application of Omnidirectional Wheelchair Robot for two scenarios bu using it as a network router and wheelchair. For these applications, the precise control of omnidirectional robot is very important issue.

The rest of this paper is structured as follows. In Sect. 2, we introduce the implementation and application of omnidirectional robot. In Sect. 3, we describe the proposed system for control of omnidirectional robot. In Sect. 4, we explain experimental results. Finally, conclusions and future work are given in Sect. 5.

2 Implementation and Application of Omnidirectional Robot

In order to control the omnidirectional robot, we designed the electric circuit for controlling three motors. In Fig. 1 is shown the image of control system for one brushless motor. Figure 2 shows interface circuit between motor driver and controller for driving simultaneously 3 motors. The wire image and 3D image of the interface is shown in Fig. 3

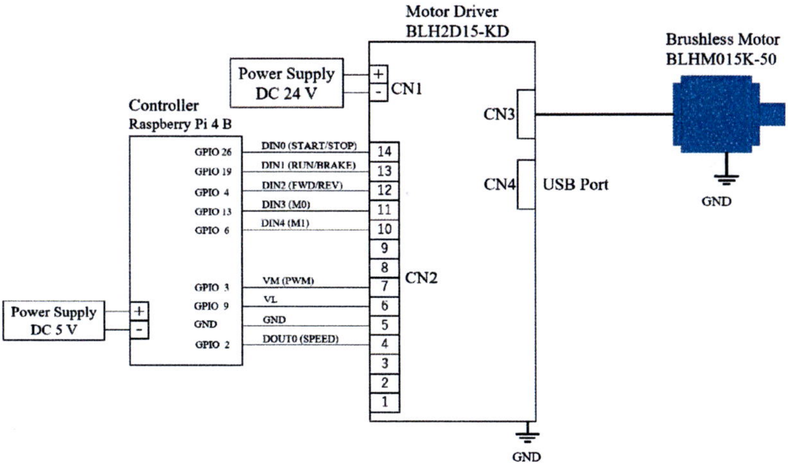

Fig. 1. Control system of brushless motor.

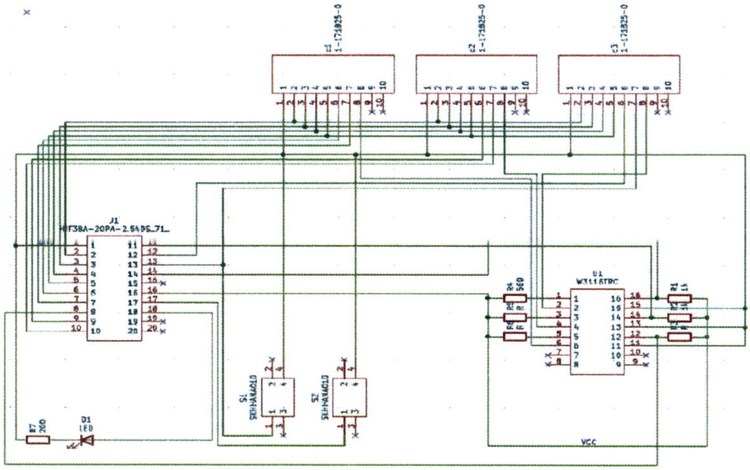

Fig. 2. Interface circuit between motor driver and controller for driving 3 motors.

2.1 Application for Moving Omnidirectional Access Point Robot

Here, we provide a description of the implemented Moving Omnidirectional Access Point (MOAP) robot shown in Fig. 4. The MOAP robot can move in all directions while maintaining a consistent orientation, allowing it to establish access points for network devices. For MOAP robot, we used omniwheels that enable omnidirectional moving in the front, back, left and right directions. Our objective is to ensure precise control for MOAP robot movement in order to create an optimal communication environment.

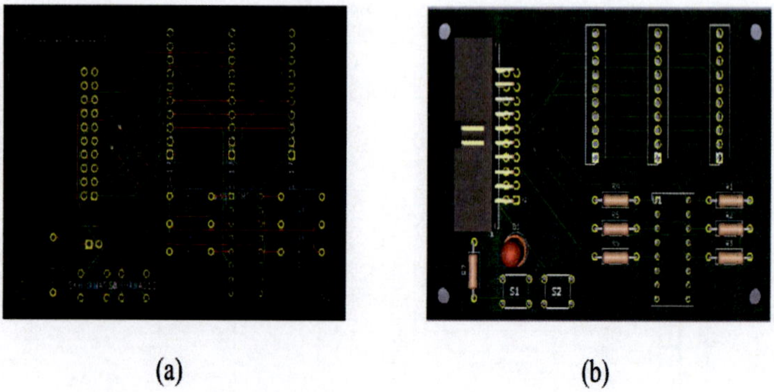

Fig. 3. Wire image and 3D image of interface circuit.

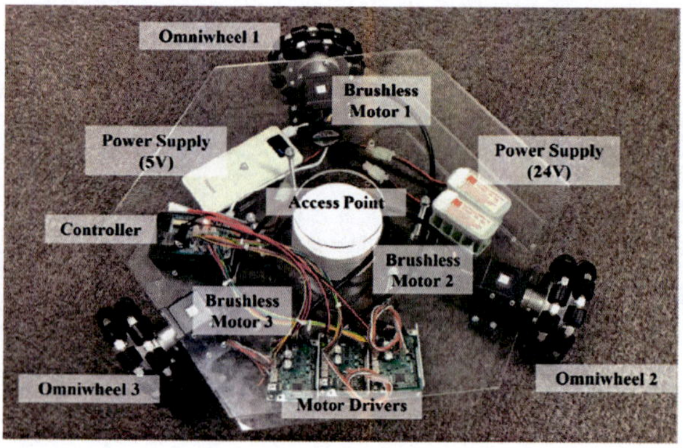

Fig. 4. MOAP robot.

The MOAP robots can cover better the clients by chaning their positions. In Fig. 5(a), the dots show clients. We deployed 200 clients on 2D space (120m $\times$ 120m). To make better communication environment for this situation, we decided optimal MOAP robot position by using K-means clustering as shown in Fig. 5(b). Then, MOAP robots will move to the optimal positions, which correspond to the centroids.

2.2 Apprication for Wheelchair Tennis

The wheelchair tennis has been utilized as a sport for Paralympics, attracting numerous athletes from various countries [12]. The popularity of wheelchair sports has significantly grown, leading to the establishment of organized events for both recreational purposes and competitive endeavors. Also, it has gained popularity for wheelchair rugby [11], basketball [2], football [1], badminton [4] and so on.

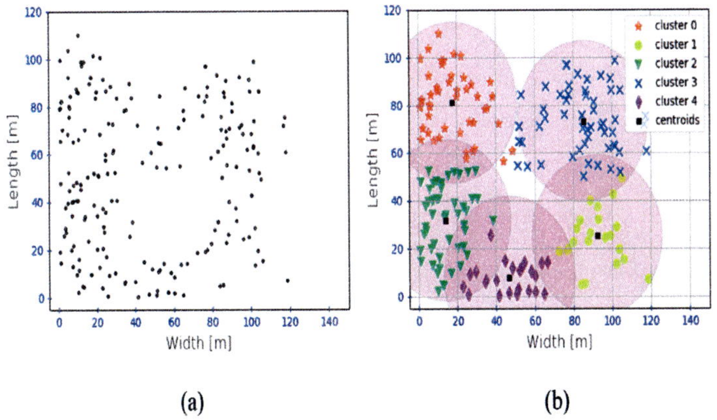

Fig. 5. Distributed MOAP robot position by using K-means clustering.

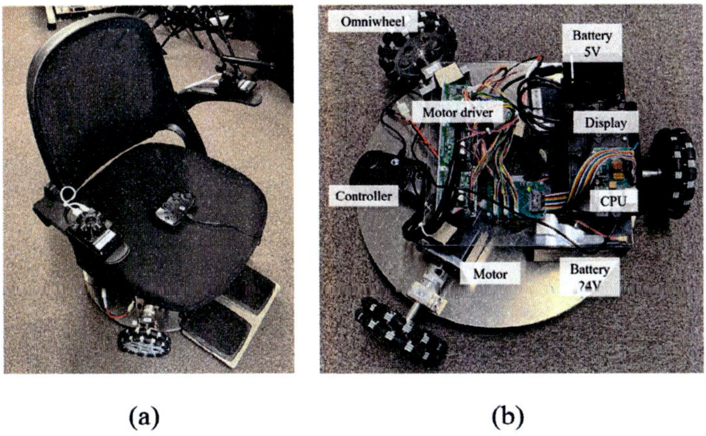

Fig. 6. Omnidirectional Wheelchair for Machine Tennis and Control Circuit.

For the wheelchair tennis is used an Omnidirectional Wheelchair Robot equipped with electric motors, enabling it to move in any direction and maintain a consistent directions. In Fig. 6(a) is shown the implemented Omnidirectional Wheelchair Robot. In Fig. 6(b) is shown the real control circuits and system.

The implemented omnidirectional wheelchair robot is designed specifically for playing tennis, offering easy usability for beginners. Nonetheless, Machine Tennis encounters certain challenges such as preventing collisions between the omnidirectional wheelchair robot and various obstacles such as walls, nets, props, or other wheelchairs during doubles matches. Therefore, it is required a precise control of MOAP robots.

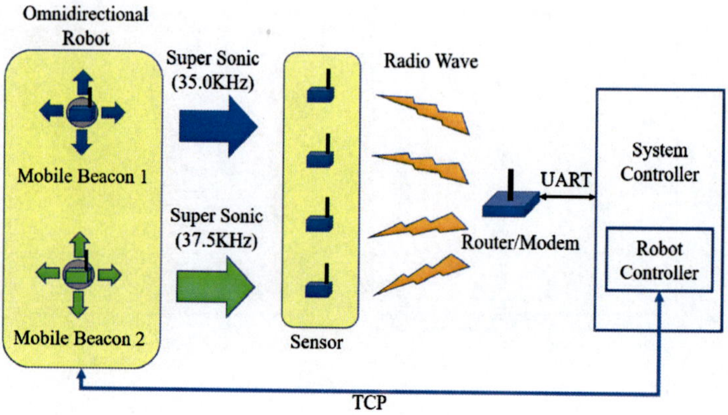

Fig. 7. Distance measurement system using supersonic signals.

3 Proposed System for Control of Omnidirectional Robot

In order to measure the position of omnidirectional robot, we use supersonic signals. In Fig. 7 is shown the distance measurement system using supersonic signals. An omnidirectional robot has two mobile beacons which will generate supersonic signals with different frequencies. For Mobile Beacon 1 is used 35.0 [KHz] and for Mobile Beacon 2 37.0 [KHz]. There are 4 sensors in the experimental environment which are able to receive these two different supersonic signals to calculate the omnidirectional robot positions. We show the real experimental environment in Fig. 8.

4 Experimental Results for Control of Omnidirectional Robot

During experiment, we moved the omnidirectional robot toward x and y directions. In Fig. 9(a) is shown the case when we moved the omnidirectional robot toward the x axis and the original position was x: 1.034 [m], y: 0.723 [m], and the target position was x: 4.034 [m], y: 0.723 [m]. Then, the robot was moved toward the target position with a free run. The result was x: 4.044 [m] and y: 0.698 [m]. The difference of distances for x direction is 0.01 [m] and for y direction is 0.025 [m].

In Fig. 9(b) is shown the case when we moved the omnidirectional robot toward the y axis and the original position was x: 4.016 [m], y: 0.706 [m], and the target position was x: 4.016 [m], y: 3.706 [m]. Then, the robot was moved toward the target position with a free run. The result was x: 4.044 [m] and y: 3.793 [m]. The difference of distances for x direction is 0.028 [m] and for y direction is 0.087 [m]. The error rates when moving toward x axis and y axis are shown in Table 1 and Table 2.

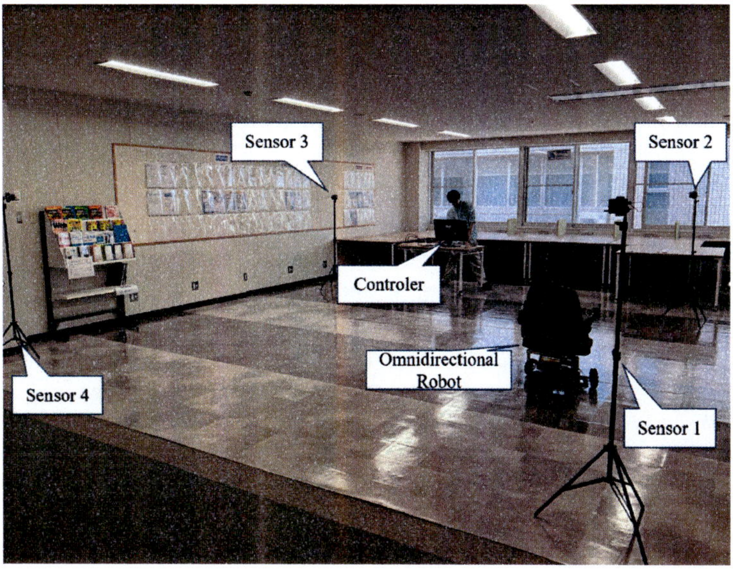

Fig. 8. Experimental environment.

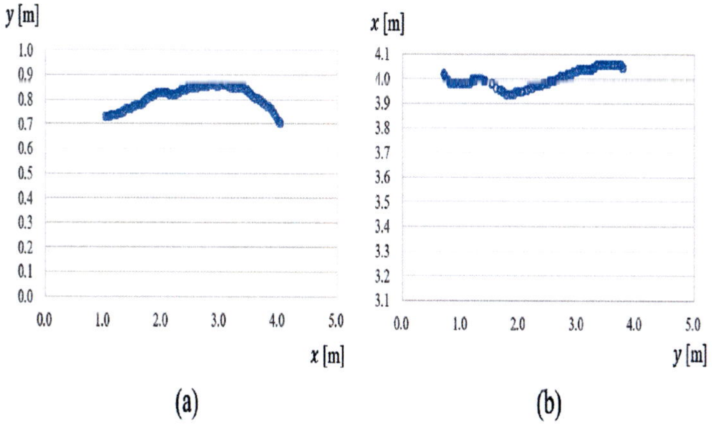

Fig. 9. Experimental results.

Table 1. Error rates when moving toward x axis.

Direction	Original Position	Target Position	Destination Position	Error Rate
x	1.034 [m]	4.034 [m]	4.044 [m]	0.30 [%]
y	0.723 [m]	0.723 [m]	0.698 [m]	0.83 [%]

Table 2. Error rates when moving toward y axis.

Direction	Original Position	Target Position	Destination Position	Error Rate
x	4.016 [m]	4.016 [m]	4.044 [m]	0.93 [%]
y	0.706 [m]	3.706 [m]	3.793 [m]	2.90 [%]

5 Conclusions and Future Work

In this paper, we introduced the implementation and application of omnidirectional robot for MOAP and wheelchair tennis. For MOAP, it is important to have an optimal control of the omidirectional robot in order to improve the network connectivity. Also, the optimal control of omnidirectional wheelchair will provide better environment for playing tennis by handicapped persons and people with disabilities.

We implemented a system for the control of omnidirectional robot using supersonic signals. We measured the accuracy of moving omnidirectional robot by using the implemented control system. The experimental results show that the omnidirectional robot was controlled correctly, because the error rate for x direction was within 1.00 [%]. However, when the robot moved toward y direction, the error rate was 2.9 [%].

In the future work, we will improve the implemented omnidirectional robot as well as conduct experiments for various situations. Also, we will consider other applications of implemented robot.

Acknowledgements. This work is supported by JSPS KAKENHI Grant Number JP22K11598.

References

1. Bragg, E., Spencer, N.L., Phelan, S.K., Pritchard-Wiart, L.: Player and parent experiences with child and adolescent power soccer sport participation. Phys. Occup. Ther. Pediat. **40**(6), 637–650 (2020)
2. Cavedon, V., Zancanaro, C., Milanese, C.: Physique and performance of young wheelchair basketball players in relation with classification. PLoS ONE **10**(11), e0143621 (2015)
3. Dahmani, M., et al.: An intelligent and low-cost eye-tracking system for motorized wheelchair control. Sensors **20**(14), 3936 (2020)
4. De Oliveira Mota Ribeiro, W., de Almeida, M.B., et al.: Performance analysis in wheelchair para-badminton matches (2020)
5. Kaye, S.A., Lewis, I., Forward, S., Delhomme, P.: A priori acceptance of highly automated cars in Australia, France, and Sweden: a theoretically-informed investigation guided by the tpb and utaut. Accid. Anal. Prevent. **137**, 105441 (2020)
6. Koester, N., Cichy, P., Antons, D., Salge, T.O.: Perceived privacy risk in the internet of things: determinants, consequences, and contingencies in the case of connected cars. Electron. Mark. **32**, 1–23 (2022)

7. Matsuo, K., Barolli, L.: Design and implementation of an omnidirectional wheelchair: control system and its applications. In: 2014 Ninth International Conference on Broadband and Wireless Computing, Communication and Applications, pp. 532–535. IEEE (2014)
8. Mitsugi, K., Matsuo, K., Barolli, L.: A comparison study of control devices for an omnidirectional wheelchair. In: Barolli, L., Amato, F., Moscato, F., Enokido, T., Takizawa, M. (eds.) WAINA 2020. AISC, vol. 1150, pp. 651–661. Springer, Cham (2020). https://doi.org/10.1007/978-3-030-44038-1_60
9. Mitsugi, K., Matsuo, K., Barolli, L.: Evaluation of a user finger movement capturing device for control of self-standing omnidirectional robot. In: Barolli, L., Woungang, I., Enokido, T. (eds.) AINA 2021. LNNS, vol. 227, pp. 30–40. Springer, Cham (2021). https://doi.org/10.1007/978-3-030-75078-7_4
10. Nauber, R., Goudu, S.R., Goeckenjan, M., Bornhäuser, M., Ribeiro, C., Medina-Sánchez, M.: Medical microrobots in reproductive medicine from the bench to the clinic. Nat. Commun. **14**(1), 728 (2023)
11. Orr, K., Malone, L.A.: Wheelchair rugby. Wheelchair sport, pp. 151–166. Human Kinetics, Champaign (2010)
12. Sánchez-Pay, A., Sanz-Rivas, D.: Wheelchair tennis, from health to competitive analysis: a narrative review. Universidad de Alicante, pp. 284–295 (2020)
13. Stankovski, S., Ostojić, G., Zhang, X., Baranovski, I., Tegeltija, S., Horvat, S.: Mechatronics, identification technology, industry 4.0 and education. In: 2019 18th International Symposium Infoteh-Jahorina (Infoteh), pp. 1–4. IEEE (2019)
14. Toyama, A., Mitsugi, K., Matsuo, K., Barolli, L.: Implementation of a moving omnidirectional access point robot and a position detecting system. In: Barolli, L., Poniszewska-Maranda, A., Park, H. (eds.) IMIS 2020. AISC, vol. 1195, pp. 203–212. Springer, Cham (2021). https://doi.org/10.1007/978-3-030-50399-4_20
15. Toyama, A., Mitsugi, K., Matsuo, K., Barolli, L.: Implementation of control interfaces for moving omnidirectional access point robot. In: Barolli, L., Takizawa, M., Enokido, T., Chen, H.-C., Matsuo, K. (eds.) BWCCA 2020. LNNS, vol. 159, pp. 436–443. Springer, Cham (2021). https://doi.org/10.1007/978-3-030-61108-8_43
16. Toyama, A., Mitsugi, K., Matsuo, K., Kulla, E., Barolli, L.: Implementation of an indoor position detecting system using mean BLE RSSI for moving omnidirectional access point robot. In: Barolli, L., Yim, K., Enokido, T. (eds.) CISIS 2021. LNNS, vol. 278, pp. 225–234. Springer, Cham (2021). https://doi.org/10.1007/978-3-030-79725-6_22

Data Augmentation Method for Improving Object Detection Accuracy of Recumbent Human in Disaster

Takahiro Uchiya(✉) and Taiga Yamada

Nagoya Institute of Technology, Gokiso-cho, Showa-ku, Nagoya 466-8555, Aichi, Japan
t-uchiya@nitech.ac.jp, t.yamada.436@nitech.jp

Abstract. For this study, we specifically examine object detection methods for finding victims at a facility who are in a lying posture because of injury or for other reasons. Most large-scale human training data available on the internet are based on the standing postures of people. Lying postures are rare. Furthermore, obtaining large numbers of images of people lying down in disaster situations and creating training data are extremely difficult. Therefore, it is necessary to learn object detection models that handle diverse postures. For this study, we assess data augmentation. Because the image of a person in a reclined posture is regarded as one in which the head and body are aligned side by side, data extension of 90° rotation to the existing image of a person in a standing posture can be used to enhance the image of a person in a reclining posture. For this study, the object detection algorithm SSD is augmented with rotational data augmentation to enhance the detection of lying human figures. However, it is unknown whether the rotation data enhancement is effective for detecting lying persons or not. For an evaluation experiment, we changed the rotation rate arbitrarily and compared the models learned with each rate. The data expansion of the rotation was evaluated using the detection accuracy for normal and recumbent human figures. We confirmed that the detection accuracy of the lying person improved at a rotation of 25%.

1 Introduction

Disasters have various phases such as prevention, response, recovery, and reconstruction. Research and development related to disaster prevention and response have been conducted to assess each phase [1]. Among them, the response phase deals most closely with lifesaving during the first 72 h after a disaster. The survival rate of victims found after this response phase is less than 25% [2]. Therefore, disaster response during the response phase directly affects the survival rate of disaster victims. To respond immediately during the response phase, disaster-resistant robots have been developed and used widely in recent years. Nevertheless, because of the special characteristics of their applications, disaster response robots are operated mainly by disaster response organizations and are not used widely in general facilities, making it difficult for general facilities to operate them immediately after a disaster. This study specifically examines disaster response during the response phase using general facility stationary robots.

L. Barolli (Ed.): BWCCA 2023, LNDECT 186, pp. 354–364, 2024.
https://doi.org/10.1007/978-3-031-46784-4_34

One earlier study [3] indicated that effective disaster prevention functions using robots in this response phase include disaster information identification, evacuation guidance, and establishment of wireless communication channels. For this study, we specifically examine the function of disaster information recognition among these disaster response functions. If disaster information such as victim conditions and locations could be obtained in before rescue operations, then it would be useful in rescue operations and assist disaster response organizations during their rescue activities. However, security cameras located at fixed points might have blind spots or break down during a disaster, leaving them unable to obtain information. Based on those and other points, this study emphasizes an investigation of the use of general facility robots to find people inside facilities who need rescue.

Although many research and development efforts have examined the use of robots to seek people needing rescue, most such efforts have been conducted outdoors. A lack of technology has existed for use inside facilities such as shopping malls or schools. Moreover, because existing techniques emphasize finding persons in distress, they are inadequate to ascertain the condition of the found person and whether or not the person can evacuate independently. Two types of person-detection methods might be used: image-based methods use cameras; infrared-based methods can detect a person's body temperature. The latter can detect a person, but it then remains difficult to ascertain the specific condition of the person.

This study defines victims of two types in a facility: persons who cannot evacuate on their own, or persons who remain in the facility. Subsequently, we aim at building a system to identify the victims using image-based method. When discriminating between victims of these two types, we consider the posture of each type of victim. Because those in need of rescue are unable to evacuate independently, they might be lying down or crawling. Therefore, if it is possible to ascertain whether or not the person in the image obtained from the robot is in a prone posture, then the state of the victim can probably be inferred. For this study, we consider judgment of whether a person detected by object detection is in a prone posture, or not. However, large-scale training data for people available on the internet include few images of people in prone postures. Training models based on such data are vulnerable to weakness in the detection of people in prone postures. The objective of this study is creation of an object detection model that is independent of the person's posture.

2 Object Detection

2.1 Overview

Object detection, which is the task of identifying object regions and object names in an image, can reveal where and what is reflected in an image. Most object detection algorithms perform the following three steps in this given order.

(1) Proposal of object region candidates.
(2) Classification of detected objects.
(3) Region adjustment (regression).

Objects are detected by finding bounding boxes that cut out multiple objects of various sizes in a single image. They are then classified into their respective classes. The Sliding Window method searches for all patterns by moving the bounding box position, but great amounts of time are necessary to search for all patterns. Therefore, various object detection algorithms have been developed, such as SSD, R-CNN, and YOLO. For this study, Single Shot Multibox Detector (SSD) is used because the operation in this study will be conducted with a camera attached to a robot. It will be used for immediate detection in the event of a disaster. Moreover, SSD is faster than the other algorithms.

2.2 Single Shot MultiBox Detector (SSD)

As a method for detecting objects in images using a single neural network, SSD [4] performs both "region candidate detection" and "classification" of objects in a single CNN operation. This rapid performance enables a faster object detection process. The use of a hierarchical feature map enables object detection in low-resolution images. Presumably, object detection is possible even with a robot camera.

2.3 Data Augmentation of SSD

In SSD, online data augmentation [5] is applied and trained.

The data augmentation process is presented below.

- Photometric Distortion: Random changes in brightness, saturation and hue.
- Random Expand: Changes the image dimensions.
- Random Crop: Crop the image so that the center of the box is in the cropped image.
- Random Mirror: Reverses the image left to right with probability of 50%.

This flow is portrayed in Fig. 1. This data augmentation improves the object detection accuracy by 8.8% when training on PascalVOC [6] 2007 and 2012 training data.

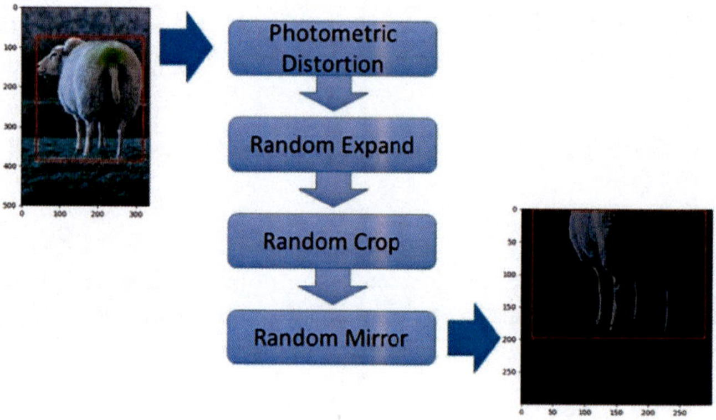

Fig. 1. Data augmentation flow in SSD.

3 Proposed Method

3.1 Outline of Proposed Method

The image of a person in a prone position is regarded as one in which the head and body are aligned horizontally. However, the majority of the training data are in the standing position, in which the head and body are aligned vertically. Therefore, we inferred that the training data can be enhanced by rotating the training data image by 90°. Herein, we propose the addition of Random Rotation processing to the SSD data augmentation flow.

- Random Rotation: Rotate the image 90° to the left or right according to probability.
- This probability is varied arbitrarily to elucidate the optimal probability. The flow of the proposed method is shown in Fig. 2. All images in the training data are subjected to the usual SSD data augmentation processes of "Photometric Distortion," "Random Expand," "Random Crop," and "Random Mirror." Subsequently, the "Random Rotation" process is performed.
- Random Rotation is not performed for all images, but for each image in each training data set. The decision to perform Random Rotation is made sequentially and by probability because learning cannot be carried out on normal training data if all images are rotated. This choice might lead to decreased object detection accuracy. For images that are rotated, a 50% probability exists that the image will be rotated to the left or right. This data augmentation is performed at each epoch. A different image is generated and trained at each epoch.

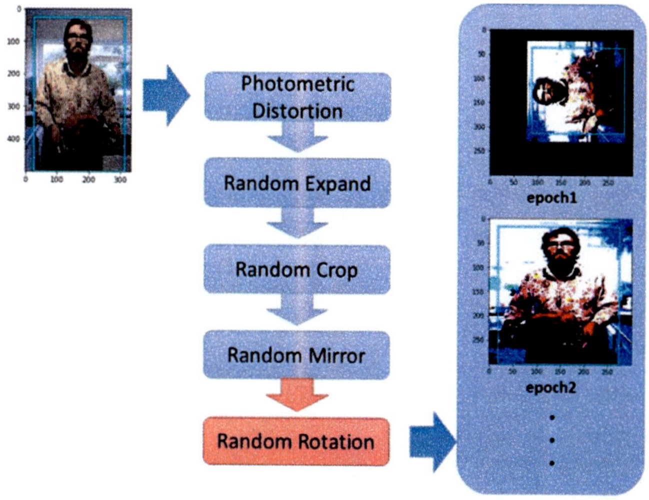

Fig. 2. Proposed method.

3.2 Learning Environment

The environment, library version, and hyperparameters used for training are presented respectively in Tables 1, 2, and 3.

Table 1. Learning environment

OS	Windows 10 Pro 64bit
CPU	Core i9-10900K
Memory	Crucial DDR4–2666 32GB
GPU	Geforce RTX2080Ti 11GB
Language	Python ver3.8

Table 2. Library version

opencv-python	4.6.0
numpy	1.20.1
matplotlib	3.3.4
pytorch	1.11.0

Table 3. Hyperparameter

Number of epoch	300
Batch size (training data)	32
Batch size (validation data)	16
Learning rate	1e-3

4 Validating the Effectiveness of the Proposed Method

4.1 Test Data

4.1.1 PascalVOC 2007 Test Data

Although the purpose of this study is improvement of the accuracy of detecting human figures in the prone posture, the accuracy of normal human images should also be ensured. For this reason, we extracted and evaluated human-only test data from the 2007 version of the PascalVOC test data, which includes a variety of images of people going about their daily lives.

4.1.2 Recumbent Figure

The images used to evaluate the accuracy of the detection of human images in the prone posture were collected from five men in their 20 s. Three lying postures (on one's side, on one's stomach, on one's back) were photographed at a distance of 2 m from the body in a room so that the entire body could be viewed from eight angles (Fig. 3). In all, 160 images were taken from eight directions in the three lying postures for each person. Annotations, which were used as test data, were created for each image. Figure 4 presents an example of the images taken.

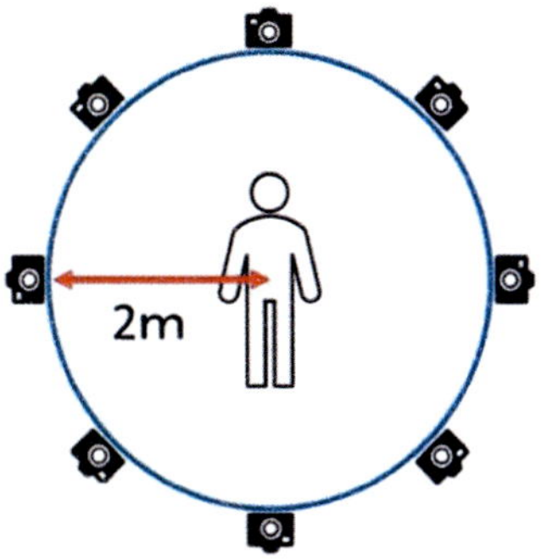

Fig. 3. Position of a person and camera during a photo shoot.

Fig. 4. Example of captured image.

4.2 Model Evaluation Metrics

4.2.1 Average Precision (AP)

Although evaluation metrics such as Precision and Recall are used to evaluate general learning models, they cannot be used to evaluate bounding boxes or confidence levels. In object detection models, evaluation metrics should behave as shown below.

- Higher when the detected object class matches the correct class.
- Higher when the detected bounding box is near the correct box.
- Lower when the detected box class is wrong despite a high confidence value.

An index called average precision (AP) is used as an evaluation index that behaves as described above.

4.2.2 Recall

For this study, failure to locate a person is the detection case which must be avoided. A missed person is the object detection model failure to detect a person who is present in an image. The final Recall value after AP calculation is calculated.

5 Evaluation Experiment

We trained 300 epochs with 5% increments of Random Rotation from 0% to 100%.

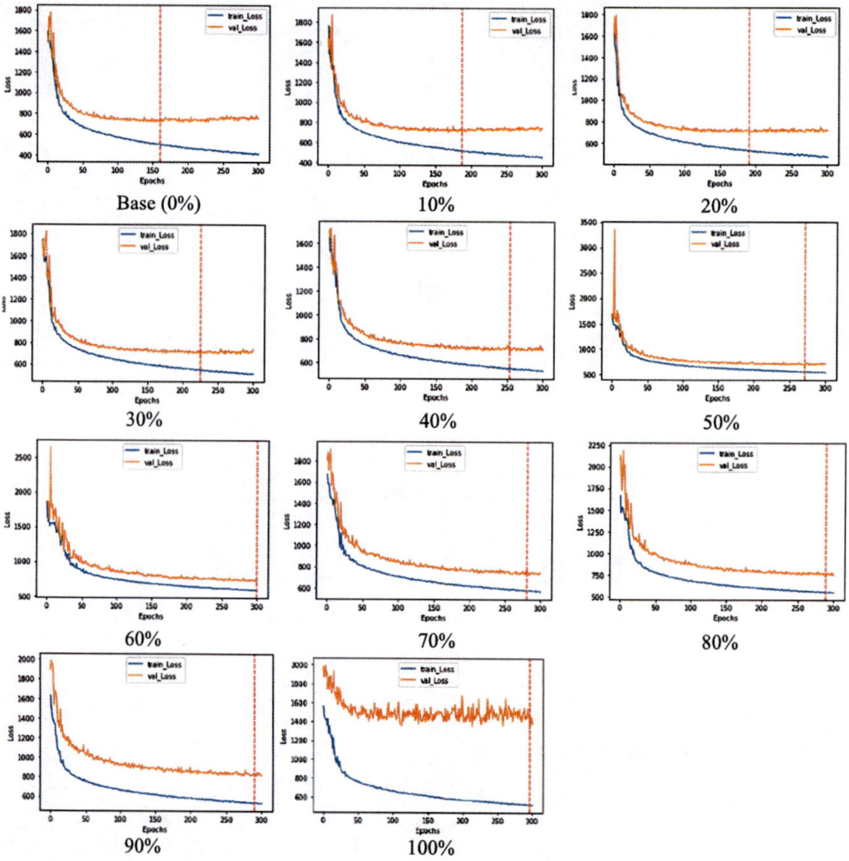

Fig. 5. Learning curve at each Random Rotation rate.

5.1 Learning Curve

Figure 5 shows a part of the learning curve of the created learning model. Figure 5 (Base (0%)) shows the learning curve with 0% Random Rotation, followed by the learning curves with increasing Random Rotation rate by 10%. Figure 5 (100%) shows the learning curve with 100% Random Rotation. The horizontal axis shows the number of epochs. The vertical axis shows the training loss. The blue and orange lines respectively represent the train loss and validation loss. The learning model evaluated for each training is the learning model with the lowest validation loss among the 300 epochs of training. The red dotted line represents the number of epochs with the lowest validation loss in the 300 epochs of training. The training model learned in these epochs is used for evaluation. In other words, learning after the red dotted line can be regarded as overlearning. The red dotted line shows the end of learning.

The 0% Random Rotation with no data augmentation completes the training at around 150 epochs. In contrast, the number of learning epochs increases as the ratio of Random Rotation increases. This finding suggests that Random Rotation can prevent overlearning, which is an important benefit of data augmentation. However, the validation loss at the end of training with 50% or less Random Rotation was around 800, whereas the validation loss with 75% or more Random Rotation was around 1,000. The value of the validation loss was nearly 1000 at 100% random rotation. The accuracy of learning was regarded as degraded. Therefore, excessive Random Rotation is inappropriate. Learning safety is regarded as compromised.

5.2 Accuracy Evaluation

Table 4 shows the AP and Recall results for each training model on the two test datasets. For Random Rotation of less than 50%, most results showed improved accuracy for both AP and Recall. For Random Rotation greater than 50%, however, the accuracy of PascalVOC on the test data decreased considerably as the percentage of Random Rotation increased.

The lowest value was obtained for 100% Random Rotation because the data augmentation was applied only to the training data. During training, the training data are split into training data and validation data. Data augmentation is applied only to the training data. For 100% Random Rotation, all training data were trained with rotated images. For that reason, the training data and validation data became completely different. Consequently, the large Random Rotation ratio increases the gap between the training data and the validation data, which hinders learning.

However, for test data in the prone posture, the training model which implemented all percentages of Random Rotation except for 100% showed higher accuracy. This finding suggests the effectiveness of Random Rotation for learning to detect people in a prone posture.

Table 4. Detection accuracy of test data of two types

Occurrence of Random Rotation	Person detection (PascalVOC)		Detection of persons in a lying position (original data)	
	AP	Recall	AP	Recall
0%	72.28%	86.45%	83.61%	98.13%
5%	74.00%	89.20%	96.94%	99.38%
10%	73.54%	88.20%	91.58%	98.75%
15%	72.10%	87.79%	90.67%	97.50%
20%	73.74%	88.76%	84.66%	96.25%
25%	73.79%	88.22%	92.84%	99.34%
30%	72.50%	87.20%	95.12%	99.38%
35%	73.41%	88.04%	86.97%	98.13%
40%	73.28%	88.88%	97.08%	100%
45%	72.87%	87.66%	87.28%	96.25%
50%	72.65%	87.97%	95.84%	100%
55%	71.51%	87.85%	99.11%	98.13%
60%	70.57%	87.09%	96.93%	100%
65%	64.98%	82.30%	99.64%	100%
70%	69.49%	85.73%	95.89%	99.38%
75%	69.64%	84.89%	96.31%	99.34%
80%	68.49%	85.00%	98.55%	100%
85%	57.79%	80.72%	99.29%	100%
90%	64.41%	81.84%	99.20%	100%
95%	50.12%	74.21%	94.52%	98.75%
100%	25.43%	55.90%	79.47%	97.50%

6 Conclusion

6.1 Summary

This study specifically examined the detection of a person lying down to assess the state of a victim remaining inside a facility. The ultimate goal of this study is construction of a system that can be operated without difficulty using only a CPU, which uses object detection to find a person in a prone posture. One difficulty with using object detection is that the training dataset includes numerous images of standing people. Object detection model trained using these images is vulnerable to weakness in detecting people in the prone posture.

For this study, we added Random Rotation to the data augmentation flow of SSD, an object detection algorithm, to train an object detection model that can guarantee detection accuracy even for recumbent human images. Results show that the Random Rotation process avoided over-learning compared to normal learning and that it improved the accuracy for learning of less than 50%. However, when training with more than 50% Random Rotation, the accuracy decreased for the PascalVOC test data, but it improved when it specifically sought the prone human figure. The accuracy decreased considerably for test data of both types at 100% Random Rotation. The results presented above suggest the effectiveness of Random Rotation, but excessive Random Rotation might lead to decreased accuracy. Because the objective of this study is creation of an object detection model that is independent of the person's posture, Random Rotation around 25% is regarded as the most effective to improve the accuracy for both test datasets.

6.2 Future Work

- Implementation of other data augmentation

  For this study, we implemented and evaluated Random Rotation as a data augmentation, but other more appropriate data augmentation methods might exist. Additionally, more advanced learning can be achieved by implementing data augmentations for a wider variety of rotations, not limited to 90°.
- Evaluation by increasing test data variety

  As described herein, we specifically examined the detection of images of people in a lying posture. The original data used for this study were such images. However, after a disaster, the posture of a person in need of rescue might not be limited to a prone posture. Postures are expected to be more diverse. Therefore, evaluating the detection accuracy in other postures is necessary to make a more multifaceted evaluation.
- Exploration and experimentation with video

  For this study, experiments were conducted to detect the lying state of a person in a still image. In a real environment, this system will be used with video from a camera mounted on an unmanned ground vehicle. Therefore, the accuracy of prone posture detection from the video obtained during a robot search must be assessed.
- Initiate a system to identify the victim condition

  The objective of this research is construction of a state discrimination system for detecting disaster victims. An evaluation is carried out to improve the detection accuracy of human figures in the prone posture. A system for discriminating victim states using the object detection model must be created based on these study results.

Acknowledgments. This work was supported by JSPS KAKENHI Grant Number JP23K11215.

References

1. Council on Competitiveness – Nippon (COCN): Disaster Response Robots and Operational Systems, Council on Competitiveness – Nippon Roundtable 2012 Project Final Report (2013)
2. Ministry of Land, Infrastructure, Transport: Learning from the Great Hanshin-Awaji Earthquake: How to Use Social Infrastructure in Times of Earthquake Disaster (2002)

3. Uchiya, T., Sugie, R., Hayashi, K., Kinoshita, T.: Proposal of disaster mitigation method using multiple robots. In: Barolli, L., Leu, F.-Y., Enokido, T., Chen, H.-C. (eds.) BWCCA 2018. LNDECT, vol. 25, pp. 659–668. Springer, Cham (2019). https://doi.org/10.1007/978-3-030-02613-4_59
4. Liu, W., et al.: SSD: single shot multibox detector. In: Leibe, B., Matas, J., Sebe, N., Welling, M. (eds.) ECCV 2016. LNCS, vol. 9905, pp. 21–37. Springer, Cham (2016). https://doi.org/10.1007/978-3-319-46448-0_2
5. Shorten, C., Khoshgoftaar, T.M.: A survey on image data augmentation for deep learning. J. Big Data **6**, 60 (2019). https://doi.org/10.1186/s40537-019-0197-0
6. PASCAL: The PASCAL Visual Object Classes Home page, http://host.robots.ox.ac.uk/pascal/VOC/index.html

Optimal Compressing and Decompressing Digital-Ink Handwriting via Sparse Gaussian Process Regression and Dynamic Programming

Jinya Yano[1] and Hiroyuki Fujioka[2(✉)]

[1] Graduate School of Engineering, Fukuoka Institute of Technology, 3-30-1 Wajiro-Higashi, Higashi-ku, Fukuoka 811-0295, Japan
mjm22103@bene.fit.ac.jp

[2] Department of System Management, Fukuoka Institute of Technology, 3-30-1 Wajiro-Higashi, Higashi-ku, Fukuoka 811-0295, Japan
fujioka@fit.ac.jp

Abstract. In this study, we delve into the challenge of compressing and decompressing digital-ink data obtained from handwritten inputs on devices such as pen tablets, etc. We present a new approach using sparse Gaussian process regression in order to compress digital-ink data into some kernel matrix consisting of a sequence of pseudo-inputs. Moreover, to enhance the accuracy of compression and decompression, we introduce a dynamic programming (DP) approach for the optimal selection of dominant pseudo-inputs. The effectiveness of our method is substantiated through a series of experimental studies.

1 Introduction

In the contemporary era of digitization, handwriting persists as an invaluable form of human expression. With advancements in technology, devices like tablets, smartphones, and digital blackboards are integrating handwriting inputs using touch or styluses, resonating with the organic nature of human communication [1]. Such handwritten trails are captured as "Digital-Ink", essentially a series of data points that chart the trajectory of the writing instrument according to a set sampling frequency.

As the density of this sampled data or the duration of the writing act grows, the sheer volume of digital-ink data escalates, prompting the need for efficient representation and storage techniques. Historically, scholars have been intrigued by this challenge. For instance, Mazalov and Watt's intriguing compression technique in [2] has leaned on orthogonal polynomial functions for its foundation. Their preliminary ideas evolved into a more streamlined, piecewise-linear compression model [3]. Moreover, the allure of B-spline methodologies became evident in related works [4], compressing digital-ink into succinct B-spline control points. However, these models may falter when faced with intricate character designs, mandating a plethora of control points or coefficients. While sparse coding avenues like Lasso hold promise [5,6], they are ensnared by their dependency on pre-defined digital-ink dictionaries [7].

This research presents a pioneering approach to the digital-ink data conundrum using sparse Gaussian process regression [8]. We usher in a transformative strategy that

L. Barolli (Ed.): BWCCA 2023, LNDECT 186, pp. 365–373, 2024.
https://doi.org/10.1007/978-3-031-46784-4_35

condenses vast digital-ink information into a manageable ensemble of pseudo-inputs, addressing the inherent complications spotlighted in prior researches [2–4]. Additionally, to further refine the precision of our compression and decompression technique, we integrate a dynamic programming (DP) methodology to optimally select dominant pseudo-inputs. This ensures the most critical data points are prioritized, enhancing both compression efficiency and data integrity. Our formulation paves the way for a tailored digital-ink representation, retaining an impressive likeness to the original manuscript.

The paper is organized as follows. As preliminaries, Sect. 2 introduces the Gaussian process (GP) regression and its application to digital-ink data compression and decompression. In Sect. 3, we introduce an approach to the digital-ink data conundrum using sparse Gaussian process regression. Then, Sect. 4 presents a dynamic programming (DP) methodology to optimally select dominant pseudo-inputs. In Sect. 5, we demonstrate the effectiveness of our proposed method. Finally, Sect. 6 concludes this work.

2 Preliminaries

As preliminaries, we here present Gaussian Process Regression (GPR) as described by [9], and its fundamental concepts when applied to digital-ink compression and decompression.

Let us consider a scenario where digital-ink data is captured using devices like a pen-tablet. From this collected data, N points are evenly chosen and represented as

$$\mathscr{D} = \{(x_i, y_i) : x_i \in [x_{\text{ini}}, x_{\text{fin}}],\ y_i \in \mathbf{R}^2,\ i = 1, 2, \cdots, N\}. \tag{1}$$

Here, $y_i \in \mathbf{R}^2$ is represented as $y_i = [X_i, Y_i]^T$. For our current context, we view the writing process as a single-stroke activity with uniform thickness. Additionally, we normalize the average of y to zero.

Given the digital-ink data $\mathscr{D}$ in (1), we model it through GPR. More specifically, our goal is to create a model that approximates a function $f(x)$ with an input x belonging to $\mathbf{R}^+$ (representing time) and digital-ink $y \in \mathbf{R}^2$ as

$$y = [X,\ Y]^T \in \mathbf{R}^2. \tag{2}$$

For ease of understanding, we assume each element in $y(x)$ of $\mathbf{R}^2$, i.e., $X(x)$ and $Y(x)$, is formulated independently. We therefore treat $y(x)$ as scalars, understanding they represent one of two components.

Introducing the function f, we suppose that the function $f(x)$ is generated from a Gaussian process with zero-mean and covariance function $k(x, x')$, i.e.

$$f \sim \text{GP}(0, k(x, x')).$$

Defining $\mathbf{y}$ as

$$\mathbf{y} = [y_1,\ y_2,\ \cdots,\ y_N]^T, \tag{3}$$

its relationship with covariance matrix $\mathbf{K}$ can be given by

$$\mathbf{y} \sim \mathscr{N}(\mathbf{0}, \mathbf{K}) \tag{4}$$

with

$$\mathbf{K} = \begin{bmatrix} k(x_1,x_1) & k(x_1,x_2) & \cdots & k(x_1,x_N) \\ k(x_2,x_1) & k(x_2,x_2) & \cdots & k(x_2,x_N) \\ \vdots & \vdots & \ddots & \vdots \\ k(x_N,x_1) & k(x_N,x_2) & \cdots & k(x_N,x_N) \end{bmatrix}. \tag{5}$$

As for $k(x_i,x_j)$, the Gaussian kernel serves as our chosen kernel function, i.e.

$$k(x_i,x_j) = \exp\left(-\frac{|x_i - x_j|^2}{\theta_1}\right) + \theta_2 \delta(x_i,x_j) \tag{6}$$

with θ_i for $i = 1,2$ being hyperparameters and $\delta(\cdot,\cdot)$ representing the Kronecker delta.

Upon deriving the covariance matrix $\mathbf{K}$ from (5) or considering the data vector $\mathbf{x} = [x_1,\cdots,x_N]^T \in \mathbf{R}^N$ that contributes to the computation of $\mathbf{K}$, we have the ability to decode the compressed digital-ink through posterior expectation calculations as follows.

Consider (x^*,y^*) with $x^* \in [x_{\text{ini}},x_{\text{fin}}]$ as the extracted digital-ink data from its compressed form, particularly, the matrix $\mathbf{K}$ referenced in (5). The combined distribution of $\mathbf{y}$, as noted in (3), and y^* following the predefined prior is characterized by a Gaussian distribution as

$$\begin{pmatrix} \mathbf{y} \\ y^* \end{pmatrix} \sim \mathcal{N}\left(\mathbf{0}, \begin{pmatrix} \mathbf{K} & \mathbf{k}_* \\ \mathbf{k}_*^T & k_{**} \end{pmatrix}\right) \tag{7}$$

with

$$\mathbf{k}_* = [k(x^*,x_1),\ldots,k(x^*,x_N)]^T, \tag{8}$$

$$k_{**} = k(x^*,x^*). \tag{9}$$

When we designate $p(y^*|x^*,\mathcal{D})$ as the posterior for y^* corresponding to a given time $x^* \in [x_{\text{ini}},x_{\text{fin}}]$, it becomes evident that

$$p(y^*|x^*,\mathcal{D}) = \mathcal{N}(\mathbf{k}_*\mathbf{K}^{-1}\mathbf{y}, k_{**} - \mathbf{k}_*^T\mathbf{K}^{-1}\mathbf{k}_*). \tag{10}$$

To revert the compressed form of $\mathbf{K}$ in (5) back to its original digital-ink representation, the requisite step involves iterative computation of the posterior expectation $p(y^*|x^*,\mathcal{D})$ from (10) as

$$E(y^*|x^*,\mathcal{D}) = \mathbf{k}_*\mathbf{K}^{-1}\mathbf{y}, \tag{11}$$

taking samples of x^* between x_{ini} and x_{fin}.

The elegance of this compression technique lies in its simplicity, needing just the storage of covariance matrix $\mathbf{K}$ from (5) or the values $x_1,\cdots,x_N$ that determine $\mathbf{K}$. However, the challenge emerges in the decompression phase, which necessitates $\mathbf{O}(N^3)$ computational time primarily due to matrix inversion in (11). This becomes a hurdle especially when N escalates to enormous scales, like in instances where there is a vast time gap in handwriting data. An intuitive resolution could be the adoption of Cholesky decomposition [10]. But, its dependency on all elements of the matrix $\mathbf{K}$ amplifies its computational intensity, especially with growing data points N.

3 Digital-Ink Compression and Decompression Using Sparse GPR

In lieu of conventional methods like Cholesky decomposition, Sparse GPR has been highlighted as an efficient technique[8]. This approach enables to treat extensive datasets in a proficient manner. Through the introduction of a limited set of inducing points, it becomes possible for Sparse GPR to make approximations to the Gaussian process' complete covariance matrix, thus considerably lowering computational complexity. In this context, we introduce an approach to compress and decompress digital-ink data utilizing Sparse GPR.

Let $\mathbf{z} \in \mathbf{R}^M$ be the vector consisting of $M(\ll N)$ pseudo-input points z_i within the range $[x_{\rm ini}, x_{\rm fin}]$, labeled as inducing points:

$$\mathbf{z} = [z_1\ z_2 \cdots z_M]^T. \tag{12}$$

While a detailed derivation will not be discussed here (refer to [8] for the details), it is shown that the kernel matrix from (5) can be represented as

$$\mathbf{K} \approx \mathbf{K}_{NM}\mathbf{K}_{MM}^{-1}\mathbf{K}_{NM}^T + \Lambda. \tag{13}$$

Here, matrices $\mathbf{K}_{NM}$ and $\mathbf{K}_{MM}$ are defined with elements $k(x_i, z_j)$ and $k(z_i, z_j)$ respectively. Additionally, Λ is a diagonal matrix outlined by $\Lambda = \mathrm{diag}\{\lambda_1, \cdots, \lambda_N\}$, where

$$\lambda_n = k(x_n, x_n) - \mathbf{k}_{nM}\mathbf{K}_{MM}^{-1}\mathbf{k}_{nM}^T,\ n = 1, \cdots, N, \tag{14}$$

and $\mathbf{k}_{nM}$ symbolizes the n-th row vector of $\mathbf{K}_{NM}$. In the Eq. (13), the computational effort for the kernel matrix is correlated with the count of pseudo-inputs M. Namely, Sparse GPR benefits from a reduction in operations as the number of pseudo-inputs diminishes.

On the other hand, let $\mathbf{u} \in \mathbf{R}^{\mathbf{M}}$ be the vector formed by collating output values corresponding to each pseudo-input point z_i:

$$u_i = f(z_i), \tag{15}$$

leading to

$$\mathbf{u} = [u_1\ u_2 \cdots u_M]^T. \tag{16}$$

Following the premise in (4), we have

$$\mathbf{u} \sim \mathcal{N}(\mathbf{0}, \mathbf{K}_{MM}). \tag{17}$$

Utilizing Bayes' theorem, the posterior distribution over $\mathbf{y}$, i.e. $p(\mathbf{u}|\mathbf{y})$, can be obtained as

$$p(\mathbf{u}|\mathbf{y}) = \mathcal{N}(\hat{\mathbf{u}}, \hat{\Sigma}_{\mathbf{u}}), \tag{18}$$

where $\hat{\mathbf{u}}$ and $\hat{\Sigma}_{\mathbf{u}}$ are given as

$$\hat{\mathbf{u}} = \mathbf{K}_{MM}\mathbf{Q}_{MM}^{-1}\mathbf{K}_{NM}^T\Lambda^{-1}\mathbf{y} \tag{19}$$

$$\hat{\Sigma}_{\mathbf{u}} = \mathbf{K}_{MM}\mathbf{Q}_{MM}^{-1}\mathbf{K}_{MM} \tag{20}$$

with

$$\mathbf{Q}_{MM} = \mathbf{K}_{MM} + \mathbf{K}_{NM}^T \Lambda^{-1} \mathbf{K}_{NM}. \tag{21}$$

Considering a new data point, denoted as (x^*, y^*) where x^* falls within $[x_{\text{ini}}, x_{\text{fin}}]$, we can execute the digital-ink decompression leveraging the posterior distribution $p(y^*|x^*, \mathscr{D})$ of y^* for a given time $x^* \in [x_{\text{ini}}, x_{\text{fin}}]$, in alignment with the conventional GPR methodology discussed in Sect. 2. The posterior distribution $p(y^*|x^*, \mathscr{D})$ is given by

$$p(y^*|x^*, \mathscr{D}) \approx \int p(y^*|\mathbf{u}) p(\mathbf{u}|\mathbf{y}) d\mathbf{u}. \tag{22}$$

From (18), we then get

$$p(y^*|x^*, \mathscr{D}, \mathbf{z}) = \mathscr{N}(\hat{y}_*, \hat{\sigma}_{y_*}), \tag{23}$$

where

$$\hat{y}_* = \mathbf{k}_{M*}^T \mathbf{K}_{MM}^{-1} \hat{\mathbf{u}}, \tag{24}$$

$$\hat{\sigma}_{y_*} = k_{**} - \mathbf{k}_{M*}^T (\mathbf{K}_{MM}^{-1} - \mathbf{Q}_{MM}^{-1}) \mathbf{k}_{M*}. \tag{25}$$

To decompress $\mathbf{K}_{MM}$ and $\mathbf{K}_{NM}$ in (13) to the digital-ink format, our primary objective is to iteratively compute the expectation of the posterior $p(y^*|x^*, \mathscr{D})$, as shown in (24), i.e.

$$E(y^*|x^*, \mathscr{D}) = \mathbf{k}_{M*}^T \mathbf{K}_{MM}^{-1} \hat{\mathbf{u}}, \tag{26}$$

where x^* is sampled from x_{ini} and x_{fin}.

4 Optimal Pseudo-input Point Selection Using Dynamic Programming

To enhance the accuracy of our compression and decompression method in Sect. 3, we introduce a dynamic programming (DP) approach based on [3] for the optimal selection of dominant pseudo-inputs. This approach ensures that the essential data points are given precedence, thereby improving both the efficiency of compression and the integrity of the data.

Now, let G be a weighted directed acyclic graph (DAG), which is a directed graph with no cycles. We then constuct G from the digital-ink data set $\mathscr{D}$ in (1). In addition, letting $V(G)$ and $E(G)$ be a set of vertices and edges of graph G, we have

$$\begin{cases} V(G) = \{v_i | 1 \leq i \leq N\} \\ E(G) = \{(v_i, v_j) | 1 \leq i < j \leq N\} \end{cases}. \tag{27}$$

Note here that $v_i,\ 1 \leq i \leq N$ corresponds to the digital-ink data $d_i,\ 1 \leq i \leq N$.

Also, let $\gamma(v_i, v_j)(\geq 0)$ be the weight of between vertices v_i and v_j. We here define as

$$\gamma(v_i, v_j) = \sum_{k=i}^{j-1} ||(x_{k+1}, y_{k+1}) - (x_k, y_k)||^2 - ||(x_j, y_j) - (x_i, y_i)||^2. \tag{28}$$

The constructed DAG, G, has distinct starting and ending points, which align with d_1 and d_N. The aim is to trace a path on G that originates from d_1 and culminates at

d_N. This path should be composed of M nodes, representing the dominant digital-ink data, and should minimize the total weight in (28). The existence of this path and its determination via DP are supported by [3].

Now, let $D_{p,j}$ be the minimum total weight of the path from v_1 and v_j, including p vertices. Then, we initially set $M = 2$ as follows:

$$D_{2,j} = \begin{cases} \infty & \text{if } j = 0 \\ \gamma(v_1, v_j) & \text{if } 1 < j \leq N \end{cases}. \tag{29}$$

For $M \geq 3$, we iteratively compute $D_{M,j}$ as

$$D_{M,j} = \begin{cases} \min_{M-2 \leq i < j}\{D_{M-1,i} + \gamma(v_i, v_j)\} & \text{if } j \geq M-1 \\ \infty & \text{otherwise} \end{cases}. \tag{30}$$

To optimally select dominant M pseudo-inputs, we thus have only to compute $D_{M,N}$, in which the minimum cumulative error of error function is found.

5 Experimental Studies

We here demonstrate the performance of digital-ink compression and decompression using the methods in Sects. 3 and 4 by some experiments.

5.1 Experimental System

Figure 1 shows an overview of the experimental system. This system includes a PC running Linux OS and a pen tablet device. Through this configuration, we can capture and save handwriting movements on the screen as digital-ink data $\mathcal{D}$, representing 2D coordinates on the XY-plane, sampled every 20 [ms]. Furthermore, the methods in Sects. 3 and 4 were developed on a PC using MATLAB (OS: Windows 10 Education, CPU: Intel Core i9-7900X, RAM: 128GB).

Fig. 1. Overview of experimental system.

5.2 Experimental Results

As an example, we present the results for the word "hello" written with a single stroke of the pen using the experimental system described in Sect. 5.1. Here, the digital-ink $\mathscr{D}$ in (1) is measured and stored by employing the experimental system with a pen-tablet device, where N is $N = 751$. Additionally, we set x_{ini} and x_{fin} as $x_{\text{ini}} = 0$ [s] and $x_{fin} = 15$ [s], respectively.

In Figs. 2 and 3, we show examples of results on the compressed and decompressed digital-ink using the method in Sect. 3. Within these illustrations, panels (a) show the decompressed digital-ink with black traces, the original measured digital-ink as red cross marks, and the pseudo-input data in blue points as chosen by the strategy from Sect. 4. For these figures, we have designated values $M = 15$ for Fig. 2 and $M = 20$ for Fig. 3. Moreover, panels (b) present results when using an equivalent number of pseudo-inputs from panels (a) but arranged at uniform intervals. The parameters θ_1 and θ_2 from (6) are assigned as $\theta_1 = 0.1$ and $\theta_2 = 0.01$ in both scenarios.

Our empirical findings demonstrate the effectiveness of the proposed compression and decompression technique. Generally, the precision was observed to augment with increasing M. However, with the error function used here, i.e. (28), pseudo-inputs may not be appropriately selected in linear sections, so there might be slightly larger errors in the decompressed areas, such as some part of the letter "h". Therefore, a different definition of the error function might be needed. Importantly, we achieved substantial compression ratios of $M/N \approx 0.019 (= 15/751)$ and $M/N \approx 0.027 (= 20/751)$ in Figs. 2 and 3, respectively.

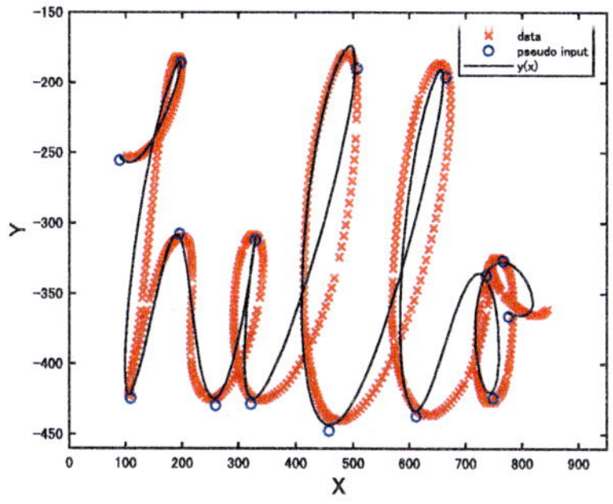

(a) with optimal selection

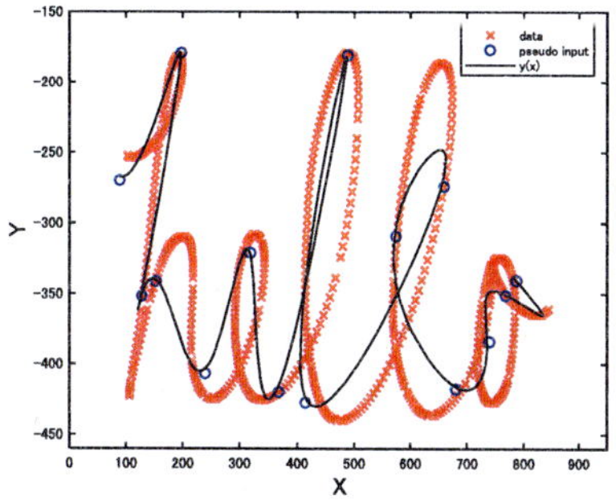

(b) with equal interval selection

Fig. 2. Compression and decompression results for the handwritten "hello" using the proposed method with $M = 15$.

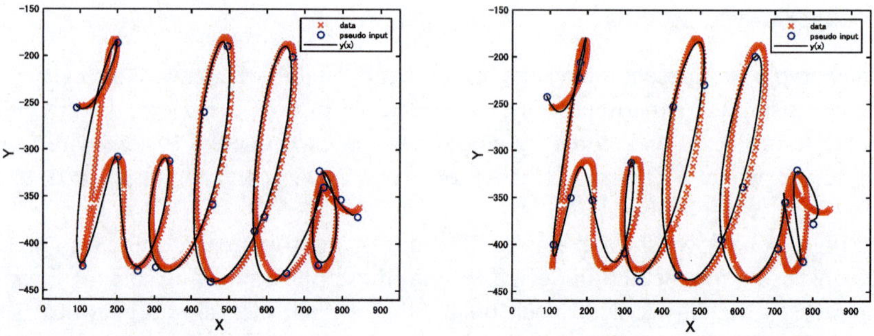

(a) with optimal selection

(b) with equal interval selection

Fig. 3. Compression and decompression results for the handwritten "hello" using the proposed method with $M = 20$.

6 Concluding Remarks

In this paper, we presented a new approach to compress and decompress handwritten digital-ink data using sparse Gaussian process regression. Our method effectively reduces the data size while preserving the intrinsic characteristics of the original digital-ink data. Furthermore, to enhance the compression performance without compromising the quality of the reconstructed digital-ink data, we incorporated a dynamic programming (DP) strategy for the optimal selection of dominant pseudo-inputs. Experimental results validate the efficacy of our methodology, showcasing a remarkable compression ratio.

While our method exhibits promising results in specific contexts, it would be beneficial to investigate its performance across diverse datasets and writing styles. The current error function may also be extended or redefined to ensure better selection of pseudo-inputs, especially in areas with linear characteristics. Additionally, the integration of machine learning techniques or the exploration of alternative regression models might further refine the compression-decompression pipeline, offering even more enhanced results. It will also be intriguing to examine the potential of our approach in real-time applications and its scalability in handling larger datasets.

References

1. Kurtenbach, G.: Pen-based computing, XRDS: crossroads. ACM Mag. Stud. - Future Interact. **16**(4), 14–20 (2010)
2. Mazalov, V., Watt, S.: Digital ink compression via functional approximation. In: Proceedings of the 12th International Conference on Frontiers in Handwriting Recognition, Kolkata, India, pp. 688–694 (2010)
3. Hu, R., Watt, S.: Optimization of point selection on digital ink curves. In: Proceedings of the 13th International Conference on Frontiers in Handwriting Recognition, Bari, Italy, pp. 525–530 (2012)

4. Mieno, Y., Fujioka, H., Kano, H.: Data compression of digital-ink with pen-slips using multi-level L1 smoothing splines. In: Proceedings of the 2015 IEEE International Conference on Systems, Man, and Cybernetics, Hong Kong, pp. 1787–1792 (2015)
5. Tibshirani, R.: Regression shrinkage and selection via the lasso. J. R. Stat. Soc. Ser. B **58**(1), 267–288 (1996)
6. Chen, S.S., Donoho, D.L., Saunders, M.A.: Atomic decomposition by basis pursuit. SIAM Rev. **43**(1), 129–159 (2001)
7. Aharon, M., Elad, M., Bruckstein, A.: The K-SVD: an algorithm for designing of overcomplete dictionaries for sparse representations. IEEE Trans. Image Process. **54**(11), 4311–4322 (2006)
8. Snelson, E., Ghahramani, Z.: Sparse gaussian process using pseudo-inputs. In: Neural Information Processing Systems, vol. 13. MIT Press (2006)
9. Bishop, C.M.: Pattern Recognition and Machine Learning. Springer, Heidelberg (2006)
10. Golub, G.H., Van Loan, C.F.: Matrix Computations: Johns Hopkins Studies in Mathematical Sciences, 3rd edn. Johns Hopkins University Press (1996)

Author Index

L. Barolli (Ed.): BWCCA 2023, LNDECT 186, pp. 375–376, 2024.
https://doi.org/10.1007/978-3-031-46784-4